MOUNTAINS: SOURCES OF WATER, SOURCES OF KNOWLEDGE

ADVANCES IN GLOBAL CHANGE RESEARCH

VOLUME 31

The titles published in this series are listed at the end of this volume.

Mountains: Sources of Water, Sources of Knowledge

Edited by

Ellen Wiegandt

Institut Universitaire Kurt Bösch,
Sion, Switzerland

Springer

A C.I.P. Catalogue record for this book is available from the Library of Congress.

ISBN 978-1-4020-6747-1 (HB)
ISBN 978-1-4020-6748-8 (e-book)

Published by Springer,
P.O. Box 17, 3300 AA Dordrecht, The Netherlands.

www.springer.com

Printed on acid-free paper

PREFACE

The idea for *Mountains: Sources of Water, Sources of Knowledge* began with the constitution of the Association Montagne 2002, whose mission was to celebrate the International Year of Mountains in the alpine canton of Valais, Switzerland. The year 2002 was the occasion for diverse activities ranging from exchanges with Bhutan and the construction of a traditional hanging bridge, to a contest for schoolchildren and roving photography exhibits. An international scientific colloquium was also organized at the University Institute Kurt Bösch. Committee members of the Association Montagne subsequently agreed that the presentations should be gathered into a book. The topic seemed particularly appropriate for the Martin Beniston-edited series, "Advances in Global Change Research," which regularly publishes the results of the annual Wengen Workshops on Global Climate Change Research as well as of other conferences.

This volume was a collective effort. The dedication of the Association Montagne committee members was unflagging. The commitment to the overall project of State Councilor Jean-Jacques Rey-Bellet and of Gabrielle Nanchen, then President of the Fondation pour le développement durable des regions de montagne, was particularly noteworthy and is gratefully acknowledged here. We appreciate the support of numerous financial sponsors of the Association Montagne 2002 and its International Year of Mountains activities, who made the conference and the book possible. Martin Beniston deserves thanks for including the volume in his series.

A book is ultimately the fruit of its authors and the reviewers who help refine arguments and clarify style. All contributors to this volume have been exemplary in revising manuscripts, responding to queries, and making corrections. They are especially thanked for their patience as this book followed its long route from conference to publication, helped in its final stages by research assistants from the Graduate Institute for International Studies in Geneva, Switzerland, Michael Flynn and Gauri Khanna.

CONTENTS

GENERAL CONCEPTS AND PROCESSES; MOUNTAIN SPECIFICITIES

FRAMING THE STUDY OF MOUNTAIN WATER RESOURCES: AN INTRODUCTION

Ellen Wiegandt
Institut universitaire Kurt Bösch
Sion, Switzerland
wiegandt@hei.unige.ch

1. ISSUES AND CHALLENGES

A crucial vulnerability the international community will face in the near future is access to fresh water in sufficient quantity and of adequate quality to meet the needs of a growing global population. As a result, mountains, which have always held a privileged relationship with water as the sources of the world's greatest rivers and the homes of huge glacier reserves, will come under increasing pressure. Already, these "water towers of the world" are threatened by major global forces. Climate change is predicted to modify the quantities of water available and shift its seasonality. Even greater challenges will come from the dynamics of human behavior. Population growth is one obvious threat to sufficient water supply, but equally important are changing norms and evolving activities. Historically dominant activities like agriculture and herding now compete with industry, leisure, domestic, and energy sectors for mountain water. These new uses are overlaid by various social constructs like pricing schemes, regulations, and property rights that shape production and consumption patterns by determining who gets how much water and when. Appropriate solutions for water use management contribute to efficient and equitable distribution. Poor administration is likely to aggravate shortages, increase social and economic disparities, and thereby make water issues a potential source of deadly local and international conflict.

A paradox of mountain water resources is that concern about deficits and competition for use coexist with the risks that come from their role as repositories. Floods, landslides, and avalanches are disasters that do not affect mountain populations uniquely, but these water-related disasters nevertheless take an especially heavy toll on mountain communities. The high volumes of water and the topography create special vulnerabilities.

E. Wiegandt (ed.), Mountains: Sources of Water, Sources of Knowledge, 3–13.

These local conditions are not the only threat, however. The wider world also impinges on the lives of mountain populations. Agricultural and industrial practices decided in national capitals or at the international level can produce water pollution that affects mountain villages. Global climate change has multiple effects on hydrological systems. Most strikingly, it has led to significant glacial retreat. Indeed, using different scenarios for greenhouse-gas emissions, Wigley and Raper (2005) predict that 73–94 percent of mountain glacier volume will have disappeared by 2400.[1] In the Alps, the area of the world for which the most detailed records are available, the basic pattern of general retreat over the past 150 years, with an accelerated pace in the past two decades, has led to an estimated 18 percent loss of area of Swiss glaciers between 1985 and 1999. From these baseline figures, some researchers predict even stronger future glacier retreat than what has been assumed so far and a probable further enhancement of glacier disintegration through positive feedbacks (Paul et al. 2004, p. 1). This evolution will ultimately affect water reserves available for all uses. It is also expected to change the thermodynamics of the atmosphere, thereby modifying temperature and precipitation regimes. Subsequent impacts on the seasonality of precipitation and water runoff will in turn affect water balance and flood regimes. The physical complexities of mountain regions are particularly sensitive and vulnerable to these kinds of changes.

Other features related to water's physical characteristics and dynamics also take on special significance because climate change is taking place in the context of rapid and widespread socioeconomic transformation. Resource management issues become primordial, with growing intricacy and contentiousness of allocation issues. Too little or too much water create equally difficult problems. While different types of use may induce relative scarcities, disasters like floods pose problems of unwanted overabundance. Historically in mountain regions, local communities took responsibility for many organizational and coordination tasks. Consolidation and integration at the national, regional, and international levels have shifted the locus of power in many areas. This institutional centralization does not necessarily reduce the overall complexity of management, however, because it has been accompanied by an increased competition for use among different sectors. Therefore, policies adopted for one group or region may be incompatible with the needs of another. For example, keeping sufficient reserves in dams for hydroelectric power generation may lessen quantities

---

[1] Such projections into the future necessarily rest on assumptions about changing levels of greenhouse-gas emissions. Their evolution is subject to great uncertainties because it will depend on energy demand, whose expected increase will be influenced by economic and political dynamics.

available during crucial seasons for agriculture or tourism; moreover, the change in flow rates in streams can also have unintended ecological consequences (Friedl and Wuest 2002). Or, it may be that national priorities will conflict with international norms. This is potentially the case with Swiss policies aimed at meeting goals of clean energy and revenues for local communities by favoring hydropower. It may be that these will be ruled incompatible with international rules of free trade negotiated under the WTO. Similar transnational confrontations arise when different national regions or countries sharing access to a lake or river have vastly different levels of use or technological capacity to access water sources (Luterbacher and Wiegandt 2005; Fisher, this volume). Putting in place arrangements for water sharing that are viewed as fair poses severe problems related to the special characteristics of water as a common pool resource. This widespread problem of common pool resources—which also include petroleum, fishing grounds, and the atmosphere—is compounded in the case of water by asymmetric access. Upstream users can seize water without consideration of downstream needs. Technological superiorrity confers similar dominance to users sharing still bodies of water because it often allows more sophisticated pumping capacity.

Addressing these kinds of problems requires analytical tools to evaluate the implications of different kinds of property rules or pricing schemes for efficient and fair use of water. It also requires deeper study of the factors leading to international conflicts over water and the international legal regimes needed to meet the goals set by the international community for universal access to an essential life resource. The authors in this volume consider all these issues, emphasizing a range of theoretical and methodological perspectives, and applying them to a set of empirical cases representing practices and policies on all continents.

Mountain regions have a long history of overseeing the resource water and thus are sources of knowledge for examining the dilemmas of managing a public good that knows no boundaries but can be diverted and traded. Political and economic forces thus intervene directly in the use of a natural resource that has its own physically induced dynamics, which were often mastered by populations who acquired know-how as they lived with glaciers or along great mountain rivers. Unique geography, particular experience, and often special institutions characterize the distinctive interactions between mountains and water. Their multifaceted relationship comprises several aspects that also relate to wider environmental issues. This volume addresses critical themes related to hydrological dynamics and human water use. Study of this varied and broad theme requires an interdisciplinary approach drawing on theory and methods from a wide range of scientific domains focused on hydrological dynamics and human water use. The volume also benefits from analysis provided by case studies in mountain areas throughout the world.

The underlying theme is the incessant struggle for water. It can be a struggle against nature that results in too much or too little water for human activities at particular moments or seasons. It can refer to competetive uses that pit economic sectors, groups of people, or nations against each other. Mountain environments highlight these fundamental issues because they are the repositories of major quantities of the world's fresh water. The volume has thus been organized around critical themes that have both special importance for mountain regions and illustrate the general problems of current water use and future challenges posed by climate change and global economic, political, and social evolution.

2. GENERAL CONCEPTS AND MOUNTAIN SPECIFICITIES

Chapters 1–4 launch the discussions by presenting an empirical overview of water resources and hydrological dynamics in mountain regions and introducing important management concepts and their theoretical underpinnings. In their chapter in this volume, Daniel Viviroli and Rolf Weingartner, "'Water Towers'—A Global View of the Hydrological Importance of Mountains," build on this perspective. They note that many uncertainties still exist concerning our ability to measure with precision the amounts of overall water resources that come from mountain regions. Such knowledge underlies allocation schemes, and thus we continue to be plagued with problems in many parts of the world. Viviroli and Weingartner predict that such problems will become more acute as climate change, growing population and changing needs alter both supply and demand.

The special qualities of water further complicate access issues. Water is a typical common pool resource because it is impossible—or extremely costly—to restrict the number of users or rates of use. This inevitably introduces a potential "tragedy of the commons," whereby users enjoy benefits without paying the costs of using the resource (Hardin 1968). Despite its abundance in many places, water is not a pure public good because use is often characterized by rivalry (one person's use deprives another of use of the same unit of the resource) due either to local problems of quantity or pollution. The way societies have confronted distribution of scarce resources is to define some sort of property rights regime defining rules of access. Two chapters in this volume—Paul Trawick's Scarcity, Equity, and Transparency: General Principles for Successfully

Governing the Water Commons" and Ellen Wiegandt's "From Principles to Action: Incentives to Enforce Common Property Water Management"—demonstrate how private property schemes are often impossible to implement for water resources because the element itself cannot be delimited. They show, however, that mountain societies in different parts of the world have devised effective forms of common property management that place various kinds of limits on the number of users and the quantities they may use. The long historical perspective adopted in Darren Crook et al.'s "The History of Irrigation and Water Control in China's Erhai Catchment: Mitigation And Adaptation To Environmental Change" shows how natural dynamics and human responses evolved synergistically over thousands of years in China, illustrating the way societies learn over time and adjust their institutions and behavior to natural cycles, with greater or less success in preventing environmental degradation and maintaining adequate water supplies. Studying shorter time spans, Trawick and Wiegandt draw similar conclusions about the importance of broad and deep knowledge about one's local environment in achieving sustainable water use.

3. MULTIPLE USES AND COMPETITION FOR MOUNTAIN WATER

New challenges for contemporary and future societies will come from the need to confront multiple demands on water resources. Even without the climate change impacts that may reduce supply in some regions, numerous analyses suggest that changing tastes and needs will require increasing amounts of water overall that may also be accompanied by changing patterns of demand throughout the seasons of the year. Often these activities will be independent of each other to some degree and subject to their own market forces, making coordination based on water availability difficult to achieve within existing political frameworks. The two chapters illustrating these phenomena are based on Swiss examples, permitting an integrated view of the general problem. In the Swiss Alps, two major and growing activities are tourism and hydroelectric production. These have replaced agriculture as the previous major source of revenue and employment. Franco Romerio's "Hydroelectric Resources between State and Market in the Alpine Countries" describes how climate change will clearly play a role at some time in the future by altering the amount, seasonality, and regularity of the supply of water as global warming affects glacier melting, snowfall, and rates of water flows, as well as the intensity and frequency of natural disasters that could imperil installations. When and to what degree these changes will occur is subject to great uncertainties and at present hydroelectric producers are far more concerned by the opening of European electricity markets. Romerio also analyzes this development and argues that mountain regions have major opportunities to profit from the flexibility of high pressure dams to respond to peak load demand. Realizing such potential will nevertheless depend on policy and regulatory choices that are currently being sharply debated.

Similar uncertainties and regulatory confusion characterize the tourism sector. Christophe Clivaz and Emmanuel Reynard ("Crans-Montana: Water Resources Management in an Alpine Tourist Resort") identify the decentralized nature of the Swiss political structure, which extends to its management of water resources, as a central cause of what they view as dysfunctional management practices. This problem is not uniquely a Swiss one, even though Switzerland is known for its strong reliance on local governance. As Saleh and Dinar (2004) show in their cross country analysis of water sector management, severe problems exist under multiple forms of ownership and regulation around the world. In evaluating ongoing institutional changes, they note the growing impact of user associations and nongovernmental agencies on policymaking. They point to the need for integrated legal and policy institutions. These become increasingly necessary as various forms of scarcity lead to greater conflict over water resource allocation, a problem also addressed in this volume.

4. WATER CONFLICTS AND CONFLICT RESOLUTION MECHANISMS

The distinctive qualities of water, particularly its open and often asymmetric access, lead to particular forms of conflict and also to a characteristic set of conflict resolution strategies. Natural resources are essential to life and therefore frequently are imbued with symbolic value irrespective of their economic or use value. Consequently, groups or nations may seek control by acting first ("first mover advantage"). As a result, conflict rather than negotiation may prevail. Numerous empirical examples permit testing various hypotheses about specificities of these kinds of resource conflicts. They also point to possible solutions that can be adapted to particular characteristics. In the Middle East, a range of competitive interactions—from tensions to outright conflict—typify relations among various states. Franklin Fisher ("Water Value, Water Management, and Water Conflict: A Systematic Approach") takes as his starting point the environmentally and politically induced asymmetric water shortages in Israel, Palestine, and Jordan. Recognizing that complex historical and strategic considerations make agreement on water allocation next to impossible under current situations, he proposes an alternative conceptualization that emphasizes the economic value of water. His model shows that this approach provides a basis under which negotiations might be separated from other issues, thereby allowing adequate water distribution to take place. Moreover, Fisher suggests that recognition of benefits from initial cooperation on water could serve to build confidence and lead to cooperation in other domains. The analysis of Turkish-Syrian relations in Serdar Güner's chapter ("Evolutionary Explanations of Syrian-Turkish Water Conflict") somewhat tempers this optimistic vision. Using game theory, Güner argues that the confrontation between "hawks" and "doves" in each population leads to stalemates. In his conception, the structure and the dynamics will remain conflictual rather than cooperative even if the situation evolves.

Confrontations do not only emerge from disagreements over resources but can stem from profoundly different perceptions of risks related to the physical properties of resources. Raymond Dacey discusses clashes between experts and lay people in his chapter on "Water Use and Risk: The Use of Prospect Theory to Guide Public Decision-Making." His analysis is embedded in the large literature on theories of risk and decision-making. Dacey applies a particular variant of these, prospect theory, to explain observations that do not correspond to the standard expected value theories of Pascal (Arnauld and Nicole 1662) or von Neumann and Morgenstern (1947). Notably, he is able to account for disputes that emerge over large scale development or conservation projects, both of which have environmental consequences and could result in water-related disasters. His models illustrate that the way different groups process information leads them to maximize different value functions and therefore arrive at different decisions. Elucidating the decision process is one step in the resolution of oppositions that otherwise paralyze decision-making or lead to catastrophes.

5. INDIGENOUS KNOWLEDGE, TECHNICAL SOLUTIONS

Studies of specific empirical cases underscore the importance of anchoring analyses in theory and applying rigorous methodologies. The chapters on Peru and Ethiopia ("Disaster, Development and Glacial Lake Control in Twentieth Century Peru" by Mark Carey and "Wetlands and Indigenous

Knowledge in the Highlands of Western Ethiopia" by Alan Dixon) provide clear evidence of the role of local expertise and indigenous perceptions in managing local environments. In Ethiopia, research on indigenous management practices reveals their deep knowledge of wetland environments and elucidates the learning processes that have allowed adaptation to small-scale environmental changes. Meeting sufficient production goals is hampered more by poverty and development considerations than by lack of understanding of the wetlands system.

The focus of Carey's Peruvian study is on the public's response to periodic catastrophes caused by glacial lake outburst floods. Knowledge about glacier hazards is present in the population but response strategies diverge from those proposed by scientists and engineers. Carey's study is a concrete illustration of the discrepancies among different groups analyzed by Dacey. Application of the methodologies Dacey proposes could help resolve confrontations that at present remain as potential hindrances to mitigation strategies. However, Carey's empirical study highlights additional complexities by identifying several stakeholder groups, which would have to be introduced into Dacey's two-actor formulation. Not only are there lay populations and experts; there are different kinds of users, including tourists, agriculturalists, and hydroelectric producers, government officials, and scientists, each with their own knowledge base and interest. The power relations among these groups evolve according to their histories and the wider political and economic context, leading to the tensions documented by Carey.

What emerges from both case studies is the importance of understanding the physical phenomena that produce particular environments. Knowledge is important to be able to maximize resources over the long term and to respond to periodic catastrophes in the short term. This knowledge can lead to institutional developments, such as property regimes or solidarity mechanisms. It can also stimulate technological developments for more efficient resource use. One technological enhancement of water resources is presented by Michel Dubas ("A New Ancient Water Mill: Remembering Former Techniques"), who describes an ancient technology developed to maximize water power. Dubas' chapter reminds us that our ability to imagine solutions to current problems often rests on our recognition that earlier populations had previously confronted them and devised responses. In the case of irrigation systems that are described by Trawick and Wiegandt in this volume, the nature of the solutions had profound and lasting impacts on institutional organization. Tracing sources of knowledge thus tells us about more than the technique itself; it also allows for reflection about how ways of life have been shaped by particular understandings of nature.

This is evident in our ongoing efforts to understand hydrological dynamics in order to master or contain their behavior at the margins. Not only are there difficulties related to water shortages, there are also negative impacts of overabundance. The most obvious examples are flood and debris flows. Two chapters deal with the special vulnerabilities of mountain populations and propose strategies to anticipate or mitigate damages. In "Water Related Natural Disasters: Strategies to Deal with Debris Flows: The Case of Tschengls, Italy," Walter Gostner et al. emphasize the importance of understanding the dynamics of debris flows. They look at the chronology of past flows and identify how people responded by building structures to contain flowing water and debris. This historical account is accompanied by a model integrating causal factors in order to anticipate events and propose timely mitigation measures.

Salvatore Manfreda and Mauro Fiorentino ("Flood Volume Estimation and Flood Mitigation: the Adige River Basin") undertake an estimation of flood volumes, arguing that it is the most important variable for the design of effective flood mitigation strategies. They complement this with probabilistic models and hydrological simulations, and introduce a model for making statistical estimates of flood volumes. They argue that their use of multiple and complementary analyses can reduce uncertainties and therefore contribute to more effective flood management strategies.

Managing water as a resource and not as a danger can also benefit from quantitative assessments, as demonstrated by Tatiana Orehova and Elena Kirilova Bojilova in their chapter, "Hydrological Assessment for Karstic Springs in the Mountain Regions of Bulgaria." They note that over the last 25 years or so, there has been a continuous decrease in rainfall and increase in air temperature that has led to reductions in river flow. Since Bulgaria's karstic springs are important sources of domestic water, these recent drought conditions may suggest future trends under global climate change and thus merit special attention. They present seasonal measurements of the impact of climate variability on spring discharge rates over an observation period and conclude that results can be used for the estimation of long-term variability.

6. POLICY IMPLICATIONS FOR EFFICIENT AND EQUITABLE WATER USE

Identifying historical patterns, refining methodologies to describe and predict hydrological dynamics, and elaborating models of human behavior toward resource use and risk are all essential components for the design of effective policy. Such scientific endeavors are also important to address current disputes over water access and to generate solutions for the predicted impacts of climate change. Any efforts to assure fair and adequate allocation of an indispensable resource, and thereby reduce potential for conflict, are predicated on an accurate assessment of water quantity and dynamics. Only rigorous scientific study of the physical resource and of human behavior and social institutions can lead to adequate policies. "Sources of water, sources of knowledge," mountain regions are critical because they physically hold much of the world's fresh water. They also contain populations that have come to understand much about water dynamics and have developed strategies to manage the resource and confront dangerous conditions that water sometimes generates. Most of the chapters in this volume illuminate different facets of this local knowledge and suggest enhancements through continued scientific study. They thus contribute indirectly to the goal of developing more effective policy to confront existing and emerging problems of supply and allocation.

The two final chapters address these issues directly. Urs Luterbacher and Dusan Mamatkanov ("Water and Mountains, Upstream and Downstream: Analyzing Unequal Relations") focus on the special asymmetries present in international relations of water use. The case they analyze is Kyrgyzstan, where tensions are acute but outright conflict has so far been avoided. This provides an opportunity to outline the conditions that create inequalities and to propose solutions that could potentially prevent escalation. Some of these solutions overlap with those proposed by other authors in the volume, including Fisher ("Water Value, Water Management, and Water Conflict: A Systematic Approach") and Karina Schoengold and David Zilberman ("Creating a Policy Environment for Sustainable Water Use"). All three contributions include cost benefit analysis as a means to provide incentives for efficient use and to give clear signals for bargaining among otherwise opposing partners. In the case of Kyrgyzstan, important advantages could be gained if both upstream and downstream users maximized their respective energy and irrigation capacities and needs. Schoengold and Zilberman propose a set of mechanisms and institutions that could improve efficiency and enhance equity by assuring that maximum resources are distributed according to transparent and accepted criteria. They point out that the inherent difficulties of an open access resource for which clear property rights are difficult to define complicate the elaboration of well-functioning schemes. In this context, better scientific understanding of hydrological conditions and their evolution—as well as of extremes of under- and oversupply—are essential to the design of effective policy.

The goal of this volume is to provide an integrated picture of the interactions between physical processes and human behavior influencing water dynamics. Broad and deep knowledge of both natural phenomena and human behavior are essential to attenuate struggles over water. Identification of several key themes served to organize the chapters in the volume. What is obvious, however, is their inseparability. There can be no efficient management if the evolution of water quantity and quality is not known. Conflict cannot be avoided if allocations are inequitable and if incentives lead to levels of use that exceed supply. The chapters in this volume bring together natural and social sciences, the past and the present, theory and practice to argue that integrated conceptualizations, methodologies, and projects can contribute to better understanding of the indispensable resource water. It is a first step in enhancing our capacity to confront shortages and attenuate struggles for this vital resource.

"WATER TOWERS"—A GLOBAL VIEW OF THE HYDROLOGICAL IMPORTANCE OF MOUNTAINS

Daniel Viviroli
Department of Geography
University of Bern, Switzerland
viviroli@giub.unibe.ch

Rolf Weingartner
Department of Geography
University of Bern, Switzerland

Abstract: Mountains and highlands are often referred to as natural "water towers" because they provide lowlands with essential freshwater for irrigation and food production, for industrial use, and for the domestic needs of rapidly growing urban populations. Therefore, better knowledge about mountain water resources in different climatic zones is essential for adequate management of these resources. In this chapter, a data-based approach is used that enables the quantification of the hydrological significance of mountains. The study reveals that the world's major water towers are found in arid and semi-arid zones and that pressure on mountain water resources in general will sharpen due to climatic change, population growth, and competing use.

Keywords: mountains, hydrology, runoff, highland, lowland, water resources, comparative assessment

1. INTRODUCTION

Most of the rivers on our planet originate in mountain regions. The discharge, which builds up in the mountains, is transported via river systems to lower lying areas where a large fraction of mountain water is used for irrigation and for food production. Although mountain regions—defined as areas more than 1,000 m above sea level—make up only 27 percent of the Earth's continental surface (Ives et al. 1997), the share of the world's population they supply with water largely surpasses this value. For this reason, mountains are often referred to as natural "water towers."

It is generally agreed that mountain regions, with their disproportionnately high discharge compared to lowlands, are of significant hydrological importance. However, as far as quantification of this significance is concerned, there is a good deal of uncertainty in the scientific world (e.g.,

E. Wiegandt (ed.), Mountains: Sources of Water, Sources of Knowledge, 15–20.

Rodda 1994). A recently published study estimated the proportion of mountain discharge to global total discharge at 32 percent (Meybeck et al. 2001), while other studies indicate figures between 40 and 60 percent (Bandyopadhyay et al. 1997). Within a region, mountain discharge can represent as much as 95 percent of the total discharge of a catchment (Liniger et al. 1998). On a global scale, few measurement series exist for discharge in mountainous regions and the periods they cover are extremely limited. This restricted data base does not adequately represent the high degree of spatial and temporal heterogeneity of discharge conditions in mountain areas. Additionally, in water-scarce regions discharge data has a high strategic value that is frequently kept secret. This makes basic scientific studies more difficult and mitigation of conflicts over water resources impossible.

2. QUANTIFICATION OF THE "WATER TOWERS"

On the basis of knowledge gained from studying the hydrology of the Alps (Viviroli 2001), a data-based approach to assessing the hydrological significance of mountains was adopted using discharge data provided by the Global Runoff Data Centre (GRDC 1999). The pattern of mean monthly discharge, changes in specific discharge with increasing catchment size, and the variation in mean monthly discharge proved to be particularly suitable parameters for assessing the hydrological significance of mountainous regions. More than twenty river basins in various parts of the world were selected for case studies on the basis of climatic and topographical criteria and availability of data. The choice of the case studies aimed at covering a wide range of climatic zones and the most important mountain ranges. The inner tropical area with the two major rivers, Amazon and Congo, was omitted because tropical rains clearly dominate the hydrograph and override mountain influences. The most restrictive criteria proved to be the presence of accessible, reliable, and representative data with gauging stations suitably distributed across the river course.

The interrelation between mountain and lowland discharge for each case study was examined through a gauging station located above an altitude of 1,000 m, which served as a "mountain station," and a second one in the vicinity of the river mouth, which served as a "lowland station." In addition, regional precipitation and temperature conditions were taken into account to incorporate the discharge regime into the climatic context of the region.

There are three particular hydrological characteristics of mountain areas. The first is a disproportionately large discharge, typically about twice the amount that could be expected from the areal proportion of the mountainous section. Mountain discharge portions ranging from 20–50 percent of total discharge are observed in humid areas, while in semi-arid and arid areas the contribution of mountains to total discharge amounts to 50–90 percent of total discharge, with extremes of over 95 percent (Figure 1). Second, discharge from mountainous areas is highly reliable and causes a significant reduction of the coefficient of variation of total discharge. Third, the source of the discharge is linked to the retarding effect of snow and glacier storage, which transforms winter precipitation into spring and summer runoff and is essential for the vegetation period in the lowlands.

These and other basic characteristics—including the extent of human utilization of mountain runoff and regional precipitation and temperature patterns—have been analyzed and quantified (Viviroli 2001), and were used to elaborate an overall assessment of the hydrological significance of mountain areas (Figure 2). The study reveals that the world's major water towers are found in arid and semi-arid zones. The drier the lowlands, the greater the importance of the more humid mountain areas (Viviroli et al. 2003). High-resolution demographic data (CIESIN/IFPRI/WRI 2000) show that—if the densely populated plains of the Indus, Yellow and Yangtze Rivers are included—about 70 percent of the world's population

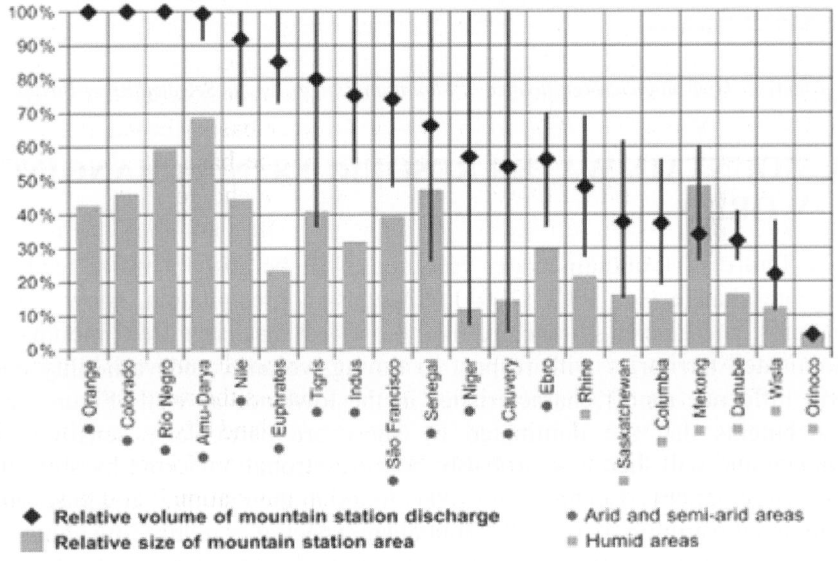

Figure 1: Mean Annual Contribution of Mountain Areas to Total Discharge and Mountain Area as Proportion of Total Catchment Area.[1]

[1] The vertical lines denote the maximum and minimum monthly contribution to total discharge.

lives between 30 degrees northern and southern latitude where drought-prone areas are frequent. This implies that mountain waters are essential and that pressure on mountain water resources will rise, especially in these zones because population growth is also more intense there.

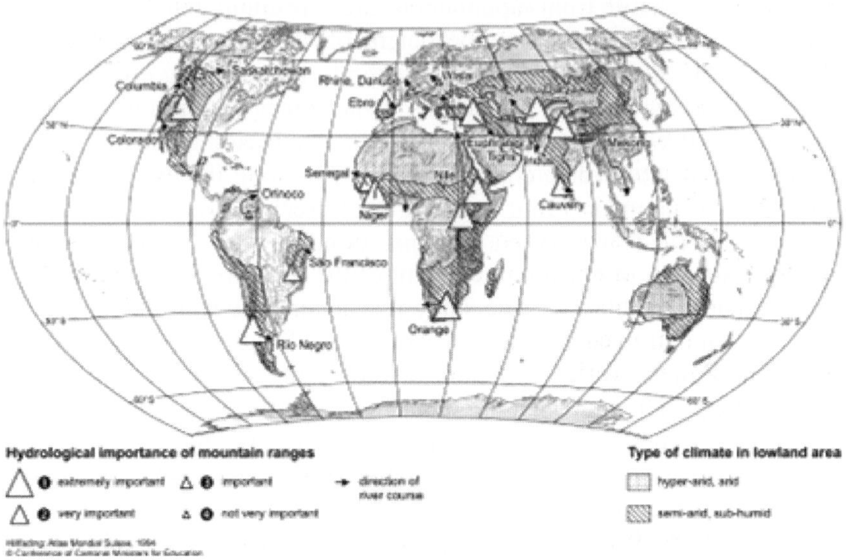

Figure 2: Hydrological Significance of Mountain Ranges for the Selected River Basins.

3. MOUNTAIN WATER RESOURCES IN A CHANGING WORLD

The future of mountain water resources will be sharply influenced by climate change (see IPCC 2001). Changes in precipitation, snow cover patterns, and glacier storage are likely to affect discharge from mountain-dominated territories with respect to timing, volume, and variability and will influence runoff characteristics in the lowlands as well (Figure 3a). Catchments that are dominated by snow are particularly sensitive to change and will therefore probably be most strongly affected by shifts in discharge patterns. Leung et al. (2004) found in their study that, along with a regional warming of 1–2.5°C, annual snowpack was reduced by about 70 percent in the western US coastal mountains by mid-century, leading to less snow accumulation and earlier snowmelt during the cold season—and thus reduced runoff and soil moisture in summer. It must be noted, however, that the reliability of current estimates dealing with discharge as influenced by climatic change is limited because of the uncertainty of regional climatic forecasts and because of the large scales used in global circulation models (GCMs), which are difficult to transform to a local scale (Arora and Boer

2001; Nijssen et al. 2001). In addition, due to the uncertainties involved in climate change scenarios, research on mountain water resources is further hindered by the problems occurring when applying hydrological models to snow-covered areas and higher altitudes, irrespective of whether meso-scale (Zappa 2002) or global-scale approaches (Döll et al. 2003) are taken.

Global climate change is not the only factor to influence mountain water resources. Population growth in critical lowland areas will strongly accentuate pressure on mountain water resources (Falkenmark and Widstrand 1992; Vörösmarty et al. 2000), especially with regard to increasing food demand and changing dietary habits (Yang and Zehnder 2002) but also because of competing use for hydropower generation and industrial use. This pressure will foster the construction of engineering works in mountainous areas, such as dams and river channels and development of large-scale irrigation schemes and energy production projects. With today's engineering skills, even large-scale river water transfers such as India's "River Link Project" (originating from the Brahmaputra and Ganges in the north to the south and west of India) or the "South-to-North Water Transfer" in China (from the Yangtze River to the Yellow River) become possible. These measures are, in turn, expected to cause changes in runoff regimes in lower lying areas with effects on water availability in the lowlands (Figure 3b) and with the potential to increase conflicts and foment political tensions. The possible constellations of power and intents of upstream and downstream users are manifold. The Southern Anatolia Project marks an example where regional development is taking place in the upstream area (Turkey) by means of a large-scale hydropower generation and irrigation project (Altinbilek 2004) and where downstream countries (Syria, Iraq) are in a weaker position. In the Nile basin, the announcements of some upstream countries (Ethiopia, Tanzania, Kenya) to make more use of the Nile flows provoke harsh retorts from the more powerful downstream country (Egypt), where the dependence on the mountain discharge is vital (Pelda 2004).

Population growth within mountain regions is also likely to result in an intensification of land-use (Figure 3b2), which in turn will lead to soil degradation and erosion. The possible consequences of this phenomenon, such as floods and poorer water quality, will affect mainly local and regional levels with only minor impacts on lower lying areas (Hofer 1998).

Although it is difficult to state specific political consequences, there are a number of issues that seem to be of importance for future policy making. With regard to climate change, an effective climate policy is imperative to prevent fundamental changes in the availability of mountain water resources. Moreover, better monitoring of vulnerable mountain areas is needed because they provide a highly sensitive warning system for environmental changes (Messerli et al. 2004). Population pressure fosters increased efforts to

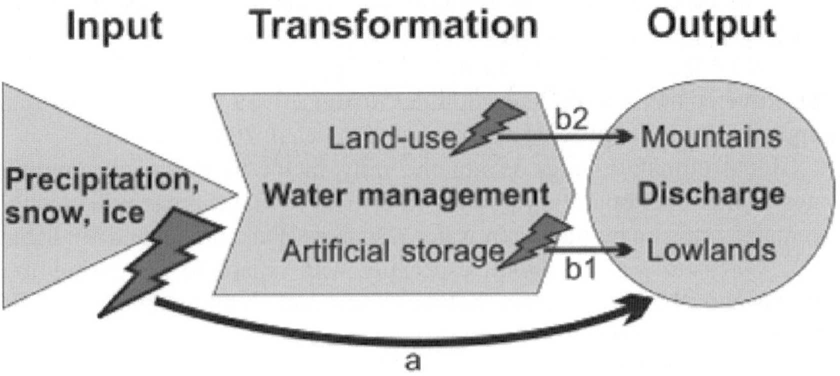

Figure 3: Future Threats to Mountain Water Resources: Climatic Change and Population Growth. (a) Climate change affects discharge in mountains as well as in lowlands; (b1) water management changes caused by population growth alter runoff in lowlands in case of extended artificial storage (b2) or in mountains in case of land-use changes. See text for further explanations.

impound mountain water resources, with dire consequences. These will have to be mitigated through more cooperative negotiation endeavors and the commitment of multilateral basin riparian organizations. Finally, it is in the interest of lowland areas to support the economic health of mountain regions so that their resources are used in a sustainable manner, which would maintain the quantity and quality of mountain runoff.

Despite the fact that global water resources are sufficient today, there exist wide spatial and inter-temporal disparities in its availability. Thus, marked increases in local and regional water scarcity can be expected in the future (OECD 2001). As the world's water towers, mountains will continue to play a paramount role in meeting the increased demands for food production, drinking water, energy supply, and industry in the twenty-first century (Viviroli and Weingartner 2002). Access to clean fresh water is a basic human right, one which is not yet available to all and will be even more difficult to guarantee in the future. Thanks to their climatic characteristics, the mountains on our planet play a special role in the water cycle, and therefore water management must start in mountain regions. There is also a complex interaction between mountains and lowlands that needs to be recognized. This interaction should be given paramount consideration in planning the development of resources (macro-scale watershed management), and the decisive role of mountains should be taken into account at future conferences on water resources and in conflict management. As a basis for deepened process knowledge and for planning, there is great need to improve the current monitoring of mountain water and climate, along with making data publicly available and for knowledge to be exchanged between neighboring countries and between highland and lowland areas.

THE HISTORY OF IRRIGATION AND WATER CONTROL IN CHINA'S ERHAI CATCHMENT: MITIGATION AND ADAPTATION TO ENVIRONMENTAL CHANGE

Darren Crook
Division of Geography & Environmental Studies
University of Hertfordshire, United Kingdom
d.crook@herts.ac.uk.

Mark Elvin
Research School of Pacific and Asian Studies
Australian National University, Australia

Richard Jones
Department of Geography
University of Exeter, United Kingdom

Shen Ji
Nanjing Department of Geography and Limnology
People's Democratic Republic of China

Gez Foster
Department of Geography
University of Liverpool, United Kingdom

John Dearing
Department of Geography
University of Liverpool, United Kingdom

Abstract: This chapter introduces an interdisciplinary methodology that combines the use of archaeological and documentary sources alongside environmental proxy indicators found in sedimentary archives to assess, on a hydrological catchment scale, historical human impacts on hydrology. The advantages and benefits of this technique are demonstrated through the results taken from ongoing work on a case study, Erhai in Yunnan province, China. This approach allows us to increase understanding of local knowledge, vulnerability, mitigation, adaptation, and resilience to local, regional, and globally derived environment and climate change.

Keywords: Erhai, human impact, catchment hydrology, environmental microvariation

E. Wiegandt (ed.), Mountains: Sources of Water, Sources of Knowledge, 21–42.

1. INTRODUCTION

Periods of environmental crisis and perturbation in China and Southeast Asia are intrinsically linked to the issue of water control (Elvin et al. 1994). Similar processes are seen throughout the world, and indeed, human activities have altered almost all of the world's river systems with major modifications to hydrological catchment areas. This global and regional problem has been exacerbated by loss of tree cover in watersheds, dam construction, reduced storage times of water in river basins, and increased severity and frequency of flooding. The massive floods (and droughts) in the lower Mekong River basin have been blamed to some extent on widespread deforestation in watershed areas, poor soil management practices, reclamation of flood plains and wetlands, and the rapid expansion of urban areas (Blake 2001). There is also growing international concern about the impacts of large dam schemes in Yunnan Province, China, on the Mekong watershed ecosystem (Lihui 2004). Such concern provides a pressing need to understand natural resource use, in particular water resources, in the large tributaries of the Mekong River in order to assess environmental vulnerability and resilience and to improve environmental security for those people living in these areas. This can only be achieved by understanding the geo-historical context in which contemporary situations have arisen.

Erhai catchment in Yunnan province, China, is a tributary of the Mekong River, with a long and fairly accessible environmental history. It is a region facing new socioeconomic and environmental challenges with large infrastructural development projects being planned, increasing permanent and transitory (tourist) populations[1] and growing concerns over water quality and quantity (UNCRD 1994). It is also a monsoon region threatened by rapid global climate change, which according to various climate models will lead to increased precipitation and more frequent and intense extreme events (Handmer et al. 1999; Arnell et al. 2001; Cubasch et al. 2001) as experienced in China during the summer of 2004 (WMO 2004). Thus, appropriate environmental knowledge and planning are essential if the long-term environmental security of this region is to be assured.

---

[1] These population increases are due mainly to the one-child policy not applying to minority nations like the Bai and to some extent to the Go West policy operated by China, which has spurred massive investment in Dali City, formerly known as Xiaguan, and led to increases in migration by Han Chinese to the area.

Our case study investigates the long-term impacts of climate and land use change on hydrology and, in particular, seeks to increase understanding of local knowledge, local community vulnerability, mitigation strategies and adaptation and resilience to local, regional, and globally derived environmental and climate change.[2] The history of irrigation and water control in the Erhai catchment is intrinsically connected to this question, and while the historical record of irrigation and other water issues in the Erhai catchment is fragmented, it nevertheless has value because its long-term perspective highlights some of the emergent issues of environmental change that are not otherwise apparent in shorter term studies. Specific aims of the project[3] relevant here include answering the following questions:

- To what extent have major shifts in hydrological processes and water quality in the historical past been triggered by human activities or climate?
- In terms of runoff and sediment generation, which parts of the landscape are most sensitive to extreme climate or human disturbance?

2. SITE DESCRIPTION OF ERHAI, NORTH AND WEST SHORES

Erhai is found in the Dali autonomous prefecture in Yunnan Province. The lake sits at ~1,970 m above sea level (as measured from Haiphong[4]) with an approximate area of 249 km^2 and a catchment of about 2,565 km^2 (Figure 1). Both the depth and area of the lake have varied during the

---

2 Our case-study is part of the IGBP/PAGES Focus 5 initiative on Human Impacts on Terrestrial Ecosystems. The project adopts an interdisciplinary methodology designed to permit global application and draws on documentary and sedimentary archives to investigate the historical impact of humans and climate on water resources in mountainous river basin environments.

3 See the LERCH website http://pcwww.liv.ac.uk/%7Edcrook/lerch/Erhai/ for details of the wider project.

4 In general, the Chinese seem to use Wusong as the benchmark for mean sea level (it is located near the mouth of the Yangzi). For this project, it was necessary for the UK research team to use Soviet maps (1970s) of China, which had discrepancies in summit and other heights from the maps used by Chinese colleagues. However, the maps seemed very similar in most ways. It was assumed that the Soviet maps were based on old French maps (the French having been dominant in the Southwest because of their occupation of Vietnam, and a sort of sphere of influence in Yunnan), and that the French had put in heights above MSL that were based on Haiphong.

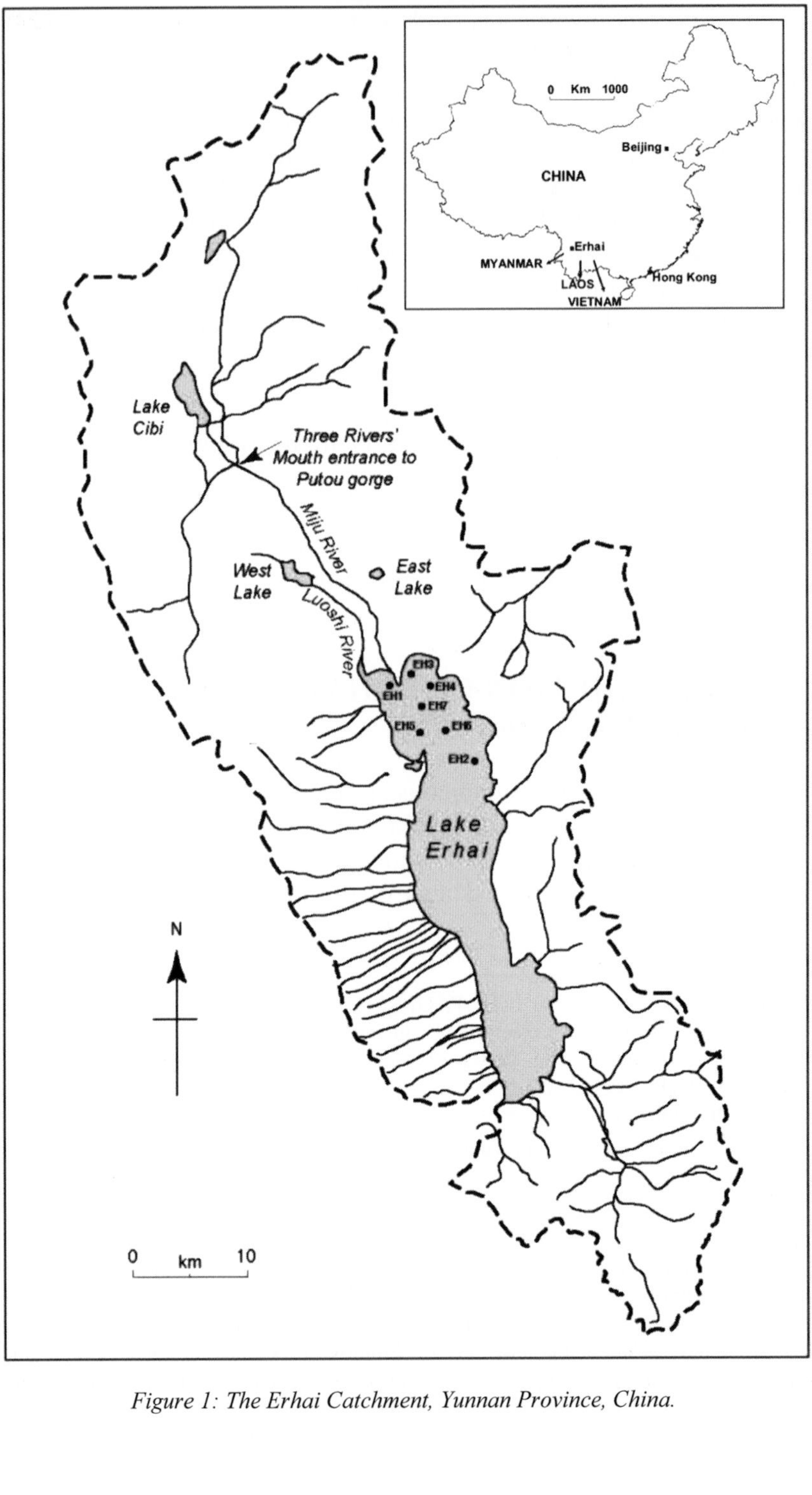

Figure 1: The Erhai Catchment, Yunnan Province, China.

Holocene period (Duan Yanxue 1989). To the north of the lake the heavily embanked and elevated Miju River draws upon an extended northern sub-catchment made up of the lower Dengchuan plain and the slightly elevated Eryuan plain, and the mountain ranges that define the northern catchment. The Miju contributes 5.18×10^8 $m^3 year^{-1}$ at the inflow into Erhai via a rapidly aggrading delta. The Miju is flanked by the smallish Luoshi River, which drains the West Lake and the Yong'an River, draining the East Lake on the Dengchuan plain. These "lakes" are not single sheets of water but interconnected groups of small lakes and ponds. The Dengchuan plain is linked to the Eryuan plain by the Putuo gorge. Lake Cibi is located on the Eryuan plain and the upper slopes of this basin form the headwaters of the Miju. The Miju River feeds an extensive river valley offtake irrigation system.

The general climatic pattern is characterized by a dry season from November to May and a rainy season from June to October, during which more than 80 percent of the precipitation falls. The west side has a wet climate with an average annual precipitation of ~1,072 mm, although rainfall rises from 1,000 mm year close to the lake shoreline to 1,800 mm near the crests of the famous Nineteen Peaks of the Diancang mountains, which rise to over 4,000 m. Seasonal snowmelt and springs supply water to an extensive slope offtake diversion irrigation system on the Dali plain, known locally as the famous Eighteen Streams (Xue Lin 1999). These steep gradient streams contribute 2.76×10^8 m^3 of water a year to the lake. Just one of these, the Wanhua Stream, has been estimated to reach a once-in-fifty-years maximum of 158.4 m^3 s^{-1}. Mean annual runoff has been calculated at above 200 mm $year^{-1}$ on the west side, but under 100 mm $year^{-1}$on most of the north side where average annual precipitation drops down to ~763 mm on the Eryuan plain. Seismic activity is frequent, and there are many documentary references to the damage done by earthquakes. The population most vulnerable to these hazards is today mainly centered on the Dali plain, which has 436,000 inhabitants and a density of 299 persons/km^2 in 1990; and on the Eryuan plain in the northern part of the catchment, which has 299,000 inhabitants and a density of 104 persons/km^2 in 1990.

3. METHODOLOGY

It was necessary to adopt an interdisciplinary approach that investigated archaeological, documentary, and sedimentary archives. This was regarded as the best way to deal with varying temporal and spatial scales of enquiry and allow for a full appraisal of a complex system. The north, east, west, and south shores of Erhai have different environmental characteristics as well as varying geomorphological, geographical, and hydrological features (see Elvin et al. 2002). The summarized story from only the north and west sides of the lake are told here because of limitations on the amount of space available. This has the advantage of matching the archaeological and documentary record to the same boundaries and spatial scale as the sedimentary analysis. It also covers the most important hydrological sectors of the catchment. Space limitations also prevented the inclusion of climate and tectonic data into this chapter.

Whilst archaeological sources survive from the Han dynasty approximately 2,000 years ago, the earliest documentary source relate to the early Tang dynasty (~ 628 CE). Documentary sources are fragmented and often not written with the objective of describing the natural milieu or to provide specific environmental information. However, with qualifications arising from a diversity of different sources (Table 1) that reflect the turbulent political history of the area (e.g., Drochon 1866; Litton 1903; Rocher 1904; Wiens 1967), this allowed us to explore social and cultural variables alongside the interactions of technical and environmental factors (see Fitzgerald 1941). The main focus lies in the subset of ecosystem reactions, usually measured as individual events, vigorous enough to be worthy of recording in the historical record, and to analyze their distinctive patterns and probable causes.

Table 1: Archaeological and Documentary Sources Used in this Study.

Type of evidence	Source
Archaeological	Lithics – steles
	Pottery
Documentary	Chinese gazetteers
	Official Chinese documents
	Lists of auspicious and uncanny events
	Minority Nation Ethnographies
	French/British Catholic and Protestant Missionary archives
	French/British colonial consular archives
	Travelogues
	NGO and GONGO reports
Cartography	Pictograms and maps

To access the longer record of prehistory and to observe longer term trends, sedimentary archives were analyzed. Sediment extraction techniques (Last and Smol 2002a) were followed by the principle analytical techniques of pollen, particle size, and mineral magnetics (see Berglund et al. 1986; Thompson and Oldfield 1986; Smol et al. 2002; Last and Smol 2002b). Sediments were collected from Erhai, the Miju River floodplain, and upland areas of the catchment. For the purpose of this chapter we concentrate our analytical attention on one of these cores known as EH2, which was taken from the deepest part of the northern basin at Erhai. It is thought to be representative of the northern part of the catchment as it is located close to the inflow of the main tributary river (Figure 1).

The core chronology is still under construction, but two range finder dates have been obtained based on Carbon-14 dating of the humin and humic organic fractions of the sediment. The two range finder dates are shown in Figure 2.

Thus, the event oriented archaeological and documentary evidence complements the theoretically continuous environmental archives provided by sedimentary archives, providing a reasonably representative chronology of environmental and hydrological change.

4. PRELIMINARY RESULTS FROM THE SEDIMENTARY ARCHIVE

A summary of the sedimentary results from EH2 are presented in Figure 2, which includes a preliminary zonation scheme based on variations in the mineral magnetic and geochemical features distinguished as sediment units 1 to 5. The zonation scheme reflects changing sediment composition up through the core, driven by shifts in the source of sediment to the lake from the catchment. The poor chronological resolution of the results is mitigated somewhat by the serendipitous finding of dateable material at the beginning of the two major inflexions in the magnetic records.

The lower zonal boundary between sediment units 1 and 2 is characterized by a marked increase in χarm (a proxy for magnetic mineral grain size) and iron (Fe) concentration in the sediment from their respective values in ~2650 BP. This trend continues through to the top of the core and appears unaffected by events after 1950 BP. It is further mirrored in the Zr/Ti ratio (a crude indicator of grain size), which appears to show a gradual increase in the clay content of the sediment. This implies that the observed trend may be due to a long-term shift in sediment source and/or supply to the lake. The cause at present is unknown but it is hoped that completion of the sediment analyses will provide the answer. Pine frequencies are seen to fluctuate whilst experiencing an overall decline, which possibly points to the onset of deforestation in the catchment. However, the low values for χfd (a proxy for the presence of magnetic material of a pedogenic origin derived from the catchment through erosion, e.g., Thompson and Oldfield 1986; Dearing 1999) indicate that the supply of topsoil material to the lake during this period was negligible, implying a degree of catchment stability.

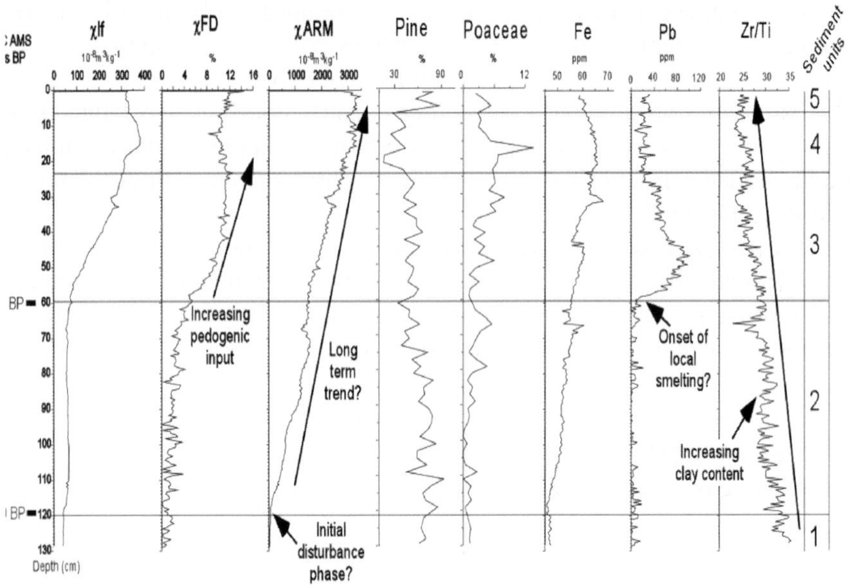

Figure 2: Results from the Sedimentary Archive.

This is clearly not the case in sediment unit 3 which is characterized by a marked and sustained increase in χlf (magnetic susceptibility) and χfd, pointing to the onset of a major period of soil erosion in the catchment at ~1950 BP. Sediment zone 3 is characterized by a marked increase in *Poaceae*, with a greater contribution of grains >37um, which may reflect the presence of cereals (data not shown). *Pinus* values remain high throughout this zone, even increasing slightly at times. Pine would certainly be expected to respond favorably to increased disturbance in the catchment allowing it to out compete previously dominant taxa particularly *Tsuga*. A significant long-distance transported pine component would also be anticipated. A more unequivocal anthropogenic signature can be seen in the geochemistry record, highlighted by the sharp increase in lead (Pb) concentration. With maximum values of ~100 ppm, too high to be simply a local bedrock signature, such a rise in Pb may reflect mining and/or smelting activity in the catchment. The earliest documentary references to silver mines and lead workings that directly relate to the Erhai catchment come from the Nanzhao period, which dates to approximately 649–902 CE.[5] However, earlier direct production of lead near Dali is recorded

[5] When there was an appropriate mixture of lead with the silver in the ore, as was usually the case, the first smelting produced an alloy of lead and silver, the second separated the two, and a third concentrated the silver further, often with the addition of a small amount of

(no clear date) for a lead mine in Heqing (Li Hao, 2002, p. 17), which is outside the border of the Erhai catchment, but nonetheless smelting activities could have taken place in the catchment as they did during the Qing dynasty (editors' note to Li Yiheng 269). The peak in silver mining in the region appears to have occurred during the period of the Dali Kingdom (960–1279 CE) (Li Hao 2002, p. 188). The limited historical data thus make a prima facie case for an upsurge in silver production in the Erhai catchment in or before the seventh century CE, a maximum in the later first millennium CE and the early part of the second, followed by a decline to the point of eventual nonexistence during the eighteenth and nineteenth centuries. This later decline may be reflected in the lead record at the transition zone between sediment units 3 and 4.

It is fascinating to note that such changes appear to be superimposed upon the longer-term trends, which commenced in sediment unit 2. Indeed, the protracted increase in χarm and Fe appears unaffected by the pronounced changes that characterize sediment unit 2, implying we are dealing with two completely separate signals operating on different temporal—maybe even spatial—scales.

A third successive shift in sediment source is highlighted in sediment unit 4, characterized by a marked increase in χlf and a slight decline in χfd. This shift coincides with a major phase of deforestation, with pine percentages falling below 30 percent for the first time in the core. This shift is paralleled by a significant rise in *Poaceae* (<12 percent), possibly pointing to the onset of a major agricultural phase in the catchment. The sustained levels of soil erosion in the catchment throughout this zone suggest that people may have expanded agriculture onto previously untouched areas of the catchment, such as upland slopes. This could also possibly represent a period of increased flood frequency. The upper most sediment unit (5) is characterized by a sharp increase in Pine values to a level comparable to those recorded in sediment unit 1. Such a rise may reflect a major period of reforestation aimed at stabilizing the catchment. Interestingly, despite this χfd values remain high; indeed, the highest values are recorded at the top of the core, which suggests that large amounts of topsoil were still being moved around the catchment at this time. A ^{210}Pb date at the top of the core provides an age of 1960 CE preliminary set at about 6 cm, indicating the observed changes in this zone may be linked to events initiated by the grandiose development campaigns of Mao Zedong, like the Great Leap Forward (see Shapiro 2001).

---

copper. The lead removed was called "silver-mine lead" and was abundant in Yunnan (Song Yingxing 238, 252).

5. ARCHAEOLOGICAL AND DOCUMENTARY EVIDENCE FOR ENVIRONMENTAL CHANGE

5.1. Early Hydrological Change on the West and North Sides of the Lake

Erhai developed in the late Pleistocene as a result of faulting in a pre-existing river valley. This increased its water levels to those higher than it is today, and which subsequently fell through the early and middle Holocene (Duan Yanxue 1989). First signs of human activity in the catchment appear in the Neolithic with a number of sites found on upper slopes of the mountains, one of which has been Carbon-14 dated to a mean of 3770 BP. The date of the initial clearance of lowland forest and dense undergrowth is not recorded, although it is reasonable to guess some time not later than the first millennium BCE. However, on the west side of the lake permanent settlements have been discovered, with a component of farming, dating back at least three millennia and with evidence of Chinese-style irrigated agriculture from at least the Han dynasty, dated at approximately 2,000 years ago (Yang Dewen 1988). The first recorded administrative unit in the catchment, Yeyu County, was established in the first century BCE by Emperor Wudi of the Western Han dynasty (206 BCE–220 CE).

Further (major) development of the region did not occur until around 745 CE when the independent[6] Southern Kingdom (Nanzhao) was established in the Erhai area (Rocher 1904). Once its lower shores were developed, the western side became the location of a number of political capitals. This was because the western shore provided the conditions for productive irrigated agriculture on a scale large enough to underpin a modest-sized polity and a thriving and urbanizing local economy (Fang Guoyu 1998). This development was accompanied by the beginnings of deforestation and the control of streams and floods. Drainage and breaching of pools and ponds helped turn upland plains into fields and a combined pastoral and irrigated agricultural economy was in operation. The first recorded irrigation system of any substantial size in the catchment, the famous 18 Streams, did not appear until the ninth century CE

[6] After the battle in 745 against the Tang forces, which ended in bloody but absolute victory for the Nanzhao forces, Nanzhao, which could be said to have been under the loose suzerainty of the Tang up to that point, became effectively independent, though for a time reliant on an alliance with the Tibetans. This independence lasted until the Mongol conquest in the thirteenth century.

(see Figure 1). This demonstrated a growing commitment of the state to hydraulic systems during a period when the hydrological regime was possibly greater than today (Elvin et al. 2002, p. 12). Integral to this system was a north-south channel 16 km in length linking 11 of the streams coming down the east-facing slopes of the Diancang Mountains.

The situation on the north shore was different with the land near Erhai, mostly an area of marshland (Li Zhengqing 1998). These marshlands were drained around 737 CE by assembling several tens of thousands of men to cut more than 5 km of channel through Bell Mountain (Yin Ren 1902), which diverted the outflow from West Lake in a different direction to the Luoshi River, eventually to debouch into Erhai (Dengchuan zhouzhi 1854). This act was said to have greatly improved soil drainage and fertility (Zhou Kang 1902). The question posed here is when did these environmental improvements turn into environmental problems?

5.2. The Ming Prelude to Environmental Crisis

An intensifying pressure on resources occurred during Ming times (1368–1644 CE), and maybe earlier (Liu Wenzheng 1991; Li Zhengqing 1998; Duan Jinlu 2000). For example, in the case of the hydraulic administration of the 18 Streams, forcible settling of Ming military colonists alongside the civilian population led to new problems, as pressure on scarce water resources grew. Yi tribesmen also unexpectedly moved into and freshly cultivated the mountains, which required the use of irrigation water. Upstream users thus began applying pressure on downstream users and introduced water quotas. In the area of the 18 Streams on the lower slopes of the Diancang Range, water quotas were allocated between soldiers and commoner–civilians in a total of 35 locations between Shangguan and Xiaguan around 1426–1435 CE (Duan Jinlu 2000). It seems likely that old registers were examined and quotas carved on stone to ensure equity. Shares were temporal, thus ensuring a proportional distribution from an annually variable water source. Except in specially noted cases, this was a daylight share. The old standard for water allocation of 10 days and nights was followed, but occasionally less than 10 shares were allocated, which probably reflects a reduction in the relative quota originally assigned to a collective group of recipients. Government officials ensured compliance and maintained peace through an interventional consultation process between soldiers (the group most under pressure) and civilians, which discussed the scheduling and temporal allocation of water rights through locations and fields as controlled by sluice gates. A ground inspection and enquiry clarified the facts under dispute and infractions of water regulations by offenders were punishable by placing a wooden neck-collar around the offender's neck.

This evidence points and leads to the conclusion that the central part of the Erhai catchment experienced a strategic seasonal shortage of water as determined by the relationship between population pressure, contemporary farming technology, and the quality of natural resources as defined by that technology. During the rice transplantation period this resulted in conflict over a scarce resource. However, human transformation of the environment did not cause acute emergencies until the second half of the eighteenth century, which at that time was part of the catchment in the north.

5.3. Crisis in the North

Late in the sixteenth century a threshold was crossed, with the first clear case of a large ecosystem reaction. The crisis occurred in the northern part of the catchment, in what was then Langqiong county, immediately south of the present-day city of Eryuan (see Figure 1), where the waters of the upper Miju River and the outflow from Lake Cibi, both from the north, the Fengyu River from the west, and the smaller Ning River from the northeast, joined together at or near the Three Rivers' Mouth to flow southwards through the narrow Putuo Gorge, and thence into what was then Dengchuan Department, and finally into the northern end of the Erhai (Elvin et al. 2002; Elvin and Crook 2003). The cross-flow of waters at the entrance of the Putuo Gorge created an obstruction, with the sediments and stones blocking the swift current (Fan Zhaoxin 1830s) and causing water in the Ning River to back up, resulting in drowned paddy and dry fields. This problem was initially resolved in 1598 by clearing and dredging the channel. By 1600 heavy rains led to further flood disasters of an increased severity, which resulted in more dredging of the river and the construction of a subsidiary river. These flooding problems continued into the mid seventeenth century, culminating in the construction of a long dyke designed to control floodwaters (Elvin and Crook 2003).

The flooding problem continued into the eighteenth century, probably because of the opening up of hill lands for cultivation and deforestation. For example, hills in front of and behind Tower Base Mountain, which were opened up between 1757 and 1758, were prone to erosion (Plate 1). Following heavy rain, sediments, stones, and flash floodwaters from these hills emerged from the Baihan Gorge, obliterating the dry cross-dyke and choking the body of the river with flood debris. The result of this was that in 1762, 1769, 1770, 1780, and 1781 disasters were reported, and remission and relief were repeatedly given (Fan Zhaoxin 1830s). Following the flood of 1762, in response to a petition, a low dry cross-dyke several thousands of feet in length was built to prevent flood damage and to replace moribund remedial work at the Three Rivers' Mouth in Langqiong County. After this event minor maintenance was carried out every year and major maintenance every third year. The problem of deforestation may also have become an issue in the western part of the catchment at this time, as it had in the southern part where programs of forest protection and reafforestation with pines were beginning to be put into effect by the 1780s. Scattered evidence for trade in wood within the catchment around this date very possibly reflected growing scarcities elsewhere, and demonstrates an intensifying pressure on resources within the catchment (Elvin et al. 2002).

Plate 1: The Remains of Tower Base on the Tower Base Mountain, Opened Up between 1757 and 1758 by Yi (© John Dearing 2003).

Above Putuo Gorge another subsidiary river was opened in 1771, but flooding remained a problem, returning in 1803, 1806, 1807, and 1808 after heavy rains resulted in the collapse of both large and small dykes, as well as ploughed hills located beside the gorge. In the case of the latter, flood debris caused massive backing up of water into the county capital. More than 1,000 men were assembled to dredge clear sands and muds and split apart huge boulders. With the old cross-dyke destroyed and the old disaster mitigation strategies clearly not working, the authorities looked into new ways of reducing the flood risk. From mountain summits the topography was scrutinized and the decision was made to rebuild a dry cross-dyke 1,100 feet long that was stabilized by willows from the mouth of the gorge to the eastern and western feet of the mountain (Plate 2). Maintenance of this dyke was an ongoing concern with further sectional repairs occurring in 1815 and in 1824. For a while the river flowed freely and peacefully.

Plate 2: Baihan Dry Dyke and the Putuo Gorge (© Darren Crook 2003).

However, there were problems with the lower reaches in Langqiong, where the earth of the cross-dyke was pared away and made thinner daily and the low willow-dyke failed to survive, thus making the possibility of rupture a real and potentially devastating threat (Fan Zhaoxin 1930s). Four options to mitigate this risk included dredging the lower reaches deeply; building up and repairing the old cross-dyke; opening another diversion channel to reduce channel velocity and diverting the point of impact onto the dyke; and forbidding digging in the mountain gorges where rivers have their sources. With respect to the last point, there were two villages in the mountains, Shachang and Baihe, inhabited by the Yi who dug up the loose, light, and unstable soil on the north facing slopes in the mountain gorges, which were vulnerable to rain splash and detachment processes leading to overland flow and rill erosion. They also initiated some deforestation in this area, driven by a new or much augmented commercial demand for timber during the Daoguang reign-period (1821–1850), which clearly demonstrates an association between development and increased erosion. There is thus a *prima facie* case that the rapid increase in the volume of sediment carried by the Miju and other rivers in the north of the catchment in the later Ming dynasty and the first two-thirds of the Qing dynasty was due to hillslope land being cleared for farming, with or without deforestation,[7] and to an increase in tree-felling.

5.4. Dengchuan Disasters

Flood disasters were not confined to the Eryuan plain. By 1552, the trouble moved south of the Putuo Gorge and continued up to 1849 (Li Zhiyang no date; Hou Yunqin no date). The sudden breaching of the dykes is evident, as is the increase in frequency of repairs. Repairs to dykes carried out by colony soldiers and civilians were first recorded during the Yongle reign (1403–1424), followed by three more records between 1506 and 1521. A local government system in charge of annual repair of dykes was first put into place between 1436 and 1469, but neglect of dyke maintenance led to further crisis in the middle of the sixteenth century (The 1563 gazetteer for Dali). At this point fixed regulations and proportional annual maintenance responsibilities were introduced according to the amount of land owned. Anyone not willing to participate was fined.

A relaxed, community-based maintenance system was replaced by tight bureaucratic control after the flood of 1552. The quantitative details of maintenance work were laid down, with the costs of repair, material and labor, remaining high. The responsibility for specific lengths of dyke was assigned on the basis of taxes paid by landowners and laborers, who normally worked for a month, under official regulation. Organization, including department overhauls, improved periodically throughout the seventeenth, eighteenth, and nineteenth centuries, ensuring a period of stability after the disasters of 1815–1817 (Elvin et al. 2002; Elvin and Crook 2003). The hydraulic engineering above the Gorge at the Baihan may also have played a part.

During the first half of the nineteenth century the population, which was recovering from various epidemics, increased (Benedict 1996), leading to increased pressure on scarce resources. People were forced into the

[7] Not all slopes were forested, but even slopes with scrub needed clearance to make them suitable for farming.

mountains to cut firewood and kindling, which they then exchanged for grain. The clearing of fragile upland slopes adversely affected hydrological conditions and incurred mounting economic burdens, though not for the protagonists. There was a displacement of responsibility. The Yi opened new mountain land above the Baihan Gorge and thus did not have to rebuild the Miju dykes downstream, which is one reason why effects of this type proved so hard to stop. It was estimated that from 1828 to 1843, the riverbed in the upper course below the Putuo Gorge rose by 10 feet in spite of dredging. This was due to land clearance and deforestation upstream, which led to increased sediment loads in the Miju River. At this time river gauges were installed to monitor dredging needs. By the early nineteenth century clearing and dredging required 60,000 men working occasionally for around three months, and half again that number to rebuild the embankments.[8]

The bed of the Miju lay above the land because turbid waters annually transported and deposited silts and sands, particularly during summer and autumn. Some of the sediment remained in suspension until it accumulated at the river débouchement into Erhai, such that with the passing of time, the mouth of the Erhai became obstructed, and the tail end of the river grew congested, preventing boat traffic. For around 30 years prior to 1854–1855, the delta grew by 2.5–3 km, a linear extension rate approaching 0.2 km/year. Thus, the disaster of sediments and stones were endured all the way from the source to the tail. A sharpening awareness of how environmental problems were interconnected occurred as new experiences arose.

5.5. West-Side Hydraulic Degradation

Irrigation systems of the west side fell into a state of administrative neglect, and in some cases allegedly irrecoverable decay during the mid-sixteenth century (*Dali Gazetteer* 1563). At this time almost none of the water supply from the 18 Streams used for farming was under proper control or receiving regular maintenance (Elvin et al. 2002). On the west side over a period of about 100 years, rich soil had turned into sands and stones because of poor water control. In periods where there was a compelling public interest, maintenance work was carried out, but rarely otherwise. Failure to dredge the disaster prevention dyke led to a huge flood some time between 1488 and 1505, which destroyed many houses in

[8] The resolution of the fan/dyke record appears to be multi-decadal and may help in reconstructing an event chronology but this constitutes work in progress.

the walled city of Dali. Thereafter arrangements were made to dredge the moat annually. The same importance was given to the outflow of the Erhai, which was cleared every third year. Failure to dredge channels north of Dali led to 100 *qing* (~ 667 ha) of rich land turning into saline waste, which provides evidence for unsustainable practices.

Increased sedimentation was potentially the result of tree cutting on the Diancang mountains as trees were being felled there at this time. There is a slight suggestion that stocks were beginning to decline, but records imply that trees were still there to be cut in the early twentieth century (*Dali County Gazetteer* 1917). Presumably these trees were even more abundant during the late Ming period.[9] These fragments of evidence leave us with several problems, which are made more difficult by an approximately 300-year gap in the available evidence. An intricate, decentralized complex of small traditional irrigation systems based on the 18 Streams, together with many smaller sources like springs, was still functioning in the early twentieth century, apparently successfully, although this success remains unexplained.

In contrast to the Miju system, the part played by the state in Dali was minimal, as their small scale made community management practicable. People seem to have been content to live with flooding problems rather than trying to control them. Allocation of water was based on what was now a long established legally binding customary rotational cycle. Other groups of villages used similar cycles, whilst still others had inherited rules, working on principles not specified. Judging by the number of names in common, the anthropogenic channel structure of the west-side complex of irrigation systems in 1917 bore little relationship to that of 1563 or that of 1426–1435, just as these latter two had at best only a limited mutual relationship. Whilst the springs, gorges, and main debouchments into the lake remained more or less constant, the manmade conveyance channels of the 18 Streams were slowly but continuously shifted over time by human action across the lower lakeside slopes of the Diancang Mountains. This was done to circumvent the problems created by sedimentation, an observation supported by the decentralized and very slightly anastomotic character of the complex. To open a new channel on the lower slopes of the Diancang channel was not an easy engineering task

[9] The main period of hill slope deforestation seems to run from the second half of the seventeenth century to at least the later part of the eighteenth (when some remedial reforestation began). This included the reign of the Yongle Emperor (1723–1735). However, we have no direct evidence that any timber from Erhai was used for palace building. (Guizhou would have been a more likely source.)

because strongly corrugated contours on steep slopes and channel seepage had to be overcome. In this difficult terrain a rate of about 33 m a day was thought possible. Thus, we conclude that work of this sort would have amounted to a substantial undertaking, though not a monumental one. In oversimplified terms, while the challenge that developed on the Miju was vertical—namely, to make the dykes higher and the river bed lower—for the 18 Streams it was horizontal, to find a way to relocate laterally. The former required an ever more onerous annual state-run bureaucratic mobilization of labor, resources, funds, and managerial skills; the second needed only intermittent, small-scale excavations.

5.6. Twentieth Century Pressures on Water Resources

The political vacillation of the twentieth century left a large legacy of environmental destruction (Shapiro 2001), which unfortunately remains hard to quantify, as the political sensitivity of this information where it survives is still high. What is clear is that the earthquake of 1925 caused massive disruption to the entire catchment, even causing a tsunami on the lake and adding to the collateral damage brought about by revolution and rebellion (Retlinger 1939). The rest of the century is characterized by increased population (in-migration and natural increase) and development pressure, which over the course of the "Mao era" led to new demands on water, including hydroelectric power (HEP) generation and periods of upland erosion associated with the construction of agricultural terraces and deforestation, which resulted from the need for fuel wood and energy for iron smelting (UNCRD 1994).

The long-term process of land reclamation and drainage on the northern plains continued, with the East Lake all but disappearing. Sediments within the Baihan Gorge cross-dyke increased and grew thicker daily, so that they were almost level with the dyke parapet. Further downstream, it has been estimated that the Miju lacustrine delta vertically accumulated at a rate of just under 50 mm $year^{-1}$ at its mouth from about 1950 to the middle 1980s, a rapid rate in comparison to that found in the Boluo delta (about 2.0 mm $year^{-1}$) in the south of the catchment. The reduced storage time of water in the Dengchuan Plain may have contributed to the development of this delta. On the western shore about a third of the way down, lakewards of the town of Xizhou and the village of Shacun (sediments village[10]), the Wanhua Stream debouches into the lake

[10] The name of this village refers to a location where sediment from the Wanhua stream regularly aggraded.

(see Figure 1). Since the 1970s this has resulted in the rapid build-up of a striking foreshore spit curved like a fishhook extending into the lake, which is visible on modern maps and remote-sensing images.

Today, water is more strongly regulated both in the Miju (HEP) and on the west shore, where water shortage led to the construction of pumping stations for irrigation from the lake in the 1980s. Water shortage in the north of the catchment led to the construction of a small reservoir (see Figure 1). More irrigation water in particular has been needed to irrigate the large amount of new land brought into production by drainage on the plains and upland terracing. In terms of pollution, the outflow of Erhai named the Xi'er River has become a major source of pollution along the Mekong. The increased use of fertilizers has also created a diffuse source of pollution leading to eutrophication of the lake. A regional development and environmental management plan for the Dali-Erhai Area was drafted in 1994 to deal with these issues. However, aggressive measures to control nutrient and sediment inputs were not taken into consideration until after the first algal bloom appeared in 1996 (Dali EPB 1997). Eutrophication still remains an ongoing concern in Erhai (Jin Xiangcan 2003).

6. DISCUSSION

Our analysis of the Erhai Catchment describes past vulnerability and adaptation to human impacts and climate change and illustrates their effects on hydrology. Using an innovative interdisciplinary methodology many questions have been answered, though some are still left unanswered and the research process has uncovered yet new questions. Thus these results are preliminary and the disentanglement process continues. Before we can come to more confident conclusions about the affects of human activities on water resources, we must integrate both climate and tectonic data to understand the importance of coupling processes and synergy in driving hydrological change.

We start this section by asking whether the initial disturbance phases of human occupation and settlement were significant enough to initiate any long-term environmental trends. Initial landscape disturbance may have started as early as 2650 BP, but methodological constraints make it difficult to draw definite conclusions here. Not until 1950 BP do we see a very strong signal for human disturbance in the catchment in the form of greater amounts of pedogenic material in the catchment, probably topsoil, entering the lake. The timing of this event is in agreement with the documentary record, which shows that around this time Chinese-style irrigated agriculture was introduced. This may not have been ubiquitous throughout the catchment, but most certainly on the western shore, where the lead levels points to a wider development process occurring at this time. In terms of the hydrology, this was a period in history when the control of water became integral to everyday life and livelihoods, producing conditions that led towards a growing commitment to regulating both supply and demand by both individuals and the state, depending on the location. Together, these data point to possible long-term processes initiated by the crossing of important environmental thresholds impacting on present-day conditions in the catchment. This means that contemporary catchment planners concerned with the amount of sedimentation and new land creation in and on the lake may find that their decisions and actions to rectify this problem driven by current Chinese environmental policy may[11] have very little impact on these longer-term trends.

Only rarely do large scale documented events potentially coincide with evidence in the sedimentary records. Indeed, this does not happen again until the eighteenth century.[12] This time it is the pollen evidence that points to a phase of deforestation in the catchment, which is demonstrated by a decrease in *pinus* and an increase in *poaceae* beginning in the transition period between sediment units 3 and 4. This evidence is potentially supported by documentary records, which suggest that people, often minority nations such as the Yi, moved into the uplands to farm partly in response to population pressure on the plains. These actions had severe environmental consequences, supported by evidence found in documentary and sedimentary archives, thus independently pointing to the eighteenth century and the early part of the nineteenth century as a critical period for the onset of rapid environmental degradation in the northern part of the Erhai catchment. This instigated a period of great flooding that initiated the development of the Erhai delta in the north of the lake, which remains an ongoing process, as sediment is slowly transported down the dendritic system. Hypothetically speaking, the dip in $\chi_{FD}\%$ at this time suggests the mobilization of a different sediment source alongside the already high soil erosion rate, which could reflect a period of high-energy floods.

Analytically, the pattern of spatial connectivity appears to be a crucial determinant in creating the Miju type of acute crisis. Water systems on the west side were marginally anastomotic in places, but predominantly independent of each other; requiring clearing of deposited sediments and

---

[11] Current government policy to prevent flooding in other parts of Yunnan calls for the land to be returned to the Lakes.

[12] Care must be taken to avoid circular argument because the exact dating of the sediment sequence is not yet available and thus it cannot be certain that these records tally with the documentary records.

restructuring, all of which entailed localized costs. In contrast, the Miju in the north was, for all intents and purposes, a single dendritic system in which perturbations in sediment loading in a multitude of inputs had a cascading effect, as they reinforced each other downstream, which in part is reflected in the different flood records of the upper and lower Miju. The large sub-catchment scale and gradients of the Miju led to this river having a far greater impact on lake hydrology than the comparable Bolou dendritic system in the south of the catchment.

More generally, within the broad framework of the concept of "traditional Chinese irrigated agriculture", it is important to draw significant distinctions between differing degrees of "sustainability," based on contrast and environmental microvariation evident in the two sides of Erhai presented here. A variety of patterns of ecosystem reaction across time characterize pre-modern Chinese-style development, which is notable given that the repertoire of pre-modern farming techniques was for all intents and purposes identical in the two sub-catchments. In the Miju River sub-catchment, a long early phase of relatively widely spaced human innovations to exploit natural opportunities or provide "solutions" to natural problems, with little trouble during the intervening periods, was followed by a dramatically intensifying sequence of crises, with other factors such as upstream stripping of mountain forest and vegetation cover adding to the complexity of causes. The long-term process of land drainage and reclamation of the shallow lakes and marshlands, most evident in the Dengchuan and Eryuan plains of the north, appears to have been an environmentally benign process unworthy of recording in documentation. What this clearly demonstrates is that this level of environmental microvariation must be considered in the formulation of present-day water policy.

Throughout the historical time frame covered by the documentary records, the perturbations and problems associated with the governance and organization of water supply and demand were numerous. The desires for equity in terms of the scheduling and allocation of water rights are apparent from an official perspective, but clearly principles and practice often failed to coincide, particularly when individual motivations and group dynamics were taken into account. Individual apathy, perhaps driven by economic malaise, resulted in periods of moribund irrigation systems and abandonment, particularly on the western shore. Some of these changes constitute part of a dynamic, but others question the long-term sustainability of these systems. The evidence for long-term flexible hydraulic adaptation on the west side of Erhai suggests that even systems with a high degree of relative "sustainability" may not have been absolutely sustainable over the long run (assuming constant technology), except at the cost of periodic restructuring in addition to ordinary maintenance. This leaves open the question of how far can "stability" be reasonably allowed to contain "dynamic" elements, as contributing to any sense of "equilibrium"?

Finally the results demonstrate the advantage of using an interdisciplinary methodology of the type applied here in that they allow an observation of events—such as changes to society and the hydrological system that impact both the competitive demand for and supply of water—over varying temporal and spatial scales. The weight of importance of short-term linear events evident in the archaeological and documentary record is sometimes challenged by longer-term trends (like cyclic and nonlinear events that are apparent in the sedimentary records), such that simple cause and effect relationships are not evident. Clearly non-stationarity is also important and any concept of vulnerability, adaptation, and resilience to changes in the hydrological regime must recognize this dynamic.

SCARCITY, EQUITY, AND TRANSPARENCY: GENERAL PRINCIPLES FOR SUCCESSFULLY GOVERNING THE WATER COMMONS*

Paul Trawick
Institute of Water and Environment
Cranfield University, United Kingdom
p.trawick@cranfield.ac.uk

Abstract: A comparative cross-cultural study of several successful farmer-operated irrigation systems in two different parts of the world—the Andes of South America and the Mediterranean coast of Spain—reveals that the same set of institutions (or rules and operating principles) produces sustainable positive outcomes in each case. Several successful irrigation systems, well-documented in the literature, are thought to be of fundamentally distinct types and known to be of widely different scales: Valencia, Alicante, and Murcia in Spain. This success can only be explained in terms of basic similarities underlying the more obvious but superficial differences noted previously by other researchers. Similarities include operating principles the author first identified in his ethnographic research on successful irrigation communities in Peru. A brief overview of the comparative literature on successful systems in other semi-arid regions—India, Nepal, the Philippines—shows that the same basic system for sharing water under conditions of scarcity has emerged independently in a great many communities throughout the world, suggesting that this system is an optimal one, and constitutes a clear and unprecedented case of parallel or convergent social evolution. The author concludes with some implications of such a general model—or "universal" schema—for both the theory and practice of sustainable local irrigation.

Keywords: water management, irrigation, common-property resources, policy, markets, Peru, Spain, Chile

---

* Fieldwork in Spain and Chile during 2003–2004, was made possible by a generous Research and Writing Grant from the John D. and Catherine T. MacArthur Foundation *Program on Global Security and Sustainability.* I express my gratitude to the Foundation and refer to Trawick 2005 for a more detailed account of fieldwork results. Sincere thanks also go to the leaders of the following water-user organizations in Spain for their generous cooperation and collaboration: the *Sindicato de Riegos de la Huerta de Alicante* (its President, Vicente de Cabo Ruíz, and especially its Secretary, Manuel José Ňigez Pérez); the *Junta de Hacendados de Murcia* (its President, Sr. Sigifredo Hernández Pérez, and especially the Vocal, Sr. Benito Avilán Cornejo), and the *Tribunal de las Aguas de la Vega de Valencia* (its President, Sr.Vicente Nácher Luz, and especially to the Tribunal's Lawyer, Sr. Alfonzo Pastor Madaleña). Special thanks are due Yolanda Lopez Vera, geography graduate student at the University of Murcia, Spain. As my Research Assistant, she was crucial to my fieldwork there.

E. Wiegandt (ed.), Mountains: Sources of Water, Sources of Knowledge, 43–61.

1. INTRODUCTION

Few works in social science have influenced policy as much as Hardin's (1968) attempt, in "The Tragedy of the Commons" to explain the tendency of people to overexploit any resources that they hold in common in terms of an irresolvable conflict between the inherently selfish interests of the individual and the cooperative needs of the group. The results of such a tragedy are of course evident today throughout much of the world in the use of common-property resources: irrigation water, pasturelands, forests, and fisheries. Yet Hardin's theory has been criticized and even refuted convincingly by many authors, based on their studies of hundreds of cases where local people have managed such resources cooperatively, and done so very effectively, over a very long period of time. This rebuttal has turned attention toward the task of devising an alternative theory to explain how people have been able to overcome their conflict of interest, escape the "commons dilemma", and pursue the common good (e.g., N.R.C. 1986; McCay and Acheson 1987; Bromley 1992).

Recent research on the management of water for irrigation, some of which will be reviewed briefly here (Trawick 2001a,b; 2002a,b; 2003a,b; 2005), lends significant new support to this effort and promises to allow a decisive revision of the conventional theory. It shows that people in a great many communities throughout the world long ago arrived, quite independently, at the same sustainable solution to the "commons dilemma", creating a set of principles for sharing scarce water in an equitable and transparent manner that minimizes social conflict. This comparative research strongly suggests that, in cases where irrigation communities have managed a scarce resource autonomously and effectively over a long period of time, the principles of distribution and use appear to be highly similar if not exactly the same, and this seems to be true regardless of whether the water is communally or privately owned.

The programs of the World Bank and the regional development banks with which it is affiliated continue to be strongly shaped by the conventional theory, and they have long advocated water privatization, one of Hardin's proposed solutions to the commons dilemma, on a massive scale (World Bank 1995). My research indicates that the creation of water markets will not solve the problems afflicting irrigation in many regions of the world today, and that such markets do not actually work in the manner that they are widely thought to work, at least not in the small-scale canal systems that typify most arid and mountainous regions. This work has revealed the existence of heretofore unrecognized but highly significant commonalities in the dynamics of successful communal and "market" systems.

Scholars and scientists have made steady and important progress in critiquing and revising Hardin's theory, most notably Ostrom (1986, 1990, 1992, 1998; also Ostrom et al. 1999, 2002) and Tang (1992), who, through comparison of a large number of case studies in different countries, have led the way in identifying basic design principles that all effective locally run irrigation systems seem to share. Their focus has tended to be on canal systems of 1,000 ha or less, the kind of "indigenous" or peasant community system found in most mountain regions (Mabry and Cleveland 1996). Such small scale, and the kind of intensive face-to-face interaction among water-users that this makes possible, seems to be a common denominator that can contribute significantly to local success (Ostrom 1987). However, most of the principles identified thus far remain rather abstract, and more suitable for predicting the general conditions under which people will be able to come up with a solution of their own than for showing them how to manage water effectively when they have failed or lost the ability to do so on their own.

The effort to revise theory and make new policy has been hindered by the limitations of the primary data in these studies, which are always highly objective and descriptive at the system level, typically presented from the point of view of a resource administrator or a water distributor, but without incorporating the more subjective and culture-bound perspective of the individual farmer and water-user. Analysts have also tended to emphasize the diversity that seems to exist among local systems while not giving enough attention to the one important feature that nearly all of them have in common, at least at certain times of the year, and that is water scarcity. All of this has obscured the fact that the keys to local success in dealing with scarcity appear to be highly similar everywhere if not exactly the same: operating principles that together instill a strong positive incentive in people to obey the rules and conserve the resource, rather than merely a negative incentive that rests entirely on punishing people for "free-riding" infractions. Once the principles are identified in a particular ethnographic case, and the way that they work together from the water-users' point of view is understood, as I will try to explain below, the parallels in other countries become evident and are striking indeed.

2. HUAYNACOTAS: AN "AUTONOMOUS" INDIGENOUS COMMUNITY

The community of Huaynacotas in the Peruvian Andes, which I studied over a period of more than two years, is located in the Cotahuasi valley of the Department of Arequipa, in the southern part of the country. It is a village of Quechua-speaking peasants (population roughly 1,080), one of only a few villages in the valley that were never directly colonized by outsiders. That is to say, the Spanish landlords who became dominant almost everywhere else in the valley never succeeded in acquiring any land there and actually residing. Although extremely remote, Huaynacotas is not a pristine community, but rather one that managed to contract its boundaries very early after losing all of its low-altitude land to these local colonial elites. The villagers thereby managed to prevent further loss of land and water to the landlords and their haciendas (private agricultural estates) and to maintain a significant kind of autonomy, having full control over their water supplies and the rules governing resource use.

The hydraulic tradition described below is also found in two other villages in this same valley, and it has been shown to exist in other indigenous communities in southern Peru (see, e.g., Treacy 1994 a,b). The scarcity of water in Huaynacotas is acute even under "normal" conditions, and this constrains the sequence and timing of distribution.[1] As in most communities, the main flows of water within the system are so small that one or two landowners at a time must take turns in using it—an arrangement known as a *mita* or *turno*. The cycle of turns is so slow that it takes two to three months to water the entire 410 ha expanse of irrigated land, even after a year of good rains. During the maize planting in September and other crop plantings thereafter (e.g., beans, potatoes), the watering frequency gradually declines, and this continues until the rains begin in early January, at which time irrigation normally ceases. This means that staple crops are watered at most three times during the agricultural year.

---

[1] The hydraulic system is a dual one with two major water sources, both of them alpine springs, a pattern that is typical of the Cotahuasi valley and the Andes as a whole. The flows are stored at night in two tanks and distributed during the daytime through two separate networks of canals. The canal system as a whole spans elevations from 3,100 to 4,100 m and encompasses roughly 410 ha of territory that is divided into named sectors of land. The sequence in which the sectors are watered is determined by micro-climatic variation making some sectors colder than others, which extends the germination time for maize, the main staple crop. This initial planting and watering sequence is then repeated for each irrigation cycle throughout the agriculture year.

The function of irrigation here is to distribute the supply equitably, efficiently, transparently and with minimal conflict, so that local people are able to make the best of a bad situation. The arrangement of land-holdings directly promotes this outcome: the highly fragmented pattern, or *minifundia*, that is characteristic of Latin America and typical of peasant villages throughout most of the world. Regardless of how much land people own, they tend to have numerous small parcels scattered in different sectors at various elevations and located along different canals. They also tend to have land in both halves of the irrigation system, which has two main water sources and two largely independent networks of canals. The pattern is so highly fragmented here that one cannot speak of the usual population of "head-enders" and "tail-enders" along each canal, since most people seem to have land in both kinds of situations.

3. THE RULES OF WATER MANAGEMENT

The canal system is operated through a system of rotating authority in which customary procedures are always followed. This is done by two elected water officials called *Kampus*, who oversee the two halves of the system. During each distribution cycle, the *Kampus* divide the flow of each main canal in half, into two standard and roughly equivalent portions (*rakis*), in the act of diverting it into the secondary canals. They then allow the water to flow down to the fields, where each share is used by a landowning family or household. This happens in both halves of the canal system at the same time according to rules that are essentially equivalent.

3.1. The Rules of Distribution

Certain procedures ensure that all parcels of land served by a given source, and all households, receive water with the same frequency, which varies with seasonal and long-term fluctuations in the supply. First, the land sectors that make up the village territory are given water consecutively in a fixed sequence based on altitude and microclimatic variation. During each cycle of the system, water passes through all the sectors currently in production, reaching every parcel before beginning again.

Second, the plots within each sector are given water in a rigid contiguous order, starting at the bottom and moving upward, in such a way that the time at which they are serviced depends only on their location, rather than on who owns them or the crops in which they are planted. Alfalfa, for example, an irrigated pasture is grown here in tiny plots; but, unlike the situation that one finds in most other local villages, here it is watered in the same way and on the same schedule as any other crop.

Third, a standard method of adjusting to drought ensures that the impact of periodic shortages is absorbed equally by all households. After a year of poor rains, as the water flows subside during the planting season, some of the sectors at the upper end of both halves of the system are taken out of production in order to prevent a further decline. This was done about thirty years ago, in response to a sustained and still ongoing drought, and the abandoned sectors have remained out of use ever since. Since everyone had land in these sectors, most people were affected by the contraction. But everyone benefited equally from it, since the goal was to keep the watering cycle from stretching out so far as to seriously jeopardize the harvest. A result of this kind of cooperative arrangement is that, even though the springs here are the most vulnerable in the entire province to droughts, conflict over water is far less prevalent in Huaynacotas than in other local villages, which are less vulnerable to drought but which in nearly all cases have far less equitable arrangements (see Trawick 2001a, 2003b, Chapters 4 and 7).

3.3. The Rules of Utilization

The entire landscape of the village is terraced into level surfaces that are designed and carefully maintained to promote the absorption and retention of water. These make it possible for the watering to be carried out through a uniform technique that ensures that the duration of irrigation, and the amount of water consumed by people in their allotments, is strictly proportional to the extent of their land. Standard water containment features (*atus*)—earthen structures of uniform height—are used by everyone. Because liquid is pooled on the surface to the same depth, irrigation time and water consumption are regulated by the technology itself. Once the pools are full, irrigation is finished and the water distributors allow no departures from this arrangement—such as returns to top up the pooling structures, or the destruction of terracing and the irrigation of slopes, practices that are common in other local villages. For the purposes of this comparative analysis, the important point here is that control over irrigation time and the amount of water consumed in each household allotment are inherent features of the local technology—control that, it should be pointed out, can be and is achieved in other kinds of irrigation systems by other means.

4. THE OPERATING PRINCIPLES

From an analysis of the rules of water management in Huaynacotasone we can derive a set of operating principles on which successful irrigation seems to depend.

BASIC PRINCIPLES OF IRRIGATION IN HUAYNACOTAS

1) *Autonomy*: The community has and controls its own flows of water.

2) *Contiguity*: Water is distributed to fields in a fixed contiguous order based only on their location along successive canals, starting at one end of the system and moving steadily across it.

3) *Uniformity*:
- Among water rights: Everyone receives water with the same frequency.
- In technique: Everyone irrigates in the same way.

4) *Proportionality* (equity):
- Among rights: No one can use more water than the amount to which the extent of their land entitles them, nor can they legally get it more often than everyone else.
- Among duties: People's contributions to maintenance must be proportional to the amount of irrigated land that they have.

5) *Transparency*: Everyone knows the rules and has the ability to confirm, with their own eyes, whether those rules are generally being obeyed, as well as to detect and denounce any violations that occur.

6) *Regularity*: Things are always done in the same way under conditions of scarcity; no exceptions are allowed, and any unauthorized expansion of irrigation is prohibited.

The control of water here is unified or centrally directed (see Hunt 1988), but the system is not directly articulated with any outside agency and is, at the present time, autonomous. Although the State theoretically owns all of Peru's irrigation water according to existing law, the government water bureaucracy has never had any presence here, as in probably the majority of highland communities. This has left the local people free to maintain their own customs. As Ostrom (1990, 1992) has noted, people are much more likely to respect the rules when it is they who set them.

The principle of proportionality, among people's rights and between their rights and duties, has been noted and discussed in many other case studies. However, without uniformity, especially in the watering frequency, no such proportionality among people's rights can exist. In Huaynacotas, uniformity is a major concern and a central theme in village social life. The right to one's proportional share of water during each distribution cycle is the basic egalitarian principle upon which life in this community has long been based. However, people do not have to worry a great deal about this uniformity or be constantly on the lookout for theft and other forms of cheating and "free-riding" that would threaten it. This is because such infractions are easily detected and therefore are, and reportedly "always" have been, extremely rare. Because of the contiguous order in which it is carried out, irrigation is a highly visible activity, so that people have the capacity to monitor the system systematically, and can easily investigate matters in this way if they have reason to. The result is a tangible and widely recognized power of families and households to protect their own water rights, which results in a very strong positive incentive to obey the rules and respect local tradition.

People's rights—de facto claimant rights (see Schlager and Ostrom 1992), otherwise known as "communal" rights—are qualitatively equal, in that everyone is subject to the same rules and procedures, which they know well. Indeed, every man in the village knows how to operate the entire system, since the male heads of household serve in the post of *Kampus* in rotation, also sponsoring and directing the yearly Water Festival, the ritual cleaning of the irrigation canals. Ostrom (1987) has observed that this kind of arrangement ensures that knowledge of the rules is evenly distributed throughout the community, rather than being concentrated in the hands of a water official.

Water rights in Huaynacotas are also quantitatively proportional to each other, varying only with the extent of a person's land. This means that no one is allowed to deprive other people of water by using more than the amount to which the extent of their land entitles them, or, as commonly happens in other places, by getting it more often than everyone else. Just as in any other stratified community, some families have more land and use more water than others, but a fundamental symmetry prevails, both in the size and frequency of household allotments, and in the corresponding duties that people must fulfill in order to preserve their rights. Because large landowners have more land and use more water, their contributions to the Water Festival, and generally to the upkeep of tanks and canals, are required to be greater, in terms of labor and other inputs, than those of the smallholder majority. Largely because of this, the infrastructure is very well maintained.

The principle of contiguity is vital for several reasons. In addition to providing a uniform frequency of irrigation, a contiguous distribution pattern limits waste due to evaporation and filtration by minimizing the total surface area of canals in use on any given day. The canals are unlined and allow a great amount of water loss. The loss decreases dramatically once a canal surface and the soil beneath it have become saturated or waterlogged. By watering the entire surrounding area before moving on, the loss is minimized.

The contiguous pattern also makes irrigation a thoroughly public affair. Since everyone knows the rules that govern distribution, and the exact order in which they are supposed to receive water, and, because the owners of adjoining parcels tend to irrigate on the same day, people are normally putting their fields in order, or simply waiting and watching, while their neighbors finish their turns. This means that monitoring is pervasive and routine, spread out among users throughout the system, rather than a specialized role that is entirely in the hands of the water distributor. The visibility, and the passive vigilance provided by neighboring landowners, helps the distributors in ensuring that traditional procedures are followed, and they have the vital effect of providing controls over theft, favoritism on the part of water officials, and other forms of corruption. The rules and the work involved, together, thus create a situation of transparency. People have a strong sense of security regarding their water rights as a result, and have a strong tendency to obey the rules and respect tradition.

Infractions and other causes of conflict are rare in Huaynacotas, but, due to the extreme scarcity of the resource, they do occur. This will happen to some extent in any irrigation system, no matter what the rule. However, such corruption cannot happen repeatedly here without soon being discovered by the other water users. When infractions are detected the penalty is severe, but graded according to the gravity of the offense. It varies from the loss of one allotment during a given cycle to the loss of one's water rights on a given field for the remainder of the year.

5. THE INCENTIVE TO COOPERATE

All of the above principles play their part in creating a transparent and equitable system, and all contribute to its effectiveness. They include regularity: the rules must be consistent at all times, in all places, for everyone involved. As I learned through my work in other valley communities, exceptions to the principle of regularity that allow some crops to be watered more often than others and special provisions that temporarily modify the watering order only promote mistrust by introducing a degree of opacity, and they open the door wide for favoritism and abuse. Allowing irrigation to expand, for example, for some people would allow them to benefit specially and disproportionately from the cooperation and frugal water use of others.

The most central principle in terms of the incentive to cooperate, however, is the uniformity of the watering frequency. When everyone irrigates their land on a single schedule, and when expansion is prohibited, the water saved through conservation and self-restraint causes the distribution cycle to run faster. Thus, by limiting watering to a fixed period of time and obeying the rules, people are able to irrigate more often, as often as possible from the long-term point of view. And, in a situation of uniformity, any "free riders"—people who ignore the rules and steal water, or who irrigate excessively—interfere with the efforts of others to keep the cycle short and cause it notably to slow down.

The feedback on such behavior is thus immediate, negative, and easily perceptible: the number of days it takes to water a given area of land. The incentive to comply is consequently remarkably strong and the tragedy of the commons, far from being inevitable, is actually rather difficult to bring about. People enjoy an extraordinarily high degree of security about their water rights because, in a transparent system, a person can maintain and protect them quite effectively him or herself.

The incentive to cooperate is thus primarily a positive one that rests on this pervasive sense of security and on the lack of any persistent threat, rather than a negative one that rests on people's constant vigilance against an ever-present danger of theft and on the resultant frequent sanctions. Cheating is truly irrational and is widely perceived in that way; and this, according to the local people, explains why such behavior is so rare. This general situation is created by the scarcity of the resource, which in this community is especially grave, and by the institutional arrangements that people have worked out for dealing with a situation that is far from ideal. That is the "commons dilemma" that the people of Huaynacotas face, but it is not one they have brought upon themselves.

6. SUCCESSFUL SYSTEMS IN OTHER PARTS OF THE WORLD

There is now a large and growing literature critiquing Hardin's explanation for the commons tragedy in examining the local use of a wide array of common-property resources. (e.g., Bromley 1992; McCay and Acheson 1987; N.R.C. 1986; Ostrom 1990, 1992, 1998; Ostrom et al. 1999, 2002; Dolsak and Ostrom 2003), and irrigation communities provide some of the most impressive examples discovered thus far of successful management. Their implications for theory and policy have never been fully appreciated, however, because these analyses have focused mainly on institutions, rather than people, on structure at the expense of agency, and they have generally not explained things clearly from the "native point of view" of the water user.

A close reading of these studies—by Coward (1979) and Siy (1982) in the Philippines, Wade (1986, 1988, 1992) in southern India, Ostrom and Gardner (1993) in Nepal, Maass and Anderson (1978), and later Ostrom (1990, 1992, 1998) in various parts of Spain—strongly suggests that the same basic principles identified above are apparently at work in many of the systems described. Where the resource is scarce relative to the area of land and the crops involved—which may be the case all of the time, as in most of the Andes, or only seasonally, as in the moister regions of Asia—these principles seem to be crucial to the success of local management traditions. Moreover, this generalization seems to include rather large-scale systems composed of many communities, such as some of the ones in Spain, and to apply to both communal and "market" systems of the peasant type. Thus neither the size of the irrigation system nor the property regime seems to matter.

Autonomous control over the resource, or over each group's collective share of it, is recognized to be present in each of the communities studied. Beyond that, the principle that has been most clearly defined in the literature is proportionality, among rights (i.e., there is a single land/water ratio for each user of each major source) and between people's rights and duties, a principle first identified by Coward (1979, p. 31) in a zanjera system in the Philippines. More recently, Ostrom and Gardner (1993, p. 100) have found this to be the basic principle governing both rights and duties in many irrigation systems in Nepal, although the exact number of examples, though large, is unclear (also see Guillet 1992, pp. 204–206). Proportionality is also shown to be present in the various systems, totaling roughly 60 communities, that Ostrom (1990, pp. 69–82) has analyzed in Valencia and Murcia-Orihuela in southern Spain, based on the earlier work of Glick (1970) and Maass and Anderson (1978, 1986). What has not been realized, however, is that proportionality among water shares can in part be a by-product of irrigation technique.

Wherever the irrigation technique is highly uniform and involves pooling water on the surface of terraced fields to a standard depth—as with the pond-field terraces used all over Southeast Asia for growing rice (Conklin 1980)—proportionality among rights exists, provided that the frequency of irrigation is the same for every user of a given water source. Taking this into account, proportionality seems to be a feature of several of the systems described in southern India by Wade—again, the exact number of cases that Wade is describing is unclear—as well as the ten zanjera communities analyzed by Siy (1982) and Ostrom (1990, pp. 82–88) in the Philippines. Although he pays little attention to this question, Wade (1986, p. 75) does note in passing that people's contributions to canal maintenance must be proportional to the land area irrigated in each case. All of this strongly suggests that equity (Hunt 1992) or proportionality among both rights and duties, the central defining principle of the Andean tradition previously described, is a key feature of self-governing irrigation communities that work well and that have lasted for a long time in other mountainous parts of the world.

Uniformity is perhaps the most crucial concept, as stated previously, since without this basic commonality—a fairly uniform technique and a frequency of irrigation that is the same for everyone—no real proportionality among people's rights can exist. One can therefore infer that this principle is present in each of the systems studied. A uniform frequency necessarily must exist in all systems where water rights are tied to the land—true in the vast majority of the cases mentioned—and where there is a comprehensive, fixed order of rotation within each community or for each major water flow, as in the seven turno systems that make up the huerta of Valencia (Ostrom 1990, pp. 71–76; Maass and Anderson 1978, p. 39), as I was able to confirm in the ethnographic research of 2003–2004. Even in situations where certain crops and certain fields are given priority during a drought emergency, so that some crops then get water in lieu of others, it nevertheless appears true that all landowners and farms receive water on the same schedule.

By the same logic, uniformity must also exist in rotation systems where the actual time of irrigation is fixed by the day and hour for each parcel of land and each landowner, as in most of the 42 canal communities[2] (employing a tanda system) that make up neighboring Murcia

[2] Four of Murcia's forty-two branch canals are always open, so that water flows into them whenever it is available. These canal communities have never had an organized tanda,

(Maass and Anderson 1978, pp. 74–79; Ostrom 1990, pp. 76–78). There, the same kind of priority system was formerly implemented in times of severe drought, so that uniformity was preserved in both of these well-known Spanish systems. Today, however, such adjustments are left up to the individual in both cases; the frequency of irrigation (in Valencia), or the amount of water that actually arrives when watering takes place (in Murcia), are cut back naturally by dwindling flows and perhaps also by administrative action, and the farmer must then decide how to adjust the mixture of crops planted and the pattern of watering according to the amount of the resource that is available.

Contiguity is somewhat easier to confirm, like the feature of transparency that derives directly from it. The importance of a fixed, contiguous order of distribution is fundamental to conservation in situations where water is in short supply, and this is recognized implicitly by most of the authors whose work is being reviewed here. Wade (1988, pp. 77–78; 1992, p. 218) implies the existence of such a fixed order for numerous villages in South India (again, the exact number is unclear). Coward (1979, pp. 32–33) describes a similar system in the Philippines, but one where water is delivered simultaneously to all parcels within a given block of land whenever there is enough available to do so, as is the case much of the time. He also speaks at length of a system for rationing water during the yearly dry season, the season of scarcity, but without ever saying explicitly what the order of rotation among adjacent fields is. Nevertheless, his account does seem to imply that the rationing order in these zanjera systems is canal-by-canal, block-by-block, and field-by-field, as does the work of Siy (1982) on nine other nearby communities. Ostrom (1990, p. 73) does not clarify this in reference to the zanjera systems, but, citing Maass and Anderson (1986, p. 28), she does explicitly note that contiguity is the rule among the turno systems of Valencia. She is less specific about Murcia and Orihuela but does speaks of those systems as being "quite rigid" and involving a regular rotation during low-water periods (Ostrom 1990, p. 76), implying a contiguous order that was conclusively confirmed in the fieldwork of 2003–2004.

Contiguity is crucial in water-rationing systems for two reasons. First, the rule is the simplest one possible and is therefore easy to understand, so that knowledge of it is evenly distributed throughout the community.

---

having instead practiced a kind of free or first-come-first-served access among the members. Not surprisingly, in these days of a prevailing chronic shortage, mitigated only by the use of local wells belonging to each user-group, these communities clearly have the highest level of water conflict, as manifested in formal complaints between neighbors about violations of the rules.

Second, we have already seen that applying it has the effect of concentrating irrigation in one area at a time and thereby making it a highly visible public activity. The result, as Ostrom (1990, pp. 73–75) points out in her analysis of Valencia—though not explicitly in the other cases—is that farmers routinely have the opportunity to observe each other irrigate and to monitor, not only each others' activity, but also that of the water distributor. Even if it is not often actually used, the availability of such potential assistance is crucial because, in the steep and convoluted terrain of the Andes and other mountainous regions, it is ultimately beyond the capacity of any distributor—even the most physically fit individual—to divert the water from the main canals, guard against theft higher upslope, and monitor the circumstances and duration of water use in the fields down below.

These observations relate to the last of the six principles as well, regularity, which also seems to be present in each of the cases at times when the resource is scarce. There are a few exceptions, as in the case of the priority that used to be given to certain crops in situations of real drought emergency in Valencia and Murcia,[3] or of occasional expansions of the system—putting new properties under irrigation—which are sometimes allowed by the communities in these two districts. But those are unusual events that require special permission by the community, and they do not alter the fact that, without them, things are always done in the same orderly way. As explained, some changes are also made in Huaynacotas during severe droughts without altering the basic equity and proportionality of the arrangement.

The realization that the same basic principles are apparently at work in at least seventy to one hundred communities worldwide and that this solution was probably arrived at independently in the many countries where it exists today has important implications. First, it is one of the most striking examples yet found of parallel or convergent cultural evolution—the emergence of similar adaptations to environmental conditions—that, despite significant differences, have in common the necessity for the social groups to solve problems of serious water scarcity. The convergence itself requires an explanation which is ultimately beyond the scope of this discussion, but it suggests the existence of a widespread, or "universal," culture of irrigation and sustainability in water-sharing, which is readily

---

[3] Fieldwork in 2003–2004 revealed that this is no longer done either in Murcia or Valencia. In the former, wells have been drilled in large numbers to attempt to compensate for the growing scarcity of water, which has been greatly exacerbated by the Tajo–Segura Pipeline, the largest hydraulic project ever carried out in Europe. Supposedly built to benefit local farmers of Murcia and the upper and lower *vegas* (plains) of the Segura, the project turned out to be a massive swindle that actually deprived them of most of the water of their own Segura river.

activated when a prevailing scarcity emerges locally among groups of farmers. Second, these revelations about successful farmer-managed irrigation systems come at a critical moment in a debate about water policy. People not only can work out effective rules and principles for governing the commons; they have done so in a great many places all over the world and arrived at the same basic rules in each case. What are the implications of this general, or universal, model during a time of crisis in thinking about water management?

7. RETHINKING POLICY

Despite the arguments of many knowledgeable people against it, and even serious disagreement within the Banks over whether the policy is advisable, "privatization", otherwise known as tradable concessionary rights in water, has until very recently been strongly promoted by the World Bank and the InterAmerican Development Bank throughout Latin America as a solution to the problems that now afflict management of the resource, particularly in mountain regions. Beginning in Peru a few years ago, but then progressively in Ecuador, Bolivia, Brazil and now in other countries too, the Banks have sponsored and overseen the drafting of essentially the same privatization law, one that is based on the 1981 Water code of Chile (see Trawick 1995, 2002a).

The results of the fieldwork of 2003–2004 strongly suggested that privatization and the buying and selling of the resource will not solve the problems that are so widespread and symptomatic of the tragedy of the commons in water management. There are several reasons for this, but the main one is quite simple. In a region where the average household irrigates less than a hectare of land—one where subsistence agriculture predominates—the amount of water that can be saved by people through more conservative use, although quite significant in the aggregate, will rarely be large enough that it could feasibly be sold to someone else, even if the infrastructure existed to make this possible (it does not in most communities in the Andes today). The transaction costs, or the amount of trouble people have to go to in order to do it reliably on a long-term basis, are too high. In the rugged terrain of the Andes, the motive for conservation can only be found in the link between the efficiency of water use—in terms of both avoiding waste and respecting the rules—and the duration of the irrigation cycle.

This is well illustrated by the situation in rural Chile, a country where private tradable water rights have existed for many years, but where no such markets have apparently emerged within the peasant communities, i.e., those with small-scale, locally run canal systems. Although there are few published studies of such cases, all indications are that little or no buying and selling of water has occurred in the communities, and many of them have sought legal recognition and protection of their communal water supplies so that these kinds of transfers between individuals have not been allowed to take place (Dourojeanni and Jouravlev 1999; Bauer 1995, 1997; Solanes 1996; Hendriks 1998; Bjornlund and McKay 2002). For the most part peasants have not wanted to be part of the national water market and they have actively prevented themselves from being incorporated into it, as regards their daily use of water for irrigation (see Castro-Lucic 2002).

There is in fact only one thorough study now available of a long-standing and fairly autonomous water market, one composed largely of peasant farmers who cultivate rather small units of land, and that study focuses on the community of Alicante in Spain, one of the three systems in that country examined by Maass and Anderson (1978, 1986) and Ostrom (1990). Alicante formerly covered about 3,700 ha and included roughly 2,400 households; yet these people together form a single user community (Maass and Anderson 1978, pp. 100–145). Alicante may have been the oldest water market in the world, dating back to the sixteenth century, and was said to be remarkably efficient, with an extremely low incidence of rule infractions and resulting social conflict. Although the water market no longer exists,[4] having been disbanded in the early 1980s, this example

[4] Although it is not widely known, the Alicante market was abolished several decades ago, precisely because it did not work nearly as well as was widely thought. The Huerta went into bankruptcy due to decades of corruption and other problems back in 1983, and was officially abolished and turned into a fully communal *tanda* system in 1987. This was one of the first facts revealed in my fieldwork there during 2003–2004, which confirmed my hypotheses about the "moral economy" model for Alicante as well as for Murcia and Valencia. Space does not permit a thorough discussion of the results here, except to say that I was able to confirm, through extensive interviews with a large number of local informants (30 in Valencia, 20 in Murcia, and 10 in Alicante [combined with a substantial amount of documentary research]), that the "moral economy" model based on the principles discussed above does indeed apply to all three of the Spanish systems (or formerly did in the case of Alicante). I was able to confirm these intuited hypotheses about how the Alicante system worked during my fieldwork with local informants (Ten farmers and former water distributors) and my documentary research in the huerta's Sindicate archives. My sincere thanks again go to the Sindicate, especially to Manuel Jose, the Secretary, without whose help and collaboration I could not have completed the research.

The Alicante "market" was basically a communal *tanda* system where a priviledged minority of wealthy feudal water owners was able to exploit the prevailing scarcity (of "new" water among peasants, by selling or renting portions of the "old" water) to peasants at auctions (the shares of which had been gradually accumulated by these elites because they were no longer attached to the land). Any water bought, however, was delivered at the

from the literature has always seemed, on the face of it, to justify and endorse the World Bank's enthusiasm for water markets, as Bank economists have long been quick to point out. A close reading of Maass and Anderson's initial description of the rules operating to distribute community-owned water (both the so-called "old water" and the "new"[5]) during times of scarcity (which is the normal situation) shows that, even though these individual shares—the core rights of the system—were available for rent or purchase by other people, both of them were still held to some extent in common. The two kinds of water were delivered together in a strict contiguous order determined by the landscape, and used on a single schedule according to a fixed set of rules. Moreover, receiving them was contingent upon the completion of certain duties, duties that in each case were proportional to rights in a basic way.

In the case of the "new" water, which dated back to the sixteenth century, these allotments were merely claimed, not purchased, because the rights were directly tied to the land and determined by, and proportional to, the amount of land a farmer owned (Maass and Anderson 1978, p. 112). But even in the case of the "old" water shares, which was by far the predominant kind of water sold (amounting to roughly half of the total local supply), and which were "private" and no longer directly tied to the land, a person had to be a huerta landowner in order to use this water, and it could not be sold to anyone outside the community.

From a practical point of view, this means that nearly everyone in the community received some of the water during each rotation. They could transfer all or part of their right to someone else—particularly the "old" or feudal water, which was essentially the only kind of water sold in the local market—but that family would, in the same way, receive the additional water during their normal turn on the same schedule as everyone else, according to the location of their farm along its canal. Thus the main distribution cycle was governed, not by the "law" of supply and demand, but by a clearly defined and fixed set of rules decided upon by the

---

same time as the buyer's regular share of new water, in the same fixed order and on the same uniform schedule. The system worked well, with very few infractions, because of the institutional foundation upon which the market rested, the set of principles discussed in this article. Contrary to the impression given in the earlier published accounts, peasants almost never rented or sold their "new" water; such trafficking could only be engaged in by the wealthy nobles and owners of "old" water. The system did not in fact work very well in the view of the vast majority of the *huerta* members (water theft was rare because the system was quite transparent and because people lived in fear of the landlords), and that is the main reason why it was ultimately abolished and made a fully communal *tanda* system.

[5] Some of Alicante's water is external, supplied separately and sold by outside utilities and delivered through pipelines.

community—institutions determining the uniform frequency with which the water was available to everyone. And people only received this water if they had paid their taxes and made the other proportional contributions to maintenance upon which their rights depended. Only direct proportionality with land was lacking, and this was the case with only about half of the community's supply (the old, essentially feudal water). Once the description is clarified in this way, it becomes apparent that this was really at heart a turno system much like any other one in Spain, a community with a fixed order of rotation and a single cycle for the two kinds of water.

What all of this means, however, is that people's motives for rule compliance and water conservation have in this case been seriously misunderstood. Even though very small quantities of water were constantly traded in this community over a period of centuries (shares of the "old water" belonging to the big landlords), what was being "maximized" by people in this community by obeying the rules was not anyone's cash income, but rather how often they had the opportunity to irrigate their fields. The fact that people could sell any extra water that they did not use, or buy more if they needed it—which the small landowners normally did because of the supply of "new water" was chronically insufficient—or that a few people had rights to water which they usually sold because they no longer had much land, did not mean that such sales or such income was their main reason for being conservative and cooperative. This was clearly revealed in the fieldwork in interviews with the few people remaining in Alicante today who are old enough and knowledgeable enough to remember in detail how the "market" system worked (a system they generally refer to simply as a "tanda"). Some monetary income was a benefit, to be sure, but this was clearly recognized as secondary to the main benefit and logic of making the irrigation round as short and rapid as possible.

The primary benefit here—a maximal frequency of availability of very scarce water and a maximal frequency or intensity of its use—was, unlike any resulting income from water sales, shared equally by everyone. Thus in Alicante, just as in all the other cases mentioned here, we can speak of a close correspondence or compatibility, even a congruity, between private self-interest and the public good, in this limited sense. Given the prevailing scarcity, it was necessary for people (at least the peasants) to cooperate, and all of them experienced the benefits of doing so, directly and immediately, as well as the negative consequences of greed, of any failure of the common good to prevail. Such behavior had little to do with money, according to people who formerly participated in the "market," and everything to do with the circumstances of water use, the simple principles upon which people had agreed for centuries that water utilization would take place.

8. CONCLUSIONS

Envy, resentment, and conflict are inevitable in any human community, mainly because people are never really equal and all communities are stratified. And of course there will probably always be some free riders in any group of people. The important question is how many, and how pervasive their antisocial attitude can become under a given set of institutions and customary rules. But all of the communities described here show that "human nature" can be, in the right institutional setting, predominately social, communal and cooperative rather than selfish. And, indeed, these two supposedly contradictory motives can under certain circumstances become entirely consistent with each other and even appear to be congruent, or essentially equivalent, in the eyes of rational people. The system of rules and principles I have described here, which I call "the moral economy of water," accomplishes this. The tradition exists in south-east Asia, where water is scarce during only a fairly short season of the year, and also in places like Valencia, Murcia, Alicante, and Huaynacotas, where the resource is scarce under all but the very best of conditions. It is based on a consensus and a spirit of cooperation that, while not perfect, are nevertheless pervasive and have clearly prevailed for a very long time.

This, I would suggest, is where we should be getting our models of, and for, human behavior, and our images of "human nature." As for policy, these irrigation systems are clearly the places from which our models for building stronger institutions and better irrigation communities should come, especially for rebuilding communities that have long been rife with water conflict. With regard to irrigation water, I think that we can begin to speak now of a universal model that emerged independently in hundreds of places and reflects a worldview that was once pervasive among the world's peasant societies. Based on the central moral concept of equity, this general type of society, I think, provides us with an unprecedented case of parallel or convergent social evolution. It could clearly serve us well as a model for the greater human community of which we are a part, far better than any story of greed and tragedy, and far better than any abstract "market." For it is a clear expression of one side of our nature, as the eminently social and cooperative species we have always been whenever circumstances required us to choose which of our innate potentials was the more valuable one in the long run, and which one should come to the fore and be activated and reinforced. The model emerged independently in a great many places throughout the world, and it has endured for hundreds, perhaps even thousands, of years.

FROM PRINCIPLES TO ACTION: INCENTIVES TO ENFORCE COMMON PROPERTY WATER MANAGEMENT

Ellen Wiegandt
Institut universitaire Kurt Bösch
Sion, Switzerland
wiegandt@hei.unige.ch

Abstract: It is now commonplace to acknowledge the long history and broad geographical spread of common property resource management schemes. At present, common property solutions are also debated in arenas such as climate change (where the atmosphere is the commons) and water allocation. These discussions raise general questions about the efficiency and equity of cooperative solutions as well about the likelihood of their implementation. Several factors are relevant in this context: how actors overcome collective action problems in order to create a community of users (i.e., how to prevent "exit"), how to regulate overuse and thus prevent dissipation of resources (i.e., how to control "entrance"), and how to maintain open and democratic decision-making about property use to assure efficient and equitable exploitation. In the high mountain communities of the Valais, common property management of some water and land resources has a long history which has continued until today. The practice is well-documented, providing a wealth of data to examine the evolution of village-level institutions developed to regulate their uncertain resource base in order to meet the needs of current and future generations. This chapter argues that individual incentives and collective control produced systems that were both efficient and equitable. These historical solutions are relevant to contemporary challenges concerning collective action and resource management.

Keywords: common property management, cooperation, equity, efficiency

1. COLLECTIVE RESOURCE MANAGEMENT: REAL BUT NOT OBVIOUS

It has now become commonplace to acknowledge the long history and broad geographical spread of common property resource management schemes. Scholars have gone beyond the "tragedy of the commons" argument (Hardin 1968), which contends that common property systems

E. Wiegandt (ed.), Mountains: Sources of Water, Sources of Knowledge, 63–79.

by their very nature cannot produce sustainable resource use. In this view, dissipation of the resource occurs because each individual maximizes the benefits from his own use of the resource but shares the costs and negative effects with all other users. This process results in an incorrect evaluation of the relation between costs and benefits, which encourages overuse of the resource and leads, ultimately, to its destruction. Numerous empirical studies throughout the world contradict this conception by showing that common property arrangements are in fact present in many regions and have existed for long periods of time—an observation also made in Paul Trawick's chapter in this volume.[1] Moreover, these studies identify factors that in fact favor collective over private property rights systems.

Current debates about sustainable development and climate change take up these arguments to highlight the relevance for today of the kinds of common property solutions described for historic communities. Dilemmas posed by regulating the global atmosphere or allocating scarce water resources invite reflection on a range of management strategies, including property rights solutions. Key questions at the heart of these discussions revolve around how to foster the cooperation necessary to implement common property arrangements and how to assure that such cooperative solutions produce efficient and equitable outcomes. These are particular manifestations of the more general problem of collective action. The first basic question is how to reconcile private interests and the public good. Related to this are the issues concerning how actors create a lasting community of users (i.e., prevent "exit"), how they regulate use and thus prevent dissipation of resources (i.e., how to control "entrance") and, once the balance of numbers is achieved, how they maintain open and democratic decision-making about property use to assure both efficiency and fairness (how to foster "voice").[2]

In the face of such challenges, the passage from principle to action is not obvious. The case of water resources, with their specificities and associated management problems, highlights some key dilemmas. To address these broad issues of vital importance to our contemporary world, it is enlightening to explore local level and historic property management schemes. The Swiss case has been invoked by numerous scholars because of its persistence and its apparent effectiveness in regulating scarce natural

---

[1] The McCay and Acheson volume published in 1987 was one of the first to bring together case studies demonstrating the robustness of common property arrangements. Today there is not only a vast and interdisciplinary literature on the subject, but also academic associations devoted to the topic, such as the International Association for the Study of Common Property (IASCP).

[2] Reference here is to the work of A.O. Hirschmann (1970).

resources (Dubuis 1999; Netting 1981; Stevenson 1991; and the special conference volume of the Société d'histoire du Valais romand 1995). The particular example of the history of water use in the Swiss Alps is presented here to describe how some mountain communities successfully solved collective action problems and developed intricate common property systems to regulate the use of scarce and vital natural resources.

In communities at high elevations in the Valais, a mountainous canton in southwestern Switzerland, water and some land resources are still managed according to common property principles. Their history is well-documented, providing us with a wealth of data to examine the evolution over several centuries of institutions developed by local communities to regulate use of their uncertain resource base in order to meet the needs of each generation without compromising the welfare of future generations. At the heart of the system were incentives that made cooperation preferable to noncooperation. We will discuss these historical patterns of allocation in small communities faced with limited and variable quantities of water to illustrate some of the general problems of collective action and resource management that bedevil today's world and show how they were solved in the Alpine context.

2. CHARACTERISTICS OF WATER RESOURCES

A brief review of some aspects of water resources is helpful to highlight the particular problems their management entails. All those who study fresh water make the obvious point that it is essential for survival and has no substitutes. Not only is water necessary to maintain the life of humans, animals, and plants, it also furnishes significant amounts of energy, making competition over types of use inevitable. These multiple uses are subject to social decisions about allocation. However, available quantities are only partly determined by human action. Challenges imposed by natural forces have therefore always influenced the supply and distribution of water, leading to very early allocation schemes. Mountain regions, although often defined as "water towers" (see Viviroli and Weingartner in this volume), are not exempt from management problems. On the one hand, water may be overabundant and cause floods and landslides; on the other, it may not have optimal seasonal distribution. Moreover, the apparent upland advantage may be transformed into conflicts with downstream neighbors, as is starkly apparent in other chapters in this book by Güner (on Turkey) and Luterbacher and Mamatkanov (on Central Asia). The notion of abundance is itself a relative concept and evolves as societies' needs and wants change. Like the atmosphere, which until recently was a pure public good, pollution and climate in effect modify the quantity of water available by affecting its quality. Management strategies must therefore address these interrelated aspects of resources.

Property rights are a fundamental mechanism to achieve sustainable resource use. As a set of entitlements, they specify the rights, privileges, obligations, and limitations associated with an entity. Property rights are, in effect, quotas that control access and intensity of use. If working correctly, they should prevent resource dissipation and ensure that exploit-tation of any given resource will not hit diminishing returns. In this way, the given resource will continue to meet needs across space and through time with greatest efficiency. Different types of rules can apply, ranging from collective forms such as open access (the object of Hardin's critique) and common property, to individual arrangements embodied by private property. Because of the wealth of existing work on this subject, we will not discuss this in any detail here but will point only to several key contrasts among perspectives.

Economic orthodoxy has often presented private property rights as superior because they lead to more efficient use of the resource. Ronald Coase notably argued that negative environmental externalities would be minimized through competitive markets for them (Coase 1960). Both theoretical and empirical work question the generality of this conclusion, however. The fundamental precondition for such markets is the existence and enforcement of property rights but, as Dasgupta and Heal point out, these may be either difficult to define or difficult to enforce (1979). Flowing water, for example, cannot be treated as a separate commodity because is in a "constant state of diffusion"—or movement (Ibid. p. 49)—and a precise unit cannot be allocated to a single individual. In the case of pools of underground water, property rights can be defined as a given area of land, but it will never be clear whether the water extracted—because of its fluid nature—comes from below the designated area or from a wider area underlying the surface property of other owners.

The characteristics of water (and other resources like oil or fishing grounds) and various factors associated with some forms of their use have created conditions under which common property management is more likely to obtain: (1) The nature of the water and similar resources distinguishes them from others. Some resources like fish or animals move and must be "captured" to claim ownership. Water, air, and petroleum, on the other hand, cannot easily be divided into clearly defined units and must be "captured" through extraction. (2) Some resources benefit from economies of scale in their use. For example, preindustrial grazing land was often owned communally. Individually owned herds were small but could benefit from large grazing areas and their owners could also share oversight duties. (3) Maintenance or capital investment often requires large groups, and collective ownership of the resource may be an effective way to mobilize the necessary capital or labor. (4) Private property rights must be protected and guaranteed. Enforcement costs may be too great to be carried by individual owners. Patrolling vast expanses of water to defend fishing zones or controlling how much water various users are withdrawing from a river or underground pool are examples of such costly monitoring or enforcement. Often these different factors are interdependent and it is the combination of the resource's inherent characteristics, technological features, and the institutional configurations related to its management that explain the existence of common property arrangements.

Specific qualities of water bring together many of the features frequently invoked as favoring common property management methods. It has fugitive aspects because it must be pumped or collected to be used. It can be a common pool resource when it is found in lakes or underground pools. It often requires large infrastructures such as irrigation systems, dams, or pumps to make it available. Finally, it is often difficult to protect from diversion. That we can identify common property water management systems through time and across space is therefore not surprising, but their presence and persistence raise further questions about how communities come together and develop rules and technologies that produce efficient and equitable water management—i.e., that avoid dissipation of the resource and free-rider problems that plague all cooperative human endeavors. The historic patterns in the Valais are a window into how general principles we have identified with common property regimes were translated into actions that guaranteed broad and long-term access to a vital resource.

3. WATER USE IN THE SWISS ALPINE AGRO-PASTORAL SYSTEM

The texture of Alpine life has been amply described by historians who tell the story of populations living primarily off their land base for much of history. The importance of adequate water resources for agriculture is evident (see, for example, the special volume of the Société d'histoire du Valais romand, 1995; Netting 1981, or Wiegandt 1980). First impressions of the rugged Alpine landscape of the Valais raise immediate questions about how people survived over generations among the steep mountains where sun was plentiful but winters were cold and long, and water scarce. Low rainfall could not guarantee sufficient seasonal or even total annual water supply, especially as the population grew and evolved, making the history of the Valais also the history of the emergence of water management strategies. A solution that acquired importance beginning in the Middle Ages was to tap into the large stocks held by mountain glaciers and to develop an intricate irrigation system that brought summer melt waters to fields and pastures. The organization and operation of this system depended on cooperative resource management schemes that solved basic collective action problems inherent in the allocation of not only water but also of other resources critical to the Alpine production system, which was based on cereal grains, potatoes (after the eighteenth century), and cattle that provided milk, meat, and cheese.

Archaeologists and historians trace the origins of the Alpine agropastoral system to the Neolithic period, when grains were the staple food, complemented by small herds for meat and milk products (Dubuis 1990). Embedded in a feudal system, there is nevertheless evidence that plots were exploited and managed by households and that there was considerable autonomy exercised by communities as they took advantage of their physical distance from overlords. Complementing these household-based resources were collectively held forests, irrigation systems, and high alpine grazing lands. Coexistence of these private and communal resources had important consequences for resource management because it progressively led to norms and rules controlling the total number of people with rights to resources while at the same time protecting these rights from seizure by small elites.

The size of cattle herds, for example, was limited by mutually reinforcing rules. The number of cows that could be wintered on village territory could not exceed the amount of fodder produced from privately owned hay meadows, and only those cows that a household could winter could be sent to the communal summer pasture (*alpage*) (Netting 1981, p. 61). Moreover, the total number of animals allowed on each summer pasture was a fixed number that was set early in village history. The distribution of these shares, which could be inherited, bought, sold, or rented, determined which owners could alp what number of cows. These combined requirements simultaneously discouraged accumulation of private resources and individual capture of the commons and prevented overuse and depletion of common resources.

Controls over population growth constituted a parallel to rules about the use and distribution of resources. Demographic mechanisms and cultural rules influenced population size and growth rates. Consistent with the Western European demographic regime, Alpine communities were characterized by late age at marriage, high celibacy rates, and high emigration rates (Netting 1981; Wiegandt 1980), keeping population size in balance with the resource base. The allocation of private resources was achieved through partible inheritance. Division of estates at each generation tended to equalize the average size of family holdings because of the observed tendency of wealthier families at any given time to have more children. As a consequence, their estates were divided into relatively more shares than those of poorer families (Wiegandt 1977, 1980).

Private holdings could be freely bought and sold as well as inherited, but communal resources, necessary to make any exploitation viable, could only be acquired by inheritance through the paternal line or sometimes through marriage. This effectively limited the number of in-migrants, who would not automatically have access to vital resources. Rules for granting or restricting *bourgeoisie* status (community membership) rights to other inhabitants, such as spouses of members or new families, were loosened or tightened at different times in history to adapt to conditions that warranted either higher or lower population levels (Wiegandt 1977).

The integrated system of population dynamics and resource use and management that evolved in the Valais prevented overexploitation and dissipation of a scarce resource base and thus avoided a tragedy of the commons. How did this come about and how were individuals or small groups prevented from gaining monopoly control over critical resources? Institutional innovations in periods of crisis shed light on factors that translated principles of solidarity into action, which then became part of the characteristic Alpine culture.

4. THE EMERGENCE OF A WATER MANAGEMENT SYSTEM

As the Alpine production system successively met challenges from the environment and from the constraints and opportunities of the evolving Swiss state (Wiegandt 2004), it elaborated strategies that resolved the tragedy of the commons dilemma without eliminating individual incentives. At each step, choices were made to address particular needs; the cumulative effect was a system that smoothed economic differences among families over time and fostered cooperative behavior. A major shift in the production system during the Middle Ages illustrates how changes in basic conditions led to a response built on existing norms, reinforcing some in ways that in turn produced the features of mixed property systems and collective commons management.

In the Middle Ages, cereal grains formed the basis of the Alpine productions system. Evidence from the historical record shows that most land for these crops was privately owned. In the fourteenth century, the Plague struck the region, decimating the population, as it did throughout Western Europe. The epidemic had many consequences that have been amply studied. One was to change the dynamics of supply and demand for foodstuffs. Le Roy Ladurie (1971) has demonstrated for other regions and sectors that the decline in population following the Plague favored new technologies and production strategies. Slowly there was a general increase in well-being, accompanied by new patterns of social relations as some individuals and groups profited more than others from new opportunities.

In a similar vein, Dubuis (1999) shows that, in the Valais, the high mortality caused by the Plague and the ensuing drop in population led to abandonment of arable land. It was neither left fallow nor later returned to grain fields, however. Through land records, he traces the shift in land use patterns and the increase in number of cattle and stable buildings as evidence of the newly emerging production system that transformed fields into hay meadows. There were two main reasons for this. Lower population levels meant decreased demand for cereal grains. This was to some degree countered by new opportunities from increased demand for meat from cities in the Piedmont and Lombardy regions of Italy. To respond to both of these changes, there was a move to increase animal production, which is less vulnerable to changes in demand because animals need not be slaughtered for their meat at any predetermined time and their by-products can be stored if milk is turned to cheese. The result of the shift in emphasis from grains to herding was a greater flexibility of the system in responding to variations both in supply and demand because of the "storage" capacity represented by live animals. At the beginning, however, it also underscored social differences because, according to Dubuis, the new system was introduced primarily by wealthier peasants, who would be most able to absorb the risks associated with the change in production strategies (Ibid., p. 85).

Despite its adaptive advantages, cattle-raising in the Alpine environment solicited the local environment in a different way and placed new burdens on the organization of labor. Grain production persisted because cereals remained an essential part of diet. There were significant advantages in producing these foodstuffs locally in this Alpine environment, where transportation was arduous. Food self-sufficiency was linked to political autonomy, given that resources were also owned and managed by users. These positive effects were nevertheless achieved at the expense of greater pressures on time and organization of work. Scheduling of tasks became more complicated in this mixed economy. Cattle require constant attention throughout the year, making summer months extremely busy. Not only were there the regular feeding and milking chores during the summer, grains had to be harvested. And, with the increased importance of cattle production, there was the additional task of cutting hay for wintering-over of the cattle.

The need to produce hay introduced competition for space as well as for time because land nearest the village was needed for several purposes—for grain fields, haying meadows, and pasture. Possibilities for territorial expansion were limited, as the earliest historical documents preserved in the majority of Valaisan communities attest. Most of the documents concern various land grants and records of arbitration between communities over boundary conflicts, thus suggesting that there was little unclaimed land even in early historical periods. The only choices were to expand into previous unused lands, such as high-mountain pastures, or to deforest. The perennial need for wood for heating, cooking, and construction made forest preservation an important priority, and therefore the area of forest that could be cleared for additional meadow or pasture was also limited. Communities early on set limits that kept the balance between forest and cultivated land relatively stable throughout the ages. Acquiring pastureland at some distance from the village (up to several days' walk) was a strategy adopted at certain times by some villages (Netting 1981, p. 51) but this put pressure on the efficient use of a small labor pool composed only of villagers. The addition to the resource base of a previously unused high summer pasture thus conferred certain advantages. It provided an additional source of nourishment for the animals. It also became a way to optimize villagers' time. All the village cows grazed together in the summer pasture and were cared for by a small team, thus freeing their owners from milking and feeding chores and enabling them to spend time to cut hay and harvest grains. This strategy was adopted throughout the region and has become an essential feature of the Valaisan agro-pastoral system.

This new and major need to produce additional animal fodder was introduced by the intensification of herding. Growing more grass could not be accomplished with the low levels of precipitation in the region, however. What we observe in the historical record is that peasants increasingly turned to the water produced by snow melt and glacial runoff. They could only efficiently exploit these sources by bringing water from high altitude streams and developing a means to distribute it over the arable lands around the village. Thus, during the course of the fifteenth century throughout the Valais local inhabitants designed and built their complex system of small, locally, and collectively managed irrigation systems (*bisses*).[3] In doing so, they were confronted with significant cooperation and collective action problems that we have outlined in our earlier theoretical discussion on water management, where we also suggested several factors that would encourage their configuration as a common property resource.

Runoff from glaciers poses an upstream-downstream problem because those at higher altitudes could divert water, leaving little or none to downstream users. In the characteristic Valaisan land use pattern, individuals held land both at higher and lower latitudes (to spread risks of microclimatic factors) and thus everyone had an interest in guaranteeing equitable distribution throughout the territory. The nature of the system also favored cooperative management. Although the irrigation canals were not major infrastructures, they nevertheless required levels of labor input and coordination among users that went beyond household capabilities. Each community faced the same limits in potential labor or capital inputs, making water trade between communities with unequal investments unlikely. These factors correspond to the general criteria outlined earlier as favoring collective management systems over private ones and helps explain their emergence in the Valais.

At the same time that collective ownership was being consolidated, private ownership of individual meadows and fields persisted, leading to the characteristic mix of common and private holdings in the Valais. The coexistence of the two forms is fundamental in explaining the puzzle of why and how principles of common property and collective management could be translated into a working system that put these principles into action. The interaction between private and collective resources facilitated the resolution of the well-known free rider problems associated with public goods and allowed Valaisan communities to achieve the level of cooperation that was necessary to create and maintain collective management of water and land resources.

Herding and the attendant upkeep of pasture and irrigation systems brought new kinds of returns that could only be achieved through greater labor investments, a different use of time, and the reallocation and reorganization of resources—particularly water. Digging irrigation canals throughout the village territory to serve each individual field, pasture, and meadow,

[3] Dubuis notes that additional research will undoubtably produce a more nuanced account of the origin of irrigation systems and their collective management systems. Even within the already particular case of the Valais, important regional differences exist, especially between the lower part, which he analyzes, and the central and upper sections of the Rhône Valley where *bisses* existed already in the thirteenth century and where cattle trade was less well developed and villages more dependent on local production.

and then maintaining the channels and assuring the flow of water in an equitable and effective way, required significant inputs of labor. One solution would have been to hire additional labor, but most inhabitants of the community, even in the Middle Ages, were property owners themselves and it is not clear how many might have been willing or able to spend time working for others rather than on their own land. Bringing in outsiders as hired laborers would have been difficult because they would have been excluded from access to all the common resources that were an essential part of each holding. Difficult terrain would preclude traveling from any great distance to work as day laborers and the nearest communities would also primarily have only members who were also landowners. Moreover, even if surplus labor had been readily available, the land use pattern of scattered plots raised problems of how to control potential shirking. Additional laborers were thus a scarcity and this granted them bargaining power, especially since most of them would have had a viable resource base. We can imagine that some might be tempted to reduce the uncertainties in supply that the shift to cattle raising brought, but only if they were granted some attractive incentives. These might have included access to new water resources in exchange for their labor to establish and maintain the system. In such an arrangement, both parties would benefit from the increasing returns to scale associated with the new resources.

This scenario is speculative but tracks the historical data, which show that small numbers of villagers initially transformed grain fields into meadows. These were then followed by larger numbers who began making this conversion during the same period that saw the construction of irrigation systems and the introduction of their collective management. This description goes some way in explaining how individuals could benefit from their decision to join the commons; however, it does not explain how the system could maintain the benefits and protect the resource from the desire of too many to join, which would result in overexploitation.

We mentioned earlier the relationship between winter and summer feeding of cows and the implicit limit that the quantity of private resources placed on the number of cattle that could have access to communal resources. This in turn provided a fixed upper limit to the number of total use rights available for each given pasture. Water rights were similarly limited by a ceiling on their numbers and their association with particular plots of individually owned land. If an individual cleared a new plot, it would not have automatic access to irrigation water. Each water right translated into the time water would flow to a particular parcel of land, which was the equivalent of setting boundaries on the quantity available for each field or pasture and for the overall cultivated territory.

The mix of common and private property was thus the mechanism that set upper bounds on the number of users. It also prevented individuals from acquiring dominant shares of communal rights. Total shares of both water and pasture rights were fixed and those not inherited had to be purchased. They would have to be bought from fellow villagers equally dependent on them. Either other villagers would not relinquish the rights because they were essential for the survival of the household, or they would only do so at a very high price. Given that the buyer was another villager with very little surplus resources or capital, his capacity to meet the price would be limited. Using one's private resources to acquire rights to common ones was self-limiting because of the interaction between the two types of resources. Water rights were linked to specific plots of land so relinquishing a plot would obviate the need for water. Seeking to winter more cows in order to acquire more alpage rights would imply intensifying hay production, which necessitated more water and more labor. Both the number of cows a domestic unit could feed and the amount of hay it could cut with its own labor pool were limited. To hire additional labor in a largely nonmonetary economy meant generating surpluses beyond subsistence levels. This again required access to communal resources of water and pasture and, moreover, would produce few benefits because surpluses would go to pay the additional laborers. These factors all served to prevent the emergence of elites.

Not only access to resources was controlled. Decision-making power over resource management and use was also shared among community members through its decentralized structure. Water was not concentrated in a single main channel but was distributed among several canal systems, each with its separate management structure. Each village also typically had more than one high alpine summer pasture. A separate (but often overlapping) slate of user-managers was elected for each association to oversee the maintenance and operation of the resource in question. The small size of the membership groups, the likelihood that most members would at one time serve as a manager, and the intimate knowledge that each user acquired about the nature of the resource and its management, meant that everyone had equal and relatively complete information about the system. Because of their deep understanding, even when they were not part of the managing committee, villagers would closely follow decisions and undertake their own surveillance to assure that they received their share of water or pasture. Control was thus highly decentralized, another factor making it difficult for individuals or small groups to make any concerted attempt to capture control of these key resources. Maintenance was also a shared task of all the users, who formed corvée labor teams to clean the canals and pastures of debris every year and to assure the free-flow of water in the irrigation canals during the watering season. The participatory decision-making that gave all users a stake in the system was accompanied by obligations to assure the persistence of the resource.

Harsh environmental conditions made it essential that a solidary group invest in and maintain resources. We have outlined the factors that tightly linked population and resources, such as late marriage, high celibacy and emigration rates, and citizenship rules that limited outsider access to common resources. In addition to these controls on population size that helped maintain a certain balance with the resource base, partible inheritance rules assured a circulation of resources throughout the community, protecting both private and public goods from overexploitation. Given the limits on private goods, the level of public goods was high to protect individuals and the community as a whole from the variability and uncertainties of the Alpine environment. Hirshleifer (1983) has categorized different forms of public good provision in different kinds of society. What we observe in the Valais corresponds to Hirshleifer's "weakest link" solution. Among the several possible variants of public good provision that he identifies, this form refers to systems in which a high degree of cooperation obtains because each individual's contribution is essential to prevent the collapse of the overall system. Thus, free riding must be kept to a minimum because it threatens the society as a whole.

Our description of Valaisan society has shown that numerous practices evolved to share risks from harsh and unpredictable climate. A single household could be devastated by an early frost, a late snow, or a rainy summer. Microclimatic effects typical of Alpine environments led to spatial variability in crop success. For this reason, a family domain spread its holdings across village territory. Additional protection came in the form of access to common resources that had to be protected for the good of individuals but also for the good of the whole. Without thriving individuals and families, the community could not survive because it relied on members to maintain the collective resources. These were in turn essential to form that buffer that kept individuals from slipping under minimal subsistence levels when crops failed. The practices associated with the use of public goods incorporated incentives for individuals to maintain them and also included rules that assured that they could not be captured for individual benefit. Information was widely diffused (there are rarely secrets in small communities) and governance decentralized. If knowledge was shared, power was more easily distributed and principles more easily became actions.

Our historical analysis suggests that the integrated system of resource use and management adopted in the Valais was based on the creation and maintenance of public goods. It required and reinforced a high level of cooperation, which contributed to the success of mountain communities in preventing overexploitation and dissipation of a scarce resource base. A central feature of the system was the combination of private and collective types of property arrangements. The coexistence of different tenure systems is longstanding in the Valais, but as we have seen from the history of water management, the precise details of property arrangements governing resources varied over time. Institutional foundations for both private and common property management coexisted, giving villagers options when faced with new circumstances and challenges. This flexibility was surely important in the successful transition to new productive strategies. Its form is characteristic of the Valaisan system but similar mechanisms exist in very different cultures, such as among the Tigray in Ethiopia (Bauer 1987). To respond to changing population dynamics in a desert environment, villagers meet to redefine land between private and common forms of tenure over the space of generations. Myriad examples of flexible tenure systems could be described here but drawing on the Ethiopian case is simply to suggest that adaptive strategies can take the form of rapid or progressive shifts in their coexistence.

Contrary to arguments that claim superiority of either private or common forms of property, the Valaisan case study suggests that what matters is that rights must be clearly defined and the institutional framework must be able to enforce the rules. It also seems that when resources are scarce or their supply unreliable, mechanisms to assure broad access favors persistence of the community. In the Valais this access was achieved through norms encouraging a high level of cooperation among all community members. When these principles do not pertain, overuse is more probable and can lead to overexploitation and environmental degradation. Subsistence production is in fact particularly vulnerable to such outcomes because workers who depend on their own labor will produce even for declining returns just to assure their minimal needs. Real costs of production are not taken into account in the price, which leads to ever-increasing intensification and, ultimately, to resource depletion. This process characterizes many developing countries whose systems operate in world markets in which other regions have well-defined property rules and cost internalization.[4]

Valaisan peasants were subsistence producers like many contemporary farmers—particularly in the developing world—but they did not enter into the spiral of intensification and exhaustion of the resource base. They benefited from the closed nature of their community, where commerce was

[4] See Chichilnisky (1994) for a detailed development of this argument.

a complement to but not an integral part of the system during many historical periods. This high level of self-sufficiency accompanied by some degree of autonomy in decision-making over resource allocation kept outside powers at bay. During the feudal period, for example, many parts of the Valais used these leveraged freedoms to resist feudal domination and subsequently continued to keep their distance from the emerging state. Surely equally important was the fact that property rights, whether private or common, were clearly specified and effectively enforced and monitored.

Highlighting the central role of property rights in shaping economic and political processes is consistent with the views of recent scholars who have also argued that the appropriate definition of property rights can solve environmental problems (Chichilnisky 1994; Dasgupta 1992; Dasgupta and Heal 1979; and Demsetz 1967) They have also claimed that property rights solutions are in fact superior to other regulatory mechanisms (Chichilnisky 1994) for attaining sustainable resource use. The coexistence and interaction of both common and private property regimes has been identified in the case study presented here as a key factor in avoidance of resource depletion and in the equitable distribution of resources and risks. Moreover, the mix of private and common property structures in the Valais was the source of valuable flexibility in institutional response. Changing production technology, exemplified by the passage from cultivation to herding, facilitated transition from one property form to another. Once the two coexisted, they were each part of an enlarged set of available strategies to respond to changing environmental, political, or economic circumstances.

5. PAST PRINCIPLES, FUTURE ACTIONS

The contemporary efforts to craft environmental and resource regimes that provide equitable access to critical resources such as water or the atmosphere also rely on existing mechanisms and draw upon the theoretical findings underlying the analysis of different property rights systems. In this context, historical studies provide empirical evidence for the complexity and variability of arrangements, as well as the strengths and weaknesses of each. Common pool resources pose problems of definition and enforcement of rights. The shortcomings of private property rules identified theoretically by Dasgupta and Heal (1979) are implicitly confirmed by the common property strategies elaborated in order to overcome the distortions that can be introduced by private property principles. Of particular interest for contemporary environmental policy is the evidence of the coexistence of common and private property systems to regulate a single resource, either at one time or over time. Both forms of management can provide valuable flexibility in institutional response and facilitate adaptation to changing production technologies. The passage from primarily grain cultivation to herding has been discussed in some detail for the Valais of the Middle Ages. More recent examples demonstrate the similar types of response. In the early part of the twentieth century, communities in the Valais used their collective control over water resources to negotiate favorable contracts with producers of hydroelectric power. They in effect rented the potential power derived from water flowing down the slopes within their territory to companies that exploited the force to generate electricity. Communities did not relinquish rights to the resource itself but sold rights to the energy implicit in its downward flow.

Currently the global population is faced with the dilemma of how to regulate public goods that are being threatened by overuse. The negotiation of the Climate Change Convention and the Kyoto Protocol address the problem of the global atmosphere as expressed by global warming. Presently there is no international agreement to regulate transboundary water use. In the context of our historical analysis, it is noteworthy that present-day environmental management strategies are based on the definition of new property rights and the mix between common and private rights systems. The mechanisms devised by Kyoto Protocol to regulate CO_2 emissions recognize the atmosphere as a common property resource but propose the creation of tradable permits that have characteristics of private property, similar to the shares in Valaisan pastures or irrigation systems. How these property principles will work out in practice will depend on governance and enforcement issues that were also foremost in historical arrangements.

The increasing inclusiveness and democratization of decision-making within local associations that characterized the introduction of new property rights to irrigation water in the Valais may also be relevant in thinking about ways to mobilize contemporary populations concerned by climate change or water shortages. For example, in the case of climate change, developing countries see no benefits in changing their emissions behavior while industrialized countries realize that no viable use of the global environmental commons is possible without developing country participation. It is imperative that incentives be devised at the individual (state) level in order to create a willingness to join collective and global efforts to regulate resources with public good aspects.

Debates similar to those over climate change are emerging in the public arena over the optimal systems to assure access to freshwater, pitting those who believe in open access to water and those who recognize the potential this has for its overuse and dissipation. Many who distrust common property solutions propose mechanisms such as pricing schemes to confer value on the resource. They claim that this will define efficiency criteria and thereby lead to optimal levels of use. Others argue that people have an intrinsic right to essential quantities of fresh water.

As we stated in the beginning of this chapter, both private and common property systems have strengths and weaknesses. Some types of common pool resources pose particular problems to pure privatization related to issues of definition and enforcement of property rights. In these cases, equity and efficiency criteria can be met through common property rules. In the absence of strong existing institutions to protect private rights, cooperative solutions can emerge from a set of mutually reinforcing incentives and controls over individual efforts to acquire exclusive possession (of private rights) or to free ride (in collective systems). The difficulty is in achieving a proper balance. To many, private property rights lead to inequity and lack of innovation (Heller 2000), while uncontrolled common access ends in resource dissipation. Both outcomes threaten not only the social system as a whole but also each individual member. Common property solutions can address both challenges to sustainable resources use and social cohesion. The historical example of the Valais is evidence that property regimes are not immutable and that societies can innovate and solve problems posed by changing relations among their environments and social needs. Exploring concepts and theories may reveal other new arrangements that would improve welfare and protect resources. There is thus a continuing need to confront the puzzles of resource management and to identify potential gains from cooperative strategies.

MULTIPLE USES AND COMPETITION FOR MOUNTAIN WATER

HYDROELECTRIC RESOURCES BETWEEN STATE AND MARKET IN THE ALPINE COUNTRIES

Franco Romerio
University Center for the Study of Energy Problems
University of Geneva, Switzerland
franco.romerio@cuepe.unige.ch

Abstract: This chapter analyzes hydroelectric exploitation in the alpine countries and provides a case study on an installation situated in Switzerland. In particular, it examines the implications of the opening of electricity markets to competition, which raises deep controversies. It takes into account the institutional, socioeconomic, environmental, and energy aspects of the problem. It outlines the historical roots of current problems, as well as the prospects for the medium and long term. As a whole, the article provides a rather positive judgment on the hydroelectric exploitation's prospects, in spite of the complexity of the problems and uncertainties. Concerning the mountain regions, it suggests taking advantage of the opportunity provided by the electricity sector's reorganization in order to improve their situation as electricity producers and traders. In this respect, they should concentrate on peak production and promote green labels.

Keywords: hydroelectric power, electricity market liberalization, alpine hydro capacity and output, costs and benefits

1. INTRODUCTION

Alpine hydroelectric power resources play a major role in energy provision in Europe, providing peak and off-peak-period energy for industry and domestic use and guaranteeing the stability of transport networks in terms of frequency and voltage. They also represent a major factor in the socio-economic development of the mountain regions, generating new infrastructures, significant economic activity, and part of local government's fiscal revenue. Their environmental effects, however, are many and contradictory: regulation of watercourses has been accompanied by their impoverishment, the Alpine scenery has been disfigured, and stunning engineering feats have been performed. The risk of disaster remains, as the collective

E. Wiegandt (ed.), Mountains: Sources of Water, Sources of Knowledge, 83–102.

memory of the tragedies at Fréjus and Vajont reminds us.[1] In many cases the harnessing of hydroelectric resources has provoked local opposition because of its environmental impact or the feeling that economic returns were inadequate.

In the past, exploitation of these resources was a monopoly controlled by regional or national companies from the public or private sectors. For some years now, however, Europe's electricity sector has been undergoing radical reorganization, notably via the introduction of free market competition.

In the first part of this article, I present an overview of the changes under way in the hydroelectric sector of the Alpine arc countries. I examine the following issues: electricity supply, market reorganization, price and time differentials, hydroelectricity production costs and alternative technologies, external costs and climatic changes, eco-marketing and water royalties. In the second part, I carry out a case study based on a hydroelectric installation situated in the Swiss Alps. More precisely, I analyze the installation's characteristics, institutions, market, external costs, water royalties, and local population opposition.

2. HYDROELECTRIC RESOURCES IN THE ALPINE ARC

2.1. Electricity Supply[2]

The hydroelectric resources of the Alpine Arc were largely explored and harnessed in the course of the twentieth century. Overall 76 percent of their technical potential is now being used, with France achieving 97 percent and Slovenia only 45 percent. In general, new projects run into both economic and ecological obstacles. The present low level of investment in this sector can be explained by construction costs and the declining returns of hydroelectric sites. In addition, major projects have been dropped in response to growing awareness of environmental and landscape issues. Table 1 provides data on the hydroelectric contribution to

[1] The rupture of the Fréjus reservoir on December 12, 1959, resulted in 400 deaths; the crumbling of ground in the Vajont reservoir on October 9, 1963, resulted in some 2,000 deaths (cf. Commission Internationale des Grands Barrages 1974, pp. 33–35; Lave and Bayvanyos 1998).

[2] Cf. Association of Power Exchanges 2001; Bartle 2002.

electricity supply in the Alpine countries and more generally in the Union for the Coordination of Transmission of Electricity (UCTE)[3]:

- Hydroelectric plants represent 24 percent of the total capacity of the Alpine countries; the proportion is 8 percent in Germany; 22–31 percent in France; Italy, and Slovenia; and 68–77 percent in Austria and Switzerland.
- For the region as a whole, 16 percent of electricity output is of hydroelectric origin: 5 percent in Germany; 13–25 percent in France, Italy, and Slovenia; and 59–71 percent in Switzerland and Austria.
- In UCTE countries as a whole, 23 percent of capacity and 13 percent of output are of hydroelectric origin; 63–64 percent of hydroelectric capacity and output are situated in the Alpine countries.

Table 1: Electricity Capacity and Output in the Alpine Countries and the UCTE, Average 2001 and 2002[4] (UCTE, Half-yearly Report, I-2003, pp. 6 and 36–40).

	Total capacity	Hydro	Total output	Hydro	Nuclear	Fossil	Other
	GW	%	TWh	%	%	%	%
Austria	17	68	53	71	–	29	–
France	111	22	514	13	79	7	–
Germany	101	8	502	5	31	63	1
Italy	76	27	268	19	–	79	2
Slovenia	3	31	13	25	39	35	2
Switzerland	17	77	68	59	38	4	–
Alpine countries	325	24	1,419	16	42	41	1
UCTE countries	516	23	2,213	13	34	52	2

The plants with daily or seasonal storage, and in some cases with pumping facilities, are of great importance in terms of electricity supply security and transport network stability in that they enable modulation of output according to daily and seasonal fluctuations in consumption. Water stored in the dams during the thaw, or during the night with the aid of

[3] The UCTE coordinates electrical systems and facilitates energy exchanges in 20 countries.

[4] In 2001, the contribution of hydropower was very important due to hydrologic conditions; in contrast, in 2002 it was below average.

pumps, enables them to meet winter and peak-hour requirements. As a rule the price of electricity is higher during these periods.

Hydropower is the basis of a large part of trade in electricity between the Alpine Arc countries and more generally in the UCTE. Austria and Switzerland export peak-hour energy generated by storage plants in the Alps and during the night import base-load energy produced by the nuclear and coal-fired power stations. France and Germany export base-load energy and import peak-hour energy. Italy buys large quantities of electricity from France, the transport networks crossing its border often being overloaded. Slovenia plays a marginal role because of the small dimension of its electric system and lack of capacity. Because of its geographical location, Switzerland is a veritable focal point for electricity trading, with its storage plants allowing frequency regulation on the interconnected networks.

2.2. Opening Electricity Markets to Competition[5]

In the past, exploitation of hydroelectricity was a monopoly controlled by regional or national companies in the public or private sectors. The European Union directives of December 1996 and June 2003 aim for complete but gradual opening of electricity markets to competition. In 2003 the level of electricity market openness in Austria, France, and Italy was less than 50 percent; in Germany it was theoretically 100 percent, but in practice probably less. Slovenia is in the process of deregulating its market under legislation passed in 1999. In Switzerland, the referendum which turned down the electricity legislation on September 22, 2002, has not stopped the reorganization process. New legislation is in preparation.

The opening up of markets to competition means increased international integration of national markets. However, a hurdle exists in the form of the bottlenecks affecting international interconnection. We are also witnessing modification—and even outright abandonment—of the autarkic policies of the past, especially with respect to electricity self-sufficiency.

The old monopoly system enabled the electricity companies to invest in the hydroelectric sector without running any great risks. This naturally meant scaling investments to fit the market or finding buyers for excess output; taking account of increasing competition from alternative energy sources; and pursuing a pricing policy consistent with their strategic objectives and directives from national and local authorities. It should be

[5] Cf. Glachant and Finon 2003; Romerio 2002b; Romerio 2003a.

pointed out here that the European market within which the big electricity companies were operating was oligopolistic in character.

Now faced with competition, the companies can no longer afford to ignore the market risks, in terms of prices and costs. Hydropower is undergoing the free market's ordeal by fire. Below we discuss the decisive short-, medium-, and long-term factors involved in keeping hydropower competitive, notably in terms of prices, the value of flexibility, production costs, alternative technologies, external costs and benefits, climatic change, eco-marketing, and the rent issue.

2.3. Electricity Prices and Time Differentials[6]

Surplus on the market has the effect of bringing prices down in a way that ultimately reflects the variable production costs of thermal power stations, and notably of their fuel. A shortfall can send prices rocketing. While waiting for fresh investments we will have "rationing prices," which are higher than the cost of developing production capacity. In a balanced situation, prices cover total costs, in particular during peak hours. In practice, however, equilibrium is the exception rather than the rule.

Figure 1 explains pricing formation on a competitive electricity market. System capacity is shown on the horizontal axis and prices on the vertical. The lower stepped curve represents variable cost and the upper one total costs. The slanting lines illustrate the demand functions:

- with D_a, the situation is one of excess capacity and prices will tend to reflect variable costs (C_V);
- with D_b, there is no excess capacity and prices will cover total costs, both fixed and variable (C_T);
- with D_c, prices will rise steeply to limit demand to available capacity (CAP); this is a "rationing price" situation.

New investment is called for to meet increased demand. This creates a zone of extreme uncertainty, especially when the system's security margin becomes too narrow.

The surplus that has characterised the European market for several years up to 2001 has pushed the spot price down and partially eroded the time differential between winter and summer, and peak and off-peak periods. Nonetheless, autumn 2000 brought a clear increase caused by rising oil prices. Data provided by the German Power Index (GPI) show

[6] Cf. Association of Power Exchanges 2001; Graves, Jenkin, and Murphy 1999; Frayer and Uludere 2001; Ferraz and Romerio 2004.

that the average price of peak period electricity was €36/MWh in 2001 and that of base-load electricity €28/MWh. Price volatility was very marked. To quote one example, because of the cold snap in Europe, on December 19, 2001, the peak-hour GPI reached €477/MWh, while the base-load GPI was €323/MWh. In reality, prices rose even higher.

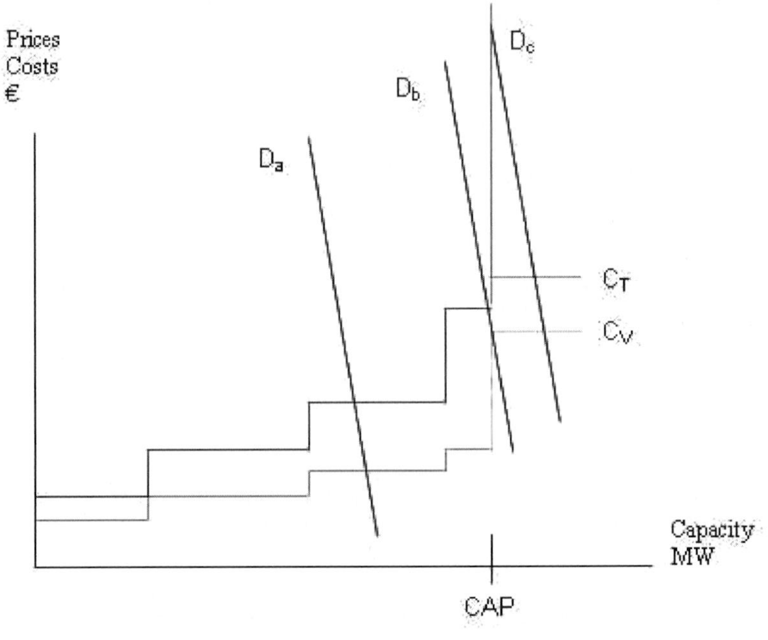

Figure 1: Determination of Electricity Prices.

Their inherent flexibility means that hydroelectric storage plants can take advantage of daily and seasonal market fluctuations, since their turbines can be activated or stopped in a matter of minutes. In this respect, the creation of spot and futures markets allows this kind of plant to optimize its potential. The regulating markets (operating reserves and grid support) offer similar opportunities.

Estimations of the value of hydropower developments fail to take this factor into account. For example, analyses based on the discounted cash flow method (DCF) give no consideration to plant flexibility or price volatility. Thus they underestimate the value of the plant. Taking the flexibility factor into account means using the real options theory, which has it that increased price volatility—and thus increased uncertainty—results in an increase in the value of the option. Seen in this light, price indices of the GPI type are of little use, in that they provide average values.

2.4. Hydroelectricity Production Costs and Alternative Technologies[7]

Unfortunately, data on production costs for hydropower developments in Europe are extremely limited and the results of an investigation we carried out were rather disappointing. Thus we shall settle for providing the following rough estimates:

- €25–35/MWh for run-of-the-river plants (base-load energy)
- €35–45/MWh for storage plants (peak-hour energy)
- €45–55/MWh for pumped storage plants

It should be pointed out, however, that spread around these averages might be quite high.

In reality, production costs for hydropower plants depend on the geographical characteristics of the site, the construction period, the degree of automation, depreciation and reserves policy, and interest rates.[8] We must not lose sight of the fact that the investment called for is massive, and that the payback period is very long. Company rationalization measures can reduce production costs as far as inefficiencies are present.

The current alternative to the run-of-the-river plants—especially when plant overhaul is an issue—is combined cycle gas turbines (CCGT), which provide base-load electricity and heat at very competitive prices. By contrast gas turbines (GT) are less worthwhile as replacements for storage plants because production costs for fewer than 1,000 h of use are relatively high.

It should nonetheless be emphasized that the hydroelectric–GT/CCGT comparison involves several unpredictable factors, in particular the price of gas, heat use, taxes on CO_2 emissions, and interest rates. Moreover, the perfect substitutability hypothesis needs to be applied with prudence in the hydroelectric–GT/CCGT context, where such other factors as availability of sites can hamper dissemination of GT/CCGT in favor of existing hydropower.

New technologies are presently being looked at, among them hydrogen-based energy storage, which could represent an alternative to hydropower storage plants some decades from now.

---

[7] Concerning alternative technologies, cf. Oud 2002.

[8] The problem of whether depreciation should be calculated according to the plant's historical value or reconstruction value has long been, and remains, a subject of vigorous debate.

2.5. External Costs and Climatic Changes[9]

Internalization of external costs generated by hydroelectricity production can penalize this energy source, but external benefits can add to its competitiveness. What should not be forgotten here, in particular, are the environmental nuisances that ensue when residual flows in rivers and streams are reduced below certain limits. Compulsory civil liability insurance aimed at more comprehensive disaster coverage than in the past is another penalizing factor for hydroelectric energy.

On the other hand, internalization of external costs due to combustion of fossil energy and nuclear fission makes hydropower generation more competitive. A tax on CO_2 can only have a positive impact. Concerning the risk of major accidents in nuclear power plants, it is not directly comparable with the risk of collapse of the reservoir or the crumbling of the ground at its interior, as is sometimes stated. In both cases, the probability of accident is very low. In the case of hydroelectric production, we are however confronted with damages, certainly very significant, but caused by an element—water—that is at the origin of life on earth. In the case of nuclear power plants, on the other hand, we are confronted with radioelements which have extremely high half-lives, and relatively little known carcinogenic, mutagenic, and teratogenic effects.[10]

Unfortunately, very little is known about the impact of climatic change on hydroelectric output. An often-cited example is that of Colombia, where El Niño apparently caused a long drought that emptied the reservoirs and pushed prices up drastically on a market already open to competition. In this respect, the changes undergone by rates of flow, snowfall, and glaciers must be looked into, together with the emergence of natural hazards—like landslides—capable of affecting hydroelectric developments. The problem is not new. In 1958, Aemmer stated the problem in these terms: "The constant rise in the average temperature has brought about a fairly general retreat of all the glaciers.... This will affect the filling of the lakes and also the availability of supplies during dry summers."[11]

---

[9] Concerning external costs, cf. Finon 1996; Krewitt, Heck, Trukenmüller, and Friedrich 1999. Concerning climatic changes, cf. Garr and Fitzharris 1994; International Panel on Climate Change 2001, pp. 399–401; Harrison and Whittington 2001.

[10] Cf. Romerio 2002a.

[11] Cf. Aemmer 1958, p. 4.

2.6. Eco-Marketing[12]

Many electricity companies have identified categories of consumers ready to buy ecologically produced power at relatively high prices. These companies are now in the process of fine-tuning their marketing strategies and organizing provision of energy via hydroelectricity, solar energy, and windmills. Special "green electron" and "blue electron" standards are being drawn up. Hydroelectric energy from small plants of less than 5–10 MW or from plants whose environmental performance is satisfactory may gain in status from this point of view. With this in mind, it would be in the operators' interest to update the norms of the hydroelectric plants from an environmental point of view, in order to obtain green labels, even if in doing so there is a slight rise in production costs.

Nakarado has clearly underlined how "the monopoly based structure [of the past] resulted in an industry culture that tended to ignore any sophisticated analysis of the electricity customer demand and the segmented dimensions of that demand", and that "rather than focussing exclusively on improving the methodological techniques of evaluation of the engineering and financial costs… the renewable industry should move toward a marketing orientation" (pp. 188–189).

2.7. Water Royalties[13]

Hydroelectric development can generate rent when electricity production costs are lower than market prices, as Figure 2 shows.[14] In theory the rent belongs to the owner of the water, usually the State or local government. It expresses the relative value of watercourses and must not be considered as a tax that internalizes a plant's external costs. In a monopoly situation the relevant authority often takes a water royalty calculated independently of the profitability of the plant, so as to collect, so to speak, the rent generated by use of the water. In a competition situation, a system is conceivable in which the royalty reflects changing electricity prices and the production costs of plants taken individually. There remains the problem of monitoring these costs. The volatility of the spot prices should also be taken into account.

---

[12] Cf. Nakarado 1996; Wüstenhagen, Markard, and Truffer 2003.

[13] Cf. Romerio 1999, p. 121; Rothman 2000; and also Amundsen, Andersen, and Sannarnes 1992; Amundsen and Tjotta 1993.

[14] Production costs include reimbursement of production factors.

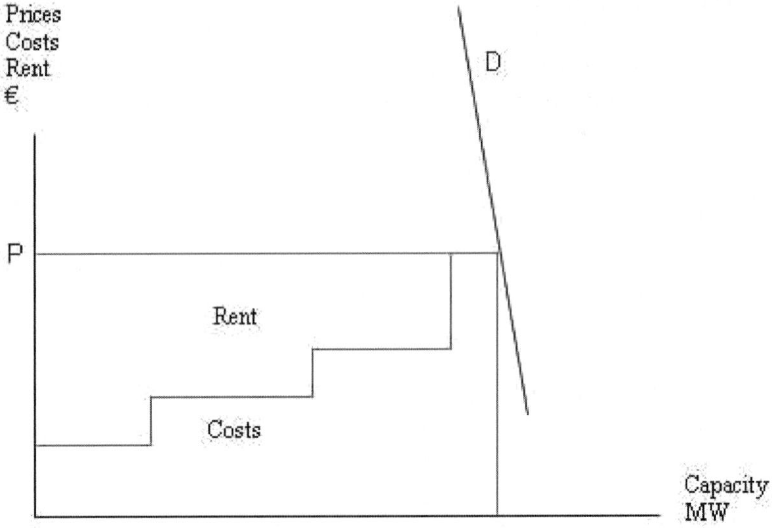

Figure 2: Determination of a Hydropower Rent.

2.8. Outlook on Hydroelectric Resources in the Alpine Arc

The hydropower industry is currently being severely jolted by the reorganization of electricity markets. The fall in wholesale prices has hit many companies quite hard, but the sector is hanging on and it seems likely that only a small number of marginal plants will fail to pay for themselves. These will become "stranded investments," which in many countries are receiving special assistance, as part of measures designed to minimize the socioeconomic costs of the transition to completely open markets.

The future of hydropower generation depends on a set of factors with positive and negative potential. Some of them—the price of fossil fuel or tensions on the electricity markets, for example—are already having a major short-term impact. Others, such as fossil fuel depletion and new technology developments, will emerge more clearly in the long-term. The effects of climatic change will make themselves felt in the medium- to long-term. Generally speaking, uncertainty rules.

There are also several factors open to influence from the authorities or market operators, via energy policy measures or business management. Achieving balance among the different existing interests is far from simple.

The opening of markets to competition is allowing the hydropower sector to stress its relative advantages, notably the flexibility of storage plants and, in many cases, comparatively low environmental impact, to the extent that green markets are created. Market opportunities and risks could be equitably shared between owners of the water and the electricity companies by establishment of an appropriate system of fixing hydropower rents.

3. CASE STUDY: THE BLENIO INSTALLATION[15]

3.1. Production

The Blenio installation is situated in the southern part of the Swiss Alps, in the canton of Ticino. Figure 3 shows the longitudinal profile of the installation. The capacity is 437 MW and the expected production is 358 GWh in winter and 300 GWh in summer.[16] Two plants are equipped with reservoirs, whereas a third is a run-of-the-river power plant. The Luzzone dam was raised in the nineties, like several others in Switzerland.[17] Today, its height is 225 m. The increase in capacity, which moves from 87 to 107×10^6 m^3, allows the transfer of production of 60 GWh from summer to

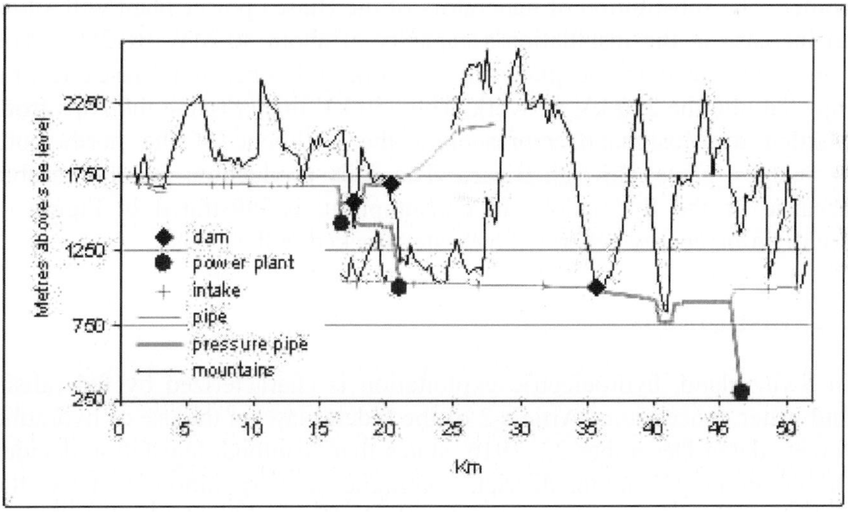

Figure 3: Longitudinal Profile of the Blenio Installation (Die Tessiner Elektrizitätswerke, Bulletin SEV/VSE, 13, 1975, p. 681).

[15] Cf. Romerio 1994; Filippini and Spreng 2001; Romerio 2003b.

[16] Cf. Office fédéral de l'économie des eaux (OFEE), *Statistiques des aménagements hydroélectriques de la Suisse.*

[17] Cf. Galli 1996.

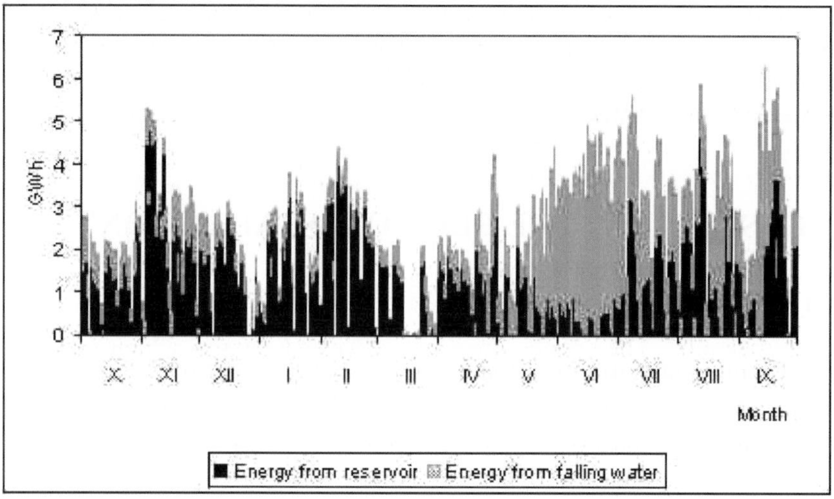

Figure 4: Daily Production of the Blenio Installation in a Year Slightly Below the Average from the Hydrological Point of View (Officine Idroelettriche della Blenio, Rendiconti).

winter. The substitution of generators of the Biasca power plant will allow an increase in the installation's capacity of about 40 MW by 2006. The power generated in the plants of Luzzone, Olivone, and Biasca is first injected into the 220 kV network. The 380 kV line crossing the Alps from Mettlen to Musignano, representing the backbone of the north-south exchanges, passes through Biasca. The daily production, greatly variable because of the fluctuations in consumption, is illustrated in Figure 4, highlighting energy generated by water stocked in the dams.

3.2. Institutions

In Switzerland, hydroelectric exploitation is characterized by federalism and water concessions. Article 2 of the federal law on the use of hydraulic forces, dated December 22, 1916, states that "cantonal legislation decides the community (canton, district, commune, or corporation) that has the right to use public water power." Article 3 affirms that "the community … can use it itself or grant it to third parties."[18]

The concessions determine the duties and the rights of the concessionary (such as the duration of the concession, right to repossess and repurchase, participation of the community in benefits). According to

[18] Cf. loi fédérale sur l'utilisation des forces hydrauliques, 22.12.1916.

the federal law, the concession cannot have a duration in excess of 80 years. At the end of this period, the conceding community has the right to take back at no cost the "wet parts" of the installation (reservoir, pressure pipes, hydraulic engines and buildings which shelter them), and to take back in return for an equitable payment the installations for the production and transport of energy. The "repurchase" of the installations can only be exercised after 2/3 of the duration of the concession, and only if it was envisaged in the concession; the "repurchase indemnity" must be "full and complete." Article 43 states that "once granted, the right of use cannot be rescinded or restricted except for reasons of public utility and upon indemnification." The concession creates "vested rights," which makes it very rigid, particularly if the repurchase option was not anticipated.[19] Opening the market to competition cannot put into question the vested rights, which will disappear only towards the middle of the twenty-first century, when the majority of the water concessions fall due.

In the case of the Blenio installation, the concession was granted by the Parliament of the canton of Ticino (according to the federal and cantonal law[20]), on November 3, 1953, to a "partner company" encompassing the canton of Ticino (20 percent of the inside capital), a private company (Aare Tessin AG für Elektrizität, ATEL, 17 percent), as well as five companies belonging to Swiss-German cantons and cities (63 percent). It is valid for 80 years, from the bringing into service of the installation (1962), and does not include repurchase right.[21]

The "partner companies," which hold the concessions of the majority of the big hydroelectric installations in Switzerland, combine electric utilities with the aim of building and operating an electric installation.[22] The partners absorb the production in proportion to their participation in the inside capital, at a price equal to the production cost. The company thus does not make a profit. The profits are realized by the partners and are taxed in the headquarter's canton, not in the mountain cantons like Ticino. In order not to remove itself completely from the tax imposition of the canton where the installations are situated, the company declares an artificial benefit, calculated by applying to the inside capital the interest

---

[19] Cf. Dubach 1979.

[20] Cf. legge cantonale riguardante l'utilizzazione delle acque del 17.5.1894.

[21] Cf. Decreto legislativo concernente la concessione per lo sfruttamento delle forze idriche della Valle di Blenio, 3.11.1953, *Bollettino ufficiale delle leggi e degli atti esecutivi del Cantone Ticino,* 79, 1953, p. 181–184, and Decreto legislativo che approva il contratto di costituzione e gli statuti delle Officine Idroelettriche di Blenio SA, 2.1.1956, *Idem,* 82, 1956, pp. 13–20.

[22] Cf. Saitzew 1950.

rate of the loans granted by the company, increased by approximately one point. Following the opening of the markets to competition, the partners redefined their strategies and today installations, such as Blenio, are controlled by companies with sometimes diverging interests.[23]

3.3. Market

Figure 5 shows the trend of the average price of electricity in Switzerland and the average production cost of Blenio related to the installation's average production as well as its observed annual production.[24] We note the growing gap between the two curves, which proves the increase in value of hydroelectric production over time. However, at the beginning, in the 1960s, the profitability of Blenio was low, even nonexistent. At that time, in fact, it represented a marginal installation in a market characterized by surplus. Moreover, since 1997 the average prices in Switzerland

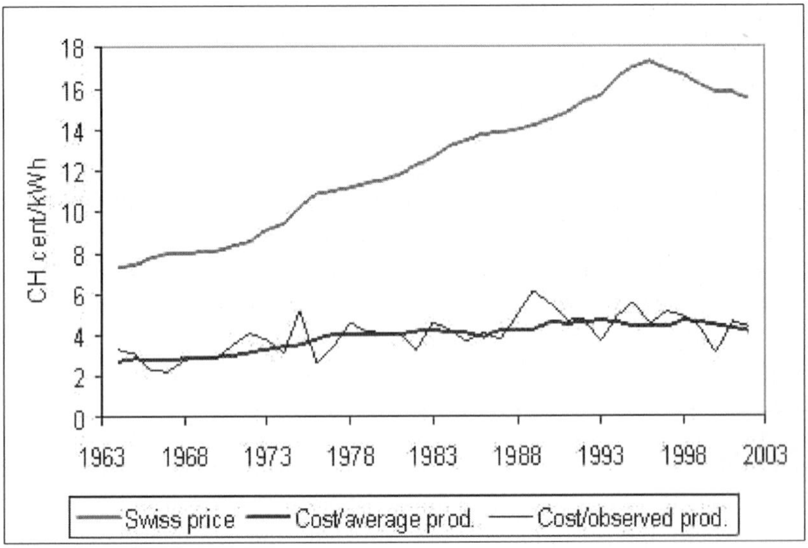

Figure 5: The Average Electricity Price in Switzerland and the Average Production Cost of Blenio (Nominal Values) [25] *(Office fédéral de l'énergie, Statistique suisse de l'électricité and Officine Idroelettriche della Blenio, Rendiconti, since 1964–1965).*

---

[23] Cf. Romerio 2003b, pp. 245–247.

[24] 1 CHF ≅ 0.66 Euro.

[25] Transport and distribution costs are included in the prices but not in the costs. At present, the costs of high-tension transportation are around 2.5 CH ¢/kWh, whereas medium- and low-tension transportation costs are on average 8.5 ¢/kWh. However, the spread around the average is very high.

have decreased (both in nominal and constant terms), because of the anticipation of the opening of the market to competition. We should also point out the great difference between the two series of costs—1989/1990 +1.85 and 2000/2001 −1.42—caused by the fluctuations of precipitations (respectively 69 percent and 152 percent of the multi-annual average).

The production cost of the Blenio installation was 4.4 CH ¢/kWh in 1999/2000. The debt financing represents 19 percent of this cost, that of equity financing 9 percent, depreciation 16 percent, staff 13 percent, miscellaneous 8.5 percent, water royalties 25 percent, and taxes 9.5 percent. The reservoir rise caused an increase in the production cost of 0.5 CH ¢/kWh for all of the production (9.0 CH ¢/kWh if the 60 GWh transferred from summer to winter is taken strictly into account). With competition, production costs could be reduced due to higher economic performance of the electric companies. On the other hand, investors' higher requirements for profitability and renovation of installations could precipitate an increase. The problem of external costs and water royalties is discussed below.

The energy produced by Blenio is used to cover the peaks and superpeaks, especially in winter, when the market prices are higher. During the winter months of 2001/2002, the Swiss market index SWEP exceeded 58 times (out of 126) 5 CH ¢/kWh, 16 times 7.5 CH ¢/kWh, and 8 times 10 CH ¢/kWh; during the winter months of 2002/2003, respectively 88 times (out of 127), 12 times, and once. The observed values can be much lower or much higher. Thus, on December 19, 2001, the SWEP was at 32.8 CH ¢, but the kWh was negotiated at about 100 CH cents. A strong reduction in the reserve margin on the European market can provoke large leaps in prices. Blenio's "blue electron" could also be valorized on the green energy markets, which are currently being developed.

3.4. External Costs

For a long time, the federal and cantonal laws did not allow for standards specific to minimal residual flows in rivers and streams. The law on water protection of January 24, 1991, filled a legal gap without, however, changing the situation appreciably because it does not apply retrospectively. While Article 31 determines the minimal residual flow, Article 33 allows the executing authorities to increase it, if the interest of landscape and nature prevails over economic interests and security of energy supply requirements. Application of the residual flow envisaged by Article 31 could cause a reduction in hydroelectric production in Switzerland of 4 percent, and even 8 percent, without counting the possibility of adjustments under consideration in Article 33.[26]

In the case of the Blenio installation, the current residual flows are the result of negotiations, sometimes quite difficult, between the canton and the company. The Blenio concession does not specifically mention the problem of minimum residual flows. At present, they are carrying out biological, hydrological, and socioeconomic studies to establish if it is advisable to increase these flows. In this respect, market pressure, as well as the green energy market requirements, should be taken into consideration. The current residual flow causes a loss of production of 2 percent, which triggers an increase in the cost of kWh of a fraction of a centime.

The civil liability insurance premiums for the big power plants with reservoirs currently vary between CHF 50 and 100×10^6 in Switzerland.[27] In the case of a serious accident, these sums of money are insufficient to cover the damage. A federal law motion aiming at reinforcing the civil liability of hydroelectric plants has been momentarily abandoned due to opposition from the parties concerned. This motion provided a compulsory coverage that could, for certain plants, reach CHF 600×10^6. In the case of a disaster—for example, when financial means are insufficient to cover the claims—the insurance scheme was no longer applicable. The Federal Assembly thus ordered an indemnity scheme. Since they are obliged to ensure themselves for considerable sums, the hydroelectric operators would have had to assume supplementary costs taking into account the extent of the risks considered. The board of directors at Blenio condemned the project because if it had been accepted the company would have had to pay insurance premiums 20 times higher.[28]

3.5. Water Royalties

In exchange for the use of hydroelectric resources, concessionaries must pay water royalties to public bodies. This represents a significant source of income for the mountain cantons. The maximum amount is fixed by federal law and is the competence of Parliament. It is the result of a political compromise between mountain cantons, lowland cantons, and the electric companies but it has failed to satisfy the stakeholders concerned. In 1916, Parliament fixed the maximum royalty at 8.20 CHF/kW. The

[26] Cf. Conseil fédéral, 1987, p. 1113.

[27] Except for the cantons of Valais and Grisons because of cantonal laws. Cf. Tercier and Roten, 2000; Office fédéral de l'économie des eaux, 2001.

[28] Cf. Officine Idroelettriche della Blenio, *Rendiconto 1998/99*, p. 12.

sixth revision was carried out in 1996. The Federal Council proposed an increase of 16 CHF/kW (from 54.– to 70.– CHF/kW), while the Union of Electric Companies argued for 6.– CHF/kW, and the mountain cantons for an increase of 26.– CHF/kW for the installations without reservoir, and of 106.– CHF/kW for the installations with reservoir. The Parliament partially met the demand of the mountain cantons by fixing the royalty at 80.– CHF/kW, but refused the supplement for the installations with reservoir.[29] With another system of rent calculation,[30] the mountain cantons will share the benefits and risks associated with hydroelectric exploitation. However, for the moment, no political party proposes any changes, as the problem is so touchy.

In the case of the Blenio installation, the company pays approximately 10^7 CHF in water royalties to the canton of Ticino, which increases production cost by approximately 25 percent.

3.6. Opposition

The mountain cantons have frequently experienced political movements that required establishment of state control over hydroelectric resources. In particular, these movements supported the creation of cantonal or communal companies that would ensure the exploitation of the hydro-electric resources by local communities. In other cases, they asked that the concessions accept norms that would allow the mountain regions to profit from hydroelectric exploitation. Exportation of production abroad and even to the urban lowland cantons was interpreted as an inadmissible distancing from the hydroelectric heritage of the cantons. Instead of exporting the electricity, they argued, it should be used in the mountain cantons themselves, by creating industries in the producing areas. If opening to competition is accompanied by privatization of hydroelectric installations, which were "cantonalized" after historical political battles, it could create renewed opposition.

The case of the Blenio installation represents an excellent illustration of these problems. The terms of the concession led to many controversies. Criticisms were related to the following points: establishment of an 80 year concession without possibility of repurchase; the limitation of cantonal participation to 20 percent of the inside capital; the absence of a norm forcing the concessionary to use part of the production *in situ* by creating

[29] Cf. Conseil fédéral, 1995.

[30] Cf. point 2.7.

new industries; the creation of a system of taxation considered to be inequitable; and the absence of standards for the minimal residual flows.

The concession's duration was defined on the basis of depreciation periods. The participation of the canton was probably decided upon judiciously, taking into account the fact that, at the time, the canton was not yet involved either in production or in the distribution of electricity. Industrialisation was not a reasonable objective because it would be irrational to use the stored energy of Blenio to cover the base-load consumption of metallurgy or electrochemistry. These industries also have a significant environmental impact. The taxation system is essentially artificial, because the profits do not reflect the electricity market value. It should, however, be recognized that it was possible to collect taxes even when profitability of the installation was low or nonexistent. With regard to the water residual flows, at the time the concession was granted, they did not represent a true worry for the affected populations.

3.7. Outlook of Blenio Installation

The Blenio power plant illustrates well the problems of hydroelectric resources in the Alpine arc. The market allows the valorization of energy from mountain reservoirs, but the costs must be meticulously controlled and market operations require a high degree of competence. Although the future holds numerous uncertainties, their resolution should not endanger the competitiveness of such plants. The environmental impact is not negligible, but relatively low if compared with the emissions from fossil power plants or when compared with the risk of a serious accident in a nuclear reactor. Updating the norms of the plant could also be profitable from an economic point of view with the introduction of green labels. The opening of electricity markets to competition creates new perspectives that mountain regions should consider without prejudice in order to enhance the value of their hydroelectric potential, such as access to spot markets via national and international interconnected networks.[31]

Around the middle of this century, when a vast majority of the big hydroelectric concessions fall due, mountain cantons will have the possibility to completely redefine the organization of the hydroelectric sector in the Swiss Alps. However, they must be well prepared. Clouds are already forming on the horizon. Unity must be established between the

[31] The electrical company of the canton of Ticino (www.aet.ch/aet), which owns 20 percent of Blenio's energy, can thus operate on the Laufenburg's spot market. The network connecting this canton to the rest of Switzerland and Italy belongs to ATEL A.G. and EGL A.G.

mountain cantons to coordinate the concessions' end and hopefully create the hydroelectric company of the Swiss Alps by 2030.

4. GENERAL CONCLUSIONS

Market reorganization creates the dynamics for change and provides opportunities that mountain regions should not miss. First, they must concentrate on peak and super-peak production and energy regulation to maximize hydroelectricity value. Recent interest shown by some companies in investment projects aimed at producing regulating energy is very significant in this respect. Furthermore, the type of peaks provoked by air-conditioning during the summer months in Southern Europe since 2001 can provide new opportunities for hydropower plants with reservoirs, as well as "demand side management" measures.[32]

Second, mountain regions ought to promote green labels in order to enhance the value of hydropower. Not only small installations should benefit, but also bigger ones, provided that their environmental impact is satisfactory. Upgrading the power plants to reduce environmental impact could then represent a good investment even from an economic point of view. Moreover, hydroelectricity should be combined with other renewable resources in order to optimize their production on green markets. Complementarities between hydropower plants and reservoir and wind energy, whose production is intermittent, deserve careful evaluation.

Finally, hydro resources must be managed taking into consideration interests and expectations of mountain-region populations. One should maximize the positive impacts of hydro resources' use while seeking to keep a certain degree of control over them. In principle, the hydro rent, which has to be linked to the market evolution, should be assigned to the local inhabitants. Decision and management centers in the Alpine arc should be promoted by maintaining or acquiring majority participation in the hydroelectric companies operating in these regions. Highly qualified jobs in engineering and finance fields are at stake. The creation of inter-regional hydroelectric companies in the Alps, operating on the European market thanks to access to networks, must be encouraged.

---

[32] Those peaks provoked some large blackouts in Italy during summer 2001.

EXAMPLE OF WATER RIGHTS: CONCESSIONS IN THE VALAIS, SWITZERLAND

Eric Wuilloud
Forces Motrices Valaisannes

The concession is an administrative act by which an authority holding a water source grants a physical or legal person (licensee) the right to use its power within certain limits. It defines:

- Extent: the watercourse(s) licensed and the amount of water that can be used.
- Term: not to exceed 80 years.
- Licensee's obligations: taxation, annual royalties, delivery of water or energy. The hydraulic fee payable to the awarding authority is one of the licensee's main obligations. These are of two types:
  - An initial tax, which according to the 1990 law is not to exceed four times the annual fee.
  - An annual royalty set in relation to the theoretical power (calculated on the basis of the head height and the usable rates of flow). This is limited by a 1996 law to no more than CHF 80.00 per kW of gross installed capacity. Facilities with power less than 1 MW are not subject to royalties; those with between 1 and 2 MW are partially subject.
- Right of reversion: The hydraulic part of facilities is to be returned free of charge to licensing communities when concession expires.
- New constraints: Affecting economic profitability of facilities are recent laws specifying residual water levels, fish ways, ecological compensations, extraction of floating refuse, among other constraints.

A sample concession

Calculation of royalties is based on the power generated by a given flow of water over time. One kW corresponds to a water flow of 1.1 liter per second over a head of 400 meters. The time factor allows calculation of kilowatts per hour (kWh). Thus a licensee who uses 10 million m^3 of water per year with a head of 1,000 meters would pay an annual royalty of CHF 240,000 based on the gross installed capacity of this facility of 27 million kWh. Spread over 8,760 hours, this is 3,000 kW. Net energy produced by the facility, taking account of losses, would be around 23 million kWh. The royalty, around one centime per kWh, thus has a considerable effect on the cost of the product (see below), even when taking into account the considerable investments required for the construction of a hydraulic facility.

	Cost [ct/kWh]
Exploitation	1 (20 percent)
Taxation and royalties	1.5 (30 percent)
Financial costs	2.5 (50 percent)

Current overcapacities on the European market have a considerable effect on producers' profit margins. In the near future we should be looking at ways of making royalties and taxes more flexible—which would be in addition to the efforts at rationalizing the exploitation that have been carried out over the last few years—in order to make hydraulic products more competitive.

CRANS-MONTANA: WATER RESOURCES MANAGEMENT IN AN ALPINE TOURIST RESORT

Christophe Clivaz
Haute Ecole Valaisanne
University of Applied Sciences of Western Switzerland,
christophe.clivaz@hevs.ch

Emmanuel Reynard
Institute of Geography
University of Lausanne, Switzerland

Abstract: Water is a key element of services supplied by Crans-Montana. It is part of the landscape and is the physical base for sport activities in its various forms (water, snow, ice). Because of the high concentration of tourist activities at this Swiss alpine resort during certain seasons—and the multiple uses of water in tourism—problems of supply sometimes emerge. When this is the case, the uses of this resource for tourism enter into competition with other types of water uses (drinking water for the resident population, irrigation, and hydroelectric production). Recent studies show that current water problems in the Crans-Montana resort are not due to water scarcity per se, but are the result of dysfunctional management. Decision makers are increasingly conscious of the need to better manage this resource. The challenge is to find how to connect "traditional" knowledge with "modern" techniques about water use and management.

Keywords: tourist uses of water, water policy, water management

1. INTRODUCTION

In spite of its particularities, the case of Crans-Montana is similar to many other situations one can find in the tourist areas of the Alps. This article details the kinds of problems that can arise in alpine tourist resorts regarding water supply and water management.[1]

[1] Data for this chapter come from two main sources: Reynard (2000a) and Clivaz (1995).

E. Wiegandt (ed.), Mountains: Sources of Water, Sources of Knowledge, 103–119.

After presenting the main characteristics and history of tourist development in Crans-Montana, we will discuss successively the tourist uses of water, the increasing competition between rival uses, the institutional regulation of water management, and the characteristics of current management of water in the resort. In conclusion, we will propose new models for improving water management that are currently being tested in the resort.

2. CRANS-MONTANA: FROM SOME "MAYENS"[2] TO A MOUNTAIN TOWN

2.1. History

Crans-Montana is a tourist resort located in the French speaking part of the canton of Valais, in south-western Switzerland (cf. Figure 1). Located on a plateau 1,500 m a.s.l.,[3] it offers an exceptional view over some of the most beautiful 4,000 m summits of the Alps, in particular the Matterhorn and Mont-Blanc. It also has an excellent climate compared to other alpine regions (good sunshine, low rainfall, and mild temperatures), which helps explain the rise in tourism there. Although today 40,000 tourists spend Christmas and New Year holidays in this famous resort, there were not any permanent inhabitants until 1892, when the first hotel was built. During the past century, the tourist development of Crans-Montana has been impressive.

Historically, golfing and skiing have been the main activities that contributed to the development of the Swiss High Plateau. In 1908, Crans-Montana inaugurated the highest 18-hole golf course in the world; three years later, one of the first genuine downhill events in the history of skiing, the Earl Robert of Kandahar Challenge Cup, was held. In 1939, the First Swiss Golf Open took place in the resort, which later became the European Masters of Golf. In 1987, Crans-Montana also hosted the World Alpine Ski Championships. Apart from these sport activities, the development of Crans-Montana was also marked by health tourism (notably for convalescence and treatment). Since the 1960s and the discovery of a cure for tuberculosis, the importance of health tourism quickly decreased, although

---

[2] The "mayens" are the huts and agricultural buildings located in the alpine pastures used in the Valais region for spring and autumn grazing.

[3] This geographical location explains why Crans-Montana is also called the "High Plateau."

it still exists to a limited extent today. In the 1970s, a new ski resort, Aminona, was built in the eastern part of the Plateau (Figure 1).

The growth of the resort of Crans-Montana, which accelerated after World War II with the rapid development of the ski industry, enabled the local population to find employment and wages without leaving their homes. However, it also generated serious problems, the effects of which are still felt today. Examples include dispersed buildings that lack architectural unity (Coppey et al. 1986; Antonietti 1993); a completely saturated road network at peak hours; air and noise pollution; social tensions between hosts and inhabitants; reduction in the profitability of the investments; and, most important for this book, water pollution and frequent water crises due to shortages (Reynard 2000b, pp. 71–72). The problems that arise from the daily activities of a mountain resort put the survival of the tourist system itself in danger. The environment has not only been damaged in its visible aspect (destruction of the landscape), but also in terms of its ability to supply natural resources, such as water, necessary to the resort (Clivaz 1995, p. 34).

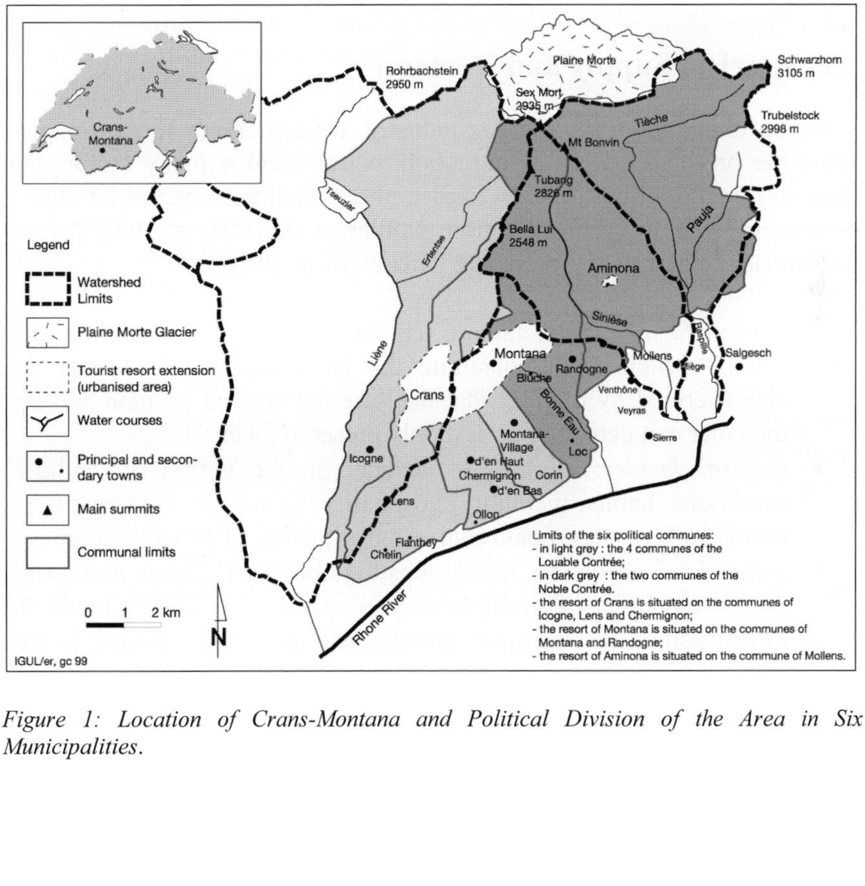

Figure 1: Location of Crans-Montana and Political Division of the Area in Six Municipalities.

To these observations, which could easily be applied to other tourist areas (see Cognat 1973), one can add a specific characteristic, which concerns Crans-Montana: the extension of the resort over six different municipalities, stretching from lowland areas (500 m a.s.l.) to the mountains (3,000 m a.s.l.) (Figure 1). Currently the resident population of these municipalities (Icogne, Lens, Chermignon, Montana, Randogne, and Mollens) is about 13,000 inhabitants, with one half living directly in the resort, and the other half in the surrounding villages on the slope. Following the development of the resort, the population of the six municipalities quadrupled in one century. The fact that there are six municipalities for one resort inevitably brings problems of coordination among the municipalities and complications in development planning, from the economic, social, and environmental points of view (Clivaz 1995, p. 35). Economic stratification is also a problem, with agriculture and industrial activities concentrated mainly in the villages situated below 1,200 m, and tourism located in the high mountainous areas. This stratification has created political tensions between the tourist area and the surrounding villages since the beginning of tourist development.

2.2. Tourism Supply and Demand

When speaking about the tourism industry, we sometimes forget to speak about the original sources of attraction, which include many factors that have no direct relationship with tourism, but which become tourist objects because of their force of attraction (Müller et al. 1993, p. 104). Among these factors, one can distinguish (Clivaz 2001, p. 105):

- *natural factors*: these factors (climate, landscape, fauna, flora, air, water), which often constitute the basic capital of tourism, are characterized by the fact that they are not created by man and that the latter can generally at best only preserve them;
- *cultural factors*: all that is included in the notion of "culture" (traditions, habits, mentality, hospitality) as well as the constructed heritage play a significant role in the attraction of a tourist area;
- *general infrastructure*: this groups together the facilities that enable the development of multiple socioeconomic activities, namely the infrastructure for transport (public and private), supplies (water,

energy, telecommunication), disposal (waste water, refuse), as well as the installations required to meet daily needs (stores, hospitals, etc.).

In the Alps, natural factors are fundamental to tourist development. This is particularly true for Crans-Montana. As mentioned earlier, this resort, in comparison to other tourist areas, enjoys an above-average amount of sunshine and a beautiful panoramic view over the Penninic Alps. The fact that Crans-Montana is located on a plateau also greatly facilitated the construction of the tourist buildings, as there were practically no physical obstacles to the multiplication of the latter. As part of the natural factors, water is a key element of the original supply. We will see in the next section that numerous tourist uses of water developed in Crans-Montana during the twentieth century.

To the original features that attract tourists we must add the derived supply, which is composed of all the facilities and services set up with the specific aim of satisfying tourists.[4] This includes 46 hotels (totaling 3,600 beds), 7,000 chalets and apartments (totaling 34,600 beds, of which only one third are rented beds), three international schools, 28 ski lifts, three golf courses (one 18-hole course and two 9-hole courses), many other sports facilities for walking, cross-country skiing, tobogganing, ice-skating, curling, mountain biking, running, tennis, swimming, etc. The derived supply also requires water for a great quantity of uses (golf course irrigation, artificial snowmaking, ice production for skating, etc.).

Tourist demand can be measured in terms of the number of overnight stays, which is about 1.5 million a year. Swiss people are the main clientele of the resort (54.7 percent) followed by French (9.5 percent), Italian (8.7 percent), German (7.5 percent), Dutch (5.3 percent), Belgian (4.6 percent), British (3 percent), and people from other countries (6.7 percent) (Crans-Montana Tourism 2004: 25). As shown in Figure 2, most of the overnight stays concern the chalets and apartments sector (56.6 percent) and only 23.5 percent concern the hotel sector, which can easily be explained by the fact that there are 10 times more beds in the first sector than in the second one. It is worth noting that Crans-Montana also counts on an important number of overnight stays at the international schools and the convalescence homes and clinics.

---

[4] To have a comprehensive view of the derived supply in Crans-Montana, see the website http://www.crans-montana.ch.

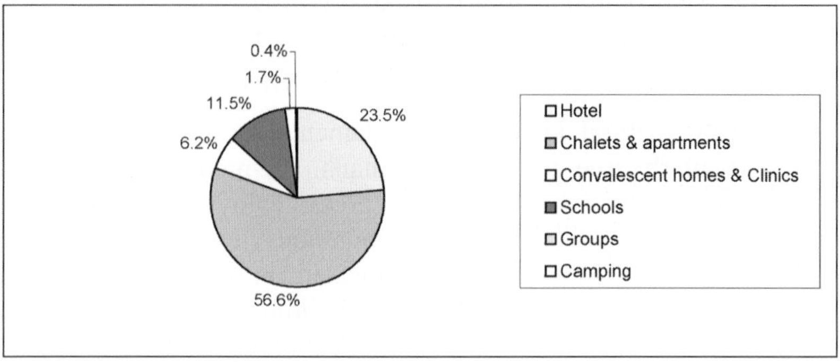

Figure 2: Distribution of Overnight Stays by Accommodation Type (2003) (Crans-Montana Tourisme 2004, p. 25).

If we look at the evolution of the overnight stays in the resort during the last decade (Figure 3), we observe a rapid decrease in the middle of the 1990s (a loss of 270,000 overnight stays between 1993 and 1996, or a decrease of 16 percent) and then a period of stagnation in demand. This means that the High Plateau today is facing the challenge of finding a way of restoring the level of overnight stays it had 10 years ago. Various measures regarding issues like destination management, the use of information and communication technologies for reservations, and better collaboration between the tourist partners were taken recently.

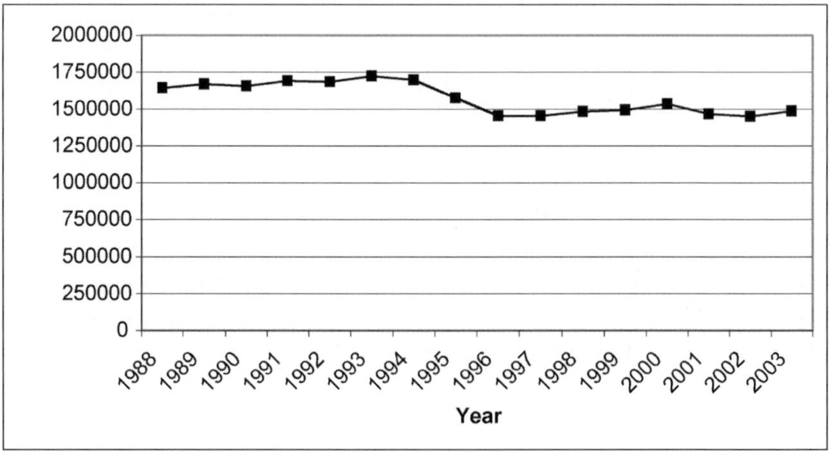

Figure 3: Evolution of the Number of Overnight Stays in the Last 15 Years (Crans-Montana Tourism 2004, p. 25).

3. TOURISM AND WATER MANAGEMENT

3.1. Tourist Uses of Water and Competition with Other Uses

This short overview shows that almost all tourist-related facilities need water in one way or another in order to function. Water is thus not only a key element of the original supply as part of the landscape or drinking water, but also a key element of the derived supply, above all as a physical base for sports activities in its various forms (water, snow, ice). Water was part of the summer tourist supply since the beginning of the tourist development in Crans-Montana and became a central element of the winter supply with the advent of skiing. Other types of tourist activities or infrastructures based on water also appeared, such as swimming pools and skating rinks, installations of artificial snowmaking, and the use of the paths along the "bisses"[5] for hiking. Additional projects of valorization of water are currently under way in the field of hydrotherapy.

While, on the whole, the tourist uses of water on the High Plateau are not very different from the ones existing in other alpine resorts, the Crans-Montana area also includes a unique atmosphere due to the presence of many lakes inside the resort itself. In this sense, the image of Crans-Montana used for tourist promotion is closely linked not only with mountains and snow, but also with lakes and water.[6]

At the same time, tourist uses of water are also linked to environmental problems, whether they are found in the High Plateau or more generally in the tourist resorts of the Alps (Clivaz 2001, pp. 115–116). Tourism contributes to an amplification of water pollution, which constitutes a difficult problem in the alpine areas where the sources are very sensitive and the capacity of regeneration of polluted water particularly weak. Moreover, this alpine water pollution has repercussions for the whole irrigation basin, thereby affecting most of the European continent. In a more specific way, the leveling or clearings undertaken in the skiing areas also have harmful consequences for the hydrological regime by supporting increased flows and an increase in erosion (Département fédéral de l'intérieur 1991, p. 25). As in other urban areas, soils are

[5] The "bisses" are mountain channels that transport water from nival or glacial rivers to cultivated fields (Société d'histoire du Valais romand, SHVR 1995).

[6] For instance, the cover page of the annual report 2001 of Crans-Montana Tourism represents a lake (Crans-Montana Tourisme 2002).

becoming much more impervious because of rapid urbanization linked to tourist development. Such imperviousness may create floods during periods of intense precipitation or during the snow melting period. These are only a few examples of the concrete implications of tourist development for water resources.

But tourism can also have other consequences for water use, particularly in the form of competition with other uses. As in the Alps in general, water is relatively plentiful in Crans-Montana (Reynard 2000a, pp. 119–179). Mean annual rainfall is about 1,000 mm at 1,500 m and 2,500–3,000 mm at 3,000 m. Evapotranspiration is about 500 mm at 1,500 m, and the available water resource is estimated to be about 100 million m^3 (Reynard 2000a, p. 148). However, because of the concentration of tourist activities in space and time and the multiple uses of water in tourism, problems of supply sometimes appear (Reynard 2000a, p. 3). When this is the case, the uses of this resource for tourism enter in competition with other types of water uses, in particular the supply of drinking water for the resident population, irrigation, and hydroelectric production. For example, during the European Masters Golf tournament, some lakes cannot be used for irrigation because they need to remain full for visual reasons. At the beginning of winter, intensive use of water for artificial snowmaking can be in competition with the drinking water supply, thereby drastically reducing the natural flow of some small water streams.

3.2. Institutional Regulation of Water Management

The Swiss political system, characterized by direct democracy and a federalist structure, imposes a three tier natural resource management structure, particularly for water management: the Confederation, the cantons, and the local municipalities (Varone et al. 2002, p. 86). Due to the historical development of the country, state affairs mostly remained in the hands of cantons until the end of the nineteenth century. Since then, tasks have been increasingly concentrated at the federal level, but the application of most public policies regulated by the Confederation is assigned to the cantons, often with considerable room for maneuver (Varone et al. 2002, p. 86). In the canton of Valais, local municipalities also have great independence.

Since the adoption of the Swiss Civil Code (SCC) in 1912, surface water, including lakes, rivers, and glaciers, are property of the state and ground water and springs are private property (Reynard et al. 2001). In Valais, the state property of water is in the hands of the local municipalities, except the Rhone River, which is the property of the canton. Many municipalities are also private owners of springs. That means that at the local level most of the water has assigned property rights and, as such, is managed by local municipalities.

Concerning the regulation of water uses and water problems, there is no general law governing water resources in Switzerland. Three general responsibilities have guided development of Swiss water-related policies (Reynard et al. 2001, pp. 118–126; Varone et al. 2002, pp. 89–90): the protection against water hazards, water exploitation (for hydroelectric production), and protection of water. The policies regulating protection against floods and erosion are very old (Federal Law on the Hydraulic Engineering Police of 1877) and the regulation is centralized at the federal level. The Federal Law on the Use of Water Power was adopted in 1916 in order to improve electricity production and supply. Hydroelectricity production is based on the system of "water concessions," which means that the owners of surface water give a concession to a hydropower company for a period—generally 80 years—against the payment of an annual fee (hydraulic fee). In Valais, most of the lateral rivers were therefore conceded to hydropower companies.

Such was the case of the Liène River in the 1940s: the concession was given by the communes of Ayent and Icogne to the Lienne SA company, which built the Tseuzier dam (capacity 50 million m^3) between 1952 and 1957. The concession will remain in force until 2037. At the time when the concession's act was signed, the main use of river water was for irrigation by communities, which had had use rights since the Middle Ages (Reynard 2002). These ancient water rights were therefore preserved. On the other hand, water used for drinking water at the time came only from spring water. For this reason, no right for the use of drinking water was written into the concession acts. Two decades later, because of the rapid development of the tourist resort, spring water was no longer sufficient to cover all the needs, especially during peak use in winter and summer. Some municipalities were reduced to buying a part of water ceded to the hydropower companies at a price much higher than the hydraulic fee (Reynard 2000a, p. 287).

The third group of water policies in Switzerland concerns the protection of the hydrosystem. Since the 1950s, the Confederation has developed a large number of technical regulations aimed at improving protection against pollution. Implementation of the Federal Laws is the responsibility of the cantons. Because of the large communal autonomy and the low financial capacity of the canton, the Canton of Valais had many difficulties in applying the federal legislation. More than 30 years after the adoption of the Federal Law against Pollution (1971), which introduced the obligation to connect the entire population to water treatment plants, several communities continue to eject effluents directly in the ecosystems without any treatment. That is not the case in Crans-Montana, where the whole population is connected to two treatment plants in the Rhone Valley, a thousand meters below. This solution resolves problems related to the rapid and seasonal population changes and to altitude (low productivity of treatment plants).

Drinking water supply is not regulated at the federal or the cantonal level. It is the task of the local municipalities to supply the entire population, including tourists and permanent residents, with sufficient water of good quality. For climatic reasons, irrigation is not, contrary to other countries, a major user of water in Switzerland; water use for irrigation is therefore very poorly regulated at the federal level. The only region where an irrigation system was developed is in the Valais because of the relative dryness of the climate (SHVR 1995). Here, irrigation networks, which gradually developed since the Middle Ages, are managed by local municipalities or by user communities, called locally "consortages," which have perpetual water rights on some rivers.

3.3. Current Water Management in the Resort

Current water management in the area is highly influenced by the institutional framework (water rights, water policies, Swiss political system) and by local factors (natural factors, political division in six municipalities, history of the region, and low cooperation between local actors).

Federal and cantonal regulations principally concern the use of water for hydroelectric production, the fight against water pollution and the management of hydrologic dangers. Drinking water supply and irrigation, on the other hand, are mainly regulated by decisions taken by the municipalities or by irrigators' associations. As most of the current water problems concern the drinking water supply, we concentrate our analysis on this use.

Drinking water is supplied independently by the six municipalities. The organization of the water network is influenced by the historical development of these communities. Since the Middle Ages, the four communities of the western part of the region were organized in a "Great Bourgeoisie" group called "Grand Lens" or "Louable Contrée" (Figure 1), which can be viewed as a federation of communities. Each section had some powers but there was no overarching central power. In 1802, the new Republic of Valais imposed a centralized power structure, which gave rise to conflicts between the four sections. In 1851, with the adoption of the cantonal law on the communal regime, a distinction was introduced between the municipal commune (or municipality) and the burger commune (run by the “bourgeoisie”). The former is composed of all the residents of the commune, whereas the latter is composed only of people that have their family origin in the commune. With this political transformation, the four sections were fused into one municipality. In 1904, after half a century of political tension, the commune was divided into the four current municipalities. With the division, some collective resources remained undivided, especially alpine meadows, forests, and some springs and water infrastructures. This explains why some springs are currently the property of the four municipalities and that the annual flows are divided with complex repartition schemes. On the eastern side of the plateau, the communes of Randogne and Mollens were also part, along with other communities, of a bourgeoisie group (“Grande Bourgeoisie de la Noble Contrée”). That might explain why most of the springs, which are property of Randogne, are situated on the territorial perimeter of Mollens (Putz 2003).

Until World War II, only spring water was used for water supply. With the rapid extension of the resort during the 1950s and 1960s, the groundwater resources were no longer sufficient and, in 1969, the municipalities of Lens and Randogne faced a severe water crisis. Both municipalities took urgent measures. Lens constructed a pipe several kilometers long for transporting water from the Tseuzier dam to the resort through a tunnel built in 1946 for irrigation purposes. Randogne also constructed a long pipeline from the Raspille River to the eastern part of the resort. Because of the peculiar distribution of property rights on this river, the result of a Bishop’s decision made in 1490 and still in use, water use depends on the consent of all other territorial communities of the watershed (Reynard 2000b, p. 72). In this instance, the municipality faced the opposition of the community of Salquenen, situated at the confluence of the river with the Rhone river, and was forced to pay an annual fee of CHF 30,000 (indexed to the living costs) (Figure 4). This is just one example of the recurrent conflicts over water rights and water allocation that have confronted several municipalities over the last 40 years (Reynard 2000a).

Reliance on surface water to meet increased demand required the construction of drinking-water treatment plants by both municipalities. Some years later, the communes of Montana and Chermignon also faced water scarcity and had to build their own water treatment plants, as they were not allowed access to the Randogne and Lens constructions. The evolution of new water needs and the absence of planning and cooperation among the various municipalities help explain the presence of four treatment plants in a small area. In 2001, the municipality of Lens renewed its 30-year-old plant and asked, without success, for the financial participation of the other municipalities in order to build a new collective plant. In the summer of 2002, the community of Chermignon renewed its own plant, situated less than 1 km from the Lens plant. This example is emblematic of the absence of global planning related to the construction of water infrastructures. It also highlights the absence of cooperation between the local municipalities, even though an intercommunal commission for water problems had been instituted in 1989. This commission does not play a key role in water-supply planning or water-use coordination, and all the important political decisions are taken at the communal level (Reynard 2000b, p. 71).

The absence of cooperation among local municipalities is illustrated by the "water market" that exists between the various political entities of the region (Figure 4). In fact, the distribution of water resources in the region is very heterogeneous (Figure 1). From a hydrological point of view, the High Plateau is divided into two main watersheds: the two main rivers, the Liène (in the west), and the Raspille (in the east). These are separated by a small catchment with very poor water resources that is the location of most of the tourist resorts and urbanized areas. This natural division explains why the municipalities rich in water resources are the communes of Icogne and Mollens, situated respectively in the Liène and Raspille watersheds at both extremities of the resort. The four most populated communes in the central part of the resort, which also face great peaks of demand during the high season, have fewer available resources. This explains the water market that developed between the various municipalities in a region that, by and large, had not faced any problem of water scarcity until recently.

These are some of the numerous cases of water competition, relative water scarcity, and water conflicts that were analyzed in the area (Reynard 2000a). Water problems in the Crans-Montana resort can be classified into three categories (Reynard 2000b, pp. 76–78; Reynard 2001a, pp. 13–14): sectoral problems that affect one type of water use (e.g., the absence of infrastructure planning for drinking water supply), intersectoral problems that affect the coordination between two or more water uses (e.g., coordination of the supply for artificial snowmaking and drinking water in early winter), and territorial problems, which are due to the absence of cooperation between the various municipalities.

In conclusion, it appears that current water problems in the Crans-Montana resort are not due to water scarcity, but are the result of dysfunctional management. This mainly includes the absence of cooperation among local municipalities, absence of planning, absence of an integrated local water policy, poor knowledge of water resource availability and water needs, and absence of cooperation between the various actors of water management in the area. The result is that water "management" is much more of an accumulation of various unconnected actions than it is a coordinated management plan. As such, water management in the High Plateau cannot be considered to be either integrated or sustainable (Reynard 2001a, p. 81). Thus, new models of water management need to be explored.

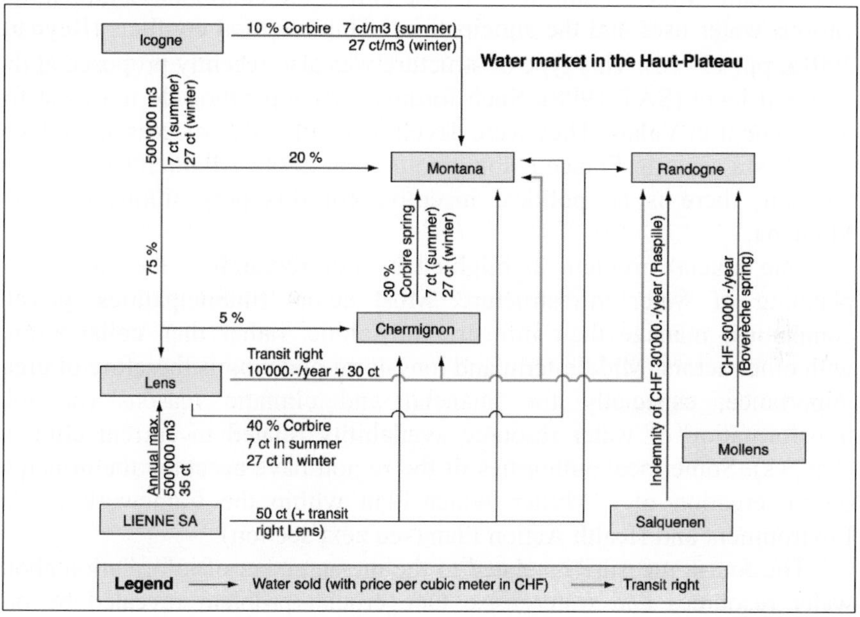

Figure 4: The Water Market between the Political Communes of the High Plateau (all prices are indexed to the living costs) (Reynard 2000a).

4. NEW MODELS OF WATER MANAGEMENT

4.1. Management Measures

In order to improve water management, we propose four types of measures (Reynard 2000a, pp. 343–356). First of all, future management should be organized at the watershed level, instead of the municipal level. However, this proposition faces natural constraints because all the populated areas in this part of the Rhone valley are situated between the rivers and not near the river banks, and political restrictions limit public activities—like transportation, urbanization, public infrastructures, and tourist infrastructures—making the logical area for political action not the watershed but the political boundaries of the six municipalities. For these reasons, an integrated water management at the water basin scale is not feasible. One solution should be the creation of an intercommunal water service, which could include the six current water services, and other water related activities.

The second measure is the creation of a new institution of management that we call "water committee," whose aim is the coordination among the various water uses and the anticipation of intersectoral conflicts (Reynard 2001a, pp. 15–16). This type of structure was also recently proposed at the cantonal level (SAT 1998). Such forums for cooperation do not exist for the moment in Valais. They were developed with some success at the local river basin scale in France (Montginoul et al. 2000; Allain 2001). For the moment, there is no political incentive for this proposition at Crans-Montana.

One crucial problem highlighted by our research is the absence of planning of water infrastructure. Most actors (municipalities, private companies) manage their infrastructure alone, rather than collaborating with other actors. Middle-term and long-term planning is therefore of great importance, especially for financial and climatic reasons (possible transformation of water resource availability related to current climatic changes). Some local authorities of the region have accepted the principle of the creation of a Master Water Plan within the framework of the Environment and Health Action Plan (see next section).

The fourth measure is related to the management of information about water resources and water uses. One crucial problem revealed by our research is the absence of systematic statistics about water, the very heterogeneous management of data, and the relatively low quality of these data. We therefore propose the creation at the regional scale of a system for collecting, structuring, managing, and upgrading water-related information (Reynard 2000a, p. 355). The system should allow the creation of thematic statistics related to water that could help to create the Water Master Plan of the High Plateau. There is also a need to organize and manage the data within a Geographical Information System (GIS). The first steps for the implementation of Water GIS have been realized within the Environment and Health Action Plan. A first study concluded that the GIS should be organized in different interconnected sub-systems reflecting the various sub-systems concerning water management (resource, drinking water, irrigation, hydroelectricity, tourist uses, waste water, hydrological hazards) and collected and organized data related to water resource (climatic data, hydrological data) (Reynard 2001b). A second study created the module related to drinking water management within the GIS software ArcView (Putz 2003). It produced varied data concerning drinking water infrastructures, characteristics of springs used for water consumption, and fountains. Local political authorities should now discuss the opportunity of continuing the implementation of other modules (see below).

4.2. Water Management and Sustainable Development of the Resort: The Environment and Health Action Plan

Water management problems are also managed within the larger framework of sustainable development actions. As a response to the different problems, including water management, that Crans-Montana has to overcome, the resort has decided to take part in the Environment and Health Action Plan as a way of prolonging and strengthening the first actions taken in the framework of a Local Agenda 21 process. This Action Plan follows from the commitments taken by Switzerland during the Earth Summit in Rio in 1992. It is supervised by the Swiss Federal Office of Public Health. The main objective of this Action Plan is to encourage people to associate environment and health in their daily lives. To reach this goal, three domains were chosen: nature and well-being, mobility and well-being, and housing conditions and well-being.[7] Due to its numerous traffic problems, Crans-Montana was chosen as a pilot region for the mobility and well-being domaine.

Since the summer of 2001, the authorities as well as numerous people involved in different fields (tourism, education, agriculture, engineering, etc.) have worked on a new approach to mobility. More precisely, 14 projects—including, for example, the development of a mobility plan for the whole resort, discovery of the regional architectural heritage by walking tours, and the promotion of local food products—have been defined. The first concrete measures were executed during Winter 2002–2003.[8] Among these 14 projects, one directly concerns the water theme and is called "Crans-Montana along the Water" ("Crans-Montana au fil de l'eau"). The aims of this project are:

1. To make the population and visitors aware of the water problems according to its principal uses: drinking water, irrigation, hydroelectric production, and tourist use. One important point in this

[7] For more information, see the following website: http://www.paes.ch.

[8] More details concerning the Action Plan Environment & Health in Crans-Montana are available at http://www.paes-crans-montana.ch.

context is to highlight the modes of water consumption and the potential of economizing.
2. To take part in the development of an "inter-municipal" master plan for a concerted management of the hydrological and water use problems of Crans-Montana.

In order to reach the first goal, several events have been organized during tourist seasons: water photography competitions in summer 2001 and water information weekends with conferences, visits of infrastructures (irrigation network, drinking water treatment plants, hydropower production plants, artificial snowmaking infrastructures) and creation of a water tourist path (2003–2004). Workshops were also organized. In 2001, the conferences focused on drinking water management problems; in 2002, the main theme was the impacts of climatic warming on water resource conservation. In order to reach the second goal, a working group composed of people representing the local political authorities, water technicians, and scientific circles are organizing the collection of all the existing information regarding water in the High Plateau. This information should be integrated in a Geographical Information System (GIS) for the six municipalities. At the moment, the water planning process is included in a much larger process of territorial planning.

As we have noted, these aims and actions in the field of water seem particularly adapted to the water management problem described in the previous sections. What is currently needed is to determine if the local authorities and their various partners in water use will be able to realize these objectives and to implement them over the long term.

4.3. Conclusion: How to Link "Traditional" with "Modern" Knowledge

Largely because of various expert studies in the field of water resources and management (Reynard 2000a) as well as the Environment & Health Action Plan, the decision-makers of Crans-Montana are now conscious that water is an important resource that should be better managed in the upcoming years. The challenge in this context is to find how to connect "traditional" knowledge with "modern" knowledge about water use and management. "Traditional" knowledge typically concerns those mechanisms aimed at developing sustainable collective management of common-pool resources, thus allowing the conservation of the resource, economic efficiency, and social equity (see Ostrom 1990). The canton of Valais has a long tradition in this type of collective management of natural resources, principally in the area of irrigation (Crook and Jones 1999) and alpine meadows.

"Modern" knowledge, on the other hand, aims to improve the management of water resources using technical and organizational solutions (such as GIS), and implementing harmonized planning (Master Water Plan) and management (local governance via Environment & Health Action Plan). The challenge is to invent new common-pool management structures (Lenhard and Rodewald 2000) that will be able to manage not only sector-based uses like irrigation, but also cross-sector problems like the coordination of various water uses.

Although we saw that the High Plateau has some specific features in terms of hydrological regime and institutional structure, we think that the challenge of connecting "traditional" and "modern" knowledge regarding water use and water management also concerns the majority of resorts in the Alps, and perhaps the majority of resorts in mountains areas of other developed countries.

WATER CONFLICTS AND CONFLICT RESOLUTION MECHANISMS

WATER VALUE, WATER MANAGEMENT, AND WATER CONFLICT: A SYSTEMATIC APPROACH*

Franklin M. Fisher
Massachusetts Institute of Technology, United States
ffisher@mit.edu

Abstract: This report on the work of an Israeli-Jordanian-Palestinian-American-Dutch project shows that water issues are best dealt with by thinking in terms of water values rather than water quantities. In this way, water conflicts can be reduced to disputes over money—in many cases, surprisingly little money. It is argued that actual free-water markets will not successfully allocate water resources, partly because water markets are unlikely to be competitive and partly because of externalities including both environmental concerns and the fact that countries place special values on the use of water in agriculture—values that exceed the returns to farmers. However, it is possible to build economic models of water use that incorporate such features and that can guide water management and infrastructure decisions. These models produce "shadow values" that can guide decisions in the same way free-market prices would if they could cope with the difficulties mentioned above. These shadow values can then be used to guide international (or other) cooperation in water. These methods are applied to Israel, Jordan, and Palestine, and the gains from cooperation are found to be much larger than the gains from reasonably large shifts in water ownership. By such means, water conflicts can be resolved.

Keywords: shadow prices, cost-benefit analysis, conflict resolution

* The project, whose results are discussed here, is the work of a great many people—too many to acknowledge them all. Chief among those are the coauthors of Fisher et al. (2002, 2005), especially Annette Huber-Lee. A shorter form of this chapter appeared as Fisher (2002). The full discussion appears in Fisher et al. (2005). Parts of the chapter have been published elsewhere and are reprinted with permission from *Environment,* April 2006, 48, 3, Heldref Publications, 1319 18th St., NW, Washington, D.C. 20036-1802. www.heldref.org © 2006 (reprinted with permission of the Helen Dwight Reid Educational Foundation.) and from "Water Casus Belli or Source of Cooperation" in *Water in the Middle East.*

E. Wiegandt (ed.), Mountains: Sources of Water, Sources of Knowledge, 123–148.

1. THINKING ABOUT WATER: THE FISHELSON EXAMPLE

So important is water that there are repeated predictions of water as a *casus belli* all over the globe. Such forecasts of conflict, however, stem from a narrow way of thinking about water.

Water is usually considered in terms of quantities only. Two (or more) parties with claims to the same water sources are seen as playing a zero-sum game. The water that one party gets is simply not available to the other, so that one party's gain is seen as the other party's loss. Water appears to have no substitute save other water.

But there is another way of thinking about water problems, a way that can lead to dispute resolution and to optimal water management. That way involves thinking about the value of water and shows that water can be traded off for other things.

The late Gideon Fishelson, an outstanding economist of Tel Aviv University, once remarked that "Water is a scarce resource. Scarce resources have value, and, no matter how much one values water, one cannot value it at more than its cost of replacement."[1] He went on to point out that desalination of seawater puts an upper bound on the value of water to any country that has a seacoast. Consider, then, the following example:[2]

A major part of the conflicting water claims of Israel and Palestine[3] consists of rival claims to the water of the so-called Mountain Aquifer. (See Figure 1) That water comes from rainfall on the hills of the West Bank and then flows underground. Most of it (even before there was a state of Israel) has always been pumped in pre-1967 Israel, in or near the coastal plain where the well depths are considerably less than in the West Bank.

Currently, the cost of desalination on the Mediterranean Coast of Israel and Palestine is between 50 and 60 US cents per cubic meter (m^3). For purposes of this example, I shall use 60 ¢/m^3. Fishelson's principle means that the value of water on the Mediterranean Coast can never exceed 60 ¢/m^3 (unless there are large changes in energy prices). But the water of the Mountain Aquifer is not on the Mediterranean Coast. To extract it and convey it to the cities of the coast[4] would cost roughly 40 ¢/m^3. But that

[1] In this chapter, "valuing water" means valuing molecules of H_2O. Particular water sources can, of course, be valued for historical or religious reasons, but such value is not the value of the water as water.

[2] In this example, I have updated Fishelson's calculation to reflect current estimates.

[3] The use of names is a sensitive subject. I do not intend here to prejudge the ultimate outcome of the Israeli-Palestinian conflict. I use the term "Palestine" out of respect for my Palestinian colleagues, and because nearly all sides now predict the existence of a Palestinian state.

[4] This example assumes that this would be the efficient use of Mountain Aquifer water. Other cases are more complicated but do not lead to qualitatively different conclusions.

means that the value of Mountain Aquifer water *in situ* cannot exceed 20 ¢/m^3 (60¢/m^3–40 ¢/m^3).

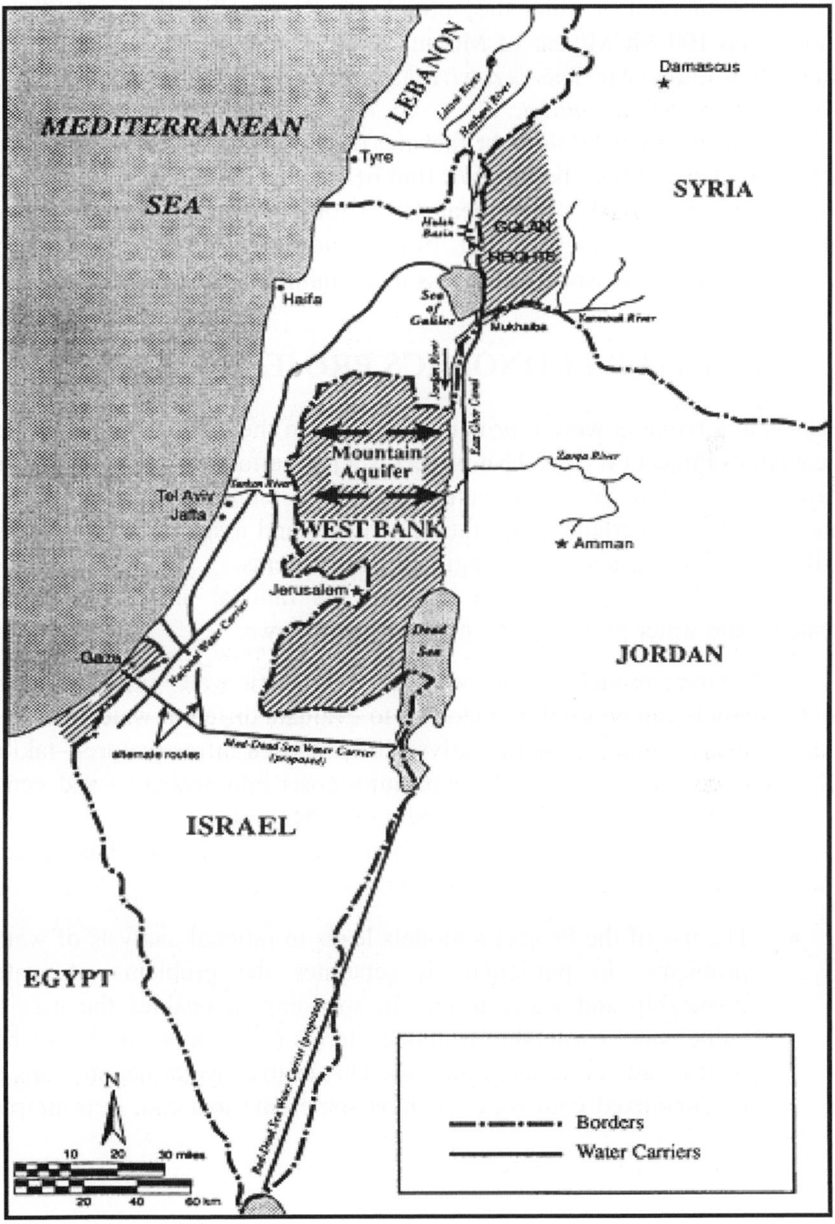

Figure 1: Simplified Map of the "Middle East" (Israel, Jordan, and Palestine), Its Major Water Resources, and Major Conveyance Infrastructure.[5]

[5] Adapted from Wolf (1994), p. 27.

To put this in perspective, observe that 100 million cubic meters (MCM) per year of Mountain Aquifer water is a very large amount in the dispute. If the Palestinians were to receive this, they would have nearly double the amount of water they now have. But the Fishelson calculation shows that 100 MCM/year of Mountain Aquifer water is not worth more than $20 million/year. *This is a trivial sum between nations. Certainly, it is not worth continued conflict.*

And it must not be thought that the desalination-cost driven numbers are more than an upper bound. We find below that desalination will not be cost-effective on the Mediterranean Coast for a number of years except in times of very substantial drought. In more normal times, the water of the Mountain Aquifer is worth much less than 20¢/m^3.

2. THE WATER ECONOMICS PROJECT

Fishelson's remarks were a principal impetus to the creation of the Water Economics Project (WEP).[6] That project is a joint effort of Israeli, Jordanian, Palestinian, Dutch, and American experts. It is facilitated by the government of The Netherlands with the knowledge and assent, but not necessarily the full agreement, of the regional governments.

The WEP has produced a tool for the rational analysis of water systems and water problems. Its goals are as follows:

1. To create models for the analysis of domestic water systems. These models can be used by planners to evaluate different water policies, to perform cost-benefit analyses of proposed infrastructure—taking systemwide effects and opportunity costs into account—and generally for the optimal management of water systems.
2. To facilitate international negotiations in water. This has several aspects:

- The use of the Project's models leads to rational analysis of water problems. In particular, it separates the problems of water ownership and water usage. In so doing, it enables the user to value water ownership in money terms (after imposing his or her own social values and policies). This enables water negotiations to be conducted with water seen as something that can, in principle,

[6] Formerly the Middle East Water Project (MEWP).

be traded. Further, since the Project shows that water values are not, in fact, very high (partly because of the availability of seawater desalination), the water problem can be made manageable. (The Project has had some success in promoting this point of view among professionals, but it is certainly far from being universally understood or accepted.)

- By using the Project's tools to investigate the water economy of the user's own country, the user can evaluate the effect of different water ownership settlements. (By making assumptions about the data, policies, and forecasts of other parties, the user can gain information about effects on the other parties as well.) This should assist in preparing negotiating positions if the ultimate agreement is to be of the standard water-ownership-division type with no further cooperation.
- Perhaps most important of all, the Project shows clearly that cooperation in water tends to be for the benefit of all parties. Such cooperation in the form of an agreement to trade water at model prices can lead to large gains to all participants (sellers as well as buyers) and is a superior solution to the standard water-quantity-division agreement. Our results show that there are large benefits to both Israel and Palestine from such an arrangement. The gains are far larger than the value of changes in the ownership of more or less of the disputed water is likely to be.
- Beyond the economic gains of such an arrangement are the gains from a flexible, cooperative water agreement in which allocations change for everyone's benefit as situations change. Such an agreement can turn water from a source of stress into a source of cooperation.

In sum, the Project hopes to promote "outside-the-box" thinking about water problems and thus to remove them as an obstacle to peace negotiations. We will show how this is possible by explaining the ideas underlying the project in greater detail and then presenting some results obtained.[7]

[7] The most extensive discussion of the WEP's methods and results is Fisher et al. (2005). See also Fisher et al. (2002). Differences in numerical results between previously published work and the present chapter are due to data revisions.

3. WATER VALUES, NOT WATER QUANTITIES

Returning to Fishelson's example, the result of the calculation of the value of the water of the Mountain Aquifer may seem surprising. But the really important insight here is that one should think about water by analyzing water values and not just water quantities. This should not come as a surprise. After all, economics is the study of how scarce resources are or should be allocated to various uses. Water is a scarce resource, and its importance to human life does not make its allocation too important to be rationally studied.

In the case of most scarce resources, free markets can be used to secure efficient allocations. This does not always work, however; the important results about the efficiency of free markets require the following conditions:

1. The markets involved must be competitive consisting only of very many, very small buyers and sellers.
2. All social benefits and costs associated with the resource must coincide with private benefits and costs, respectively, so that they will be taken into account in the profit and loss calculus of market participants.

Neither of these conditions is generally satisfied when it comes to water, partly because water markets will not generally be competitive with many small sellers and buyers, and partly because water in certain uses—for example, agricultural or environmental uses—is often considered to have social value in addition to the private value placed on it by its users. The common use of subsidies for agricultural water, for example, implies that the subsidizing government believes that water used by agriculture is more valuable than the farmers themselves consider it to be.

This does not mean, however, that economic analysis has no role to play in water management or the design of water agreements. One can build a model of the water economy of a country or region that explicitly optimizes the benefits to be obtained from water, taking into account the issues mentioned above.[8] Its solution, in effect, provides an answer in which the optimal nature of markets is restored and serves as a tool to guide policy makers.

Such a tool does not itself make water policy. Rather it enables the user to express his or her priorities and then shows how to implement them optimally. While such a model can be used to examine the costs and benefits of different policies, it is not a substitute for, but an aid to the policy maker.

---

[8] The pioneering version of such a model (although one that does not explicitly perform maximization of net benefits) is that of Eckstein et al. (1994).

It would be a mistake to suppose that such a tool only takes economic considerations (narrowly conceived) into account. The tool leaves room for the user to express social values and policies through the provision of low (or high) prices for water in certain uses, the reservation of water for certain purposes, and the assessment of penalties for environmental damage. These are, in fact, the ways that social values are usually expressed in the real world.

I first briefly describe the theory behind such tools applied to decisions within a single country. I then consider the implications for water negotiations and the structure of water agreements. I give examples drawn from the analysis of water in the Middle East.

4. THE "WAS" TOOL

The tool is called WAS, or the "Water Allocation System." At present, it is a single year, annual model, although the conditions of the year can be varied and different situations evaluated. (Since this chapter was written, a multi-year version has been developed.)

The country or region to be studied is divided into districts. Within each district, demand curves for water are defined for household, Industrial, and agricultural use of water. Extraction from each water source is limited to the annual renewable amount. Allowance is made for treatment and reuse of wastewater and for interdistrict conveyance. This procedure is followed using actual data for a recent year and projections for future years.

Environmental issues are handled in several ways. Water extraction is restricted to annual renewable amounts; an effluent charge can be imposed; the use of treated wastewater can be restricted; and water can be set aside for environmental (or other) purposes. Other environmental restrictions can also be introduced.

The WAS tool permits experimentation with different assumptions as to future infrastructure. For example, the user can install treatment plants, expand or install conveyance systems, and create seawater desalination plants.

Finally, the user specifies policies toward water. Such policies can include: specifying particular price structures for particular users; reserving water for certain uses; imposing ecological or environmental restrictions, and so forth.

Figure 2 shows an example of the main menu that the user sees when using WAS.

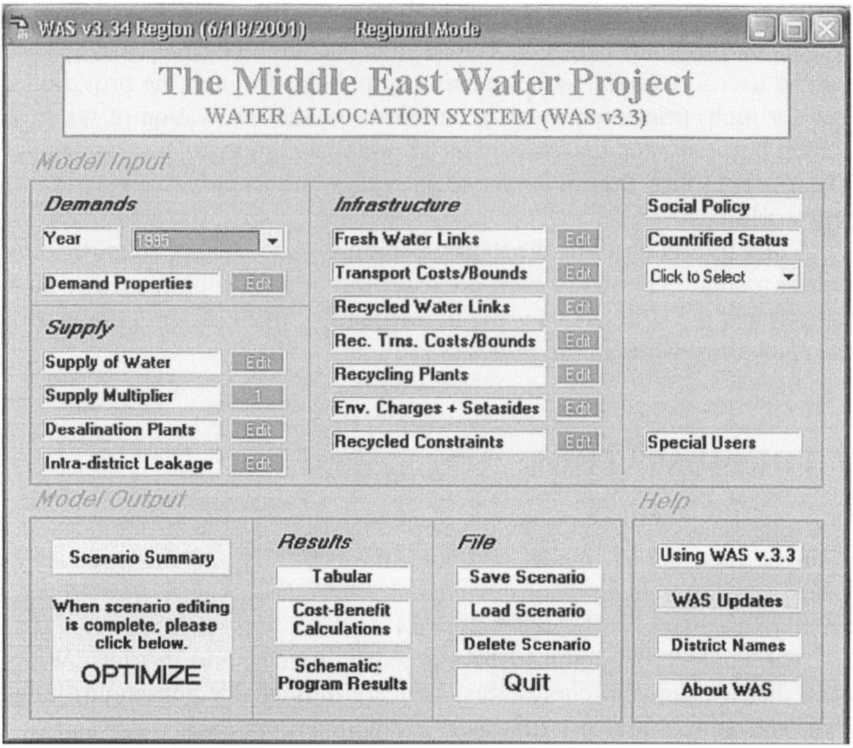

Figure 2: Water Allocation System: WAS Main Menu.

Given the choices made by the user, the model allocates the available water so as to maximize total net benefits from water. These are defined as the total amount that consumers are willing to pay for the amount of water provided, less the cost of providing it.[9]

Along with the optimal allocation of water, WAS generates a *shadow value* for water in each district. The shadow value of water in a district shows the amount by which net benefits would increase if there were an additional cubic meter of water available there. It is the true value of additional water in that district. Similarly, the shadow value of water at the source is the *scarcity rent* of the water in that source—the true measure of what water is worth at the margin.

---

[9] The total amount that consumers are willing to pay for an amount of water, Q*, is measured by the area up to Q* under their aggregate demand curve for water. Note that "willingness to pay" includes ability to pay. The provision of water to consumers that are very poor is taken to be a matter for government policy embodied in the pricing decisions made by the user of WAS.

One should not be confused by such use of marginal valuation. The fact that water is necessary for human life is taken into account in WAS by assigning large benefits to the first relatively small quantities of water allocated. But the fact that the benefits derived from the first units are greater than the marginal value does not distinguish water from any other economic good. It merely reflects the fact that demand curves slope down and that water would be (even) more valuable if it were scarcer.

It is the scarcity of water and not merely its importance for existence that gives water its value. Where water is not scarce, it is not valuable.

WAS provides a powerful tool for the analysis of the costs and benefits of various infrastructure projects. For example, if one runs the model without assuming the existence of seawater desalination facilities, then the shadow values in coastal districts provide a cost target that seawater desalination must meet to be economically viable. Alternatively, by running the model with and without a proposed conveyance line, one can find the increase in annual benefits that the line in question would bring. Taking the present discounted value of such increases gives the net benefits that should be compared with the capital cost of plant construction. Note that such calculations take into account the system-wide effects that result from the projected infrastructure.

5. INFRASTRUCTURE ANALYSIS: SOME RESULTS

The first WAS-generated results concern Israel, Palestine, and Jordan, and are presented for each party separately, assuming each of them only has access to the water it had at the end of 2003. Results involving cooperation are given later.

5.1. Desalination: Israel

Figure 3 shows the shadow values obtained for 2010, both in a situation of normal availability of natural resources, "normal hydrology" (the upper numbers), and in a severe drought, when that availability is reduced by 30 percent (the lower numbers). Israel's price policy ("Fixed Price Policies") of 1995 are assumed to remain in effect. These policies heavily subsidize water for agriculture while charging higher prices to household and industrial users. Note that Israel's practice of reducing the quantity of subsidized agricultural water in times of drought has not been modeled, so the results are *more* favorable to the need for desalination than would be the case in practice.[10]

The important result with which to start can be seen in the upper shadow values for the coastal districts: Acco, Hadera, Raanan, Rehovot, and Lachish. The highest shadow value is at Acco and is only $\$.319/m^3$—well below the cost of desalination. This means that desalination plants would not be needed in years of normal hydrology.

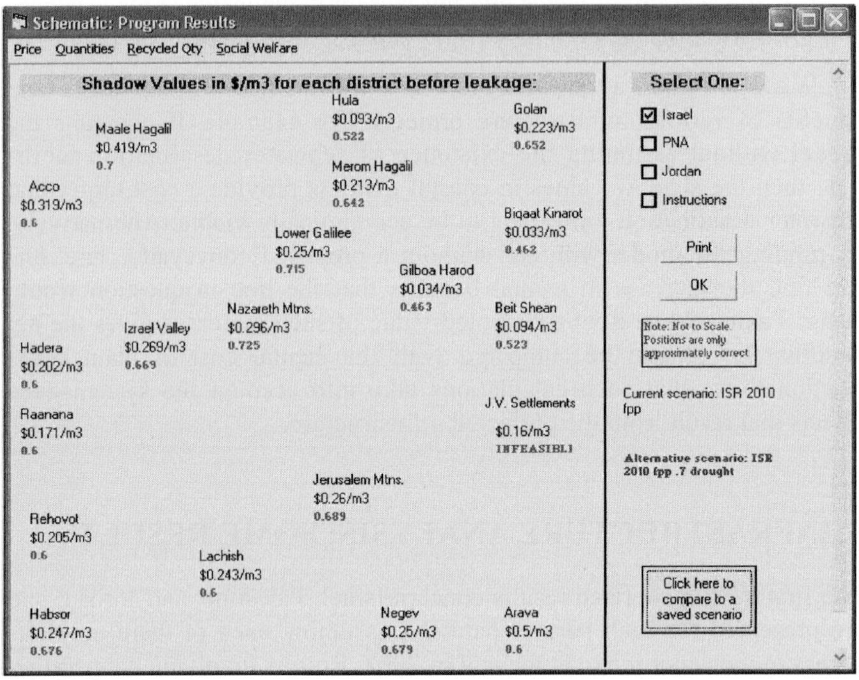

Figure 3: 2010 Shadow Values with Desalination: Normal Hydrology vs. 30 Percent Reduction in Naturally Occurring Fresh Water Sources; Fixed-Price Policies in Effect.

On the other hand, such plants would be desirable in severe drought years. In the lower numbers in Figure 3, desalination plants operate in all the coastal districts at an assumed cost of $\$.60/m^3$. The required sizes of such plants (obtained by running WAS without restricting plant capacity and observing the resulting plant output) are given in Table 1.

[10] The infeasibility listed for the Jordan Valley Settlements in the drought case reflects the fact that the full amount of subsidized water demanded by agriculture cannot be delivered there at the fixed prices cannot be delivered.

Results for 2020 are similar, although, as one should expect, it does not take so severe a drought to make desalination efficient, and the required plant sizes in each district are larger.

Of course, much of the costs of desalination consist of capital costs—included here in the price (or target price) per m^3. Such costs are largely incurred when the plant is constructed. After that, the plants would be used in normal years unless the operating costs were above the upper shadow values in Figure 3 (highest $.319/$m^3$). Israel therefore needs to consider whether the insurance for drought years provided by building desalination plants is worth the excess capital costs.[11] (Note that the system of Fixed Price Policies contributes substantially to the need for desalination; without such policies, the plants required for severe drought would be far smaller than shown in Table 1, and some would not be required at all.)

Table 1: Desalination (or Import) Requirements in Mediterranean Coastal Districts in 2010 with 30 Percent Reduction in Natural Fresh Water Sources and Fixed-Price Policies in Effect.

District	**Water requirements (MCM/Year)**
Acco	80
Hadera	64
Raanana	17
Rehovot	51
Lachish	29
TOTAL	241

5.2. Desalination: Palestine

A similar analysis for Palestine produces quite a surprising result. Palestine can desalinate seawater only on the seacoast of the Gaza Strip (See Figure 1). Consider Figure 4 on the following page. Here results for 2010 are presented on the assumption that Palestine builds recycling plants and conveyance lines.

[11] Note that a multi-year version of WAS (discussed below) could be of substantial aid in such a calculation.

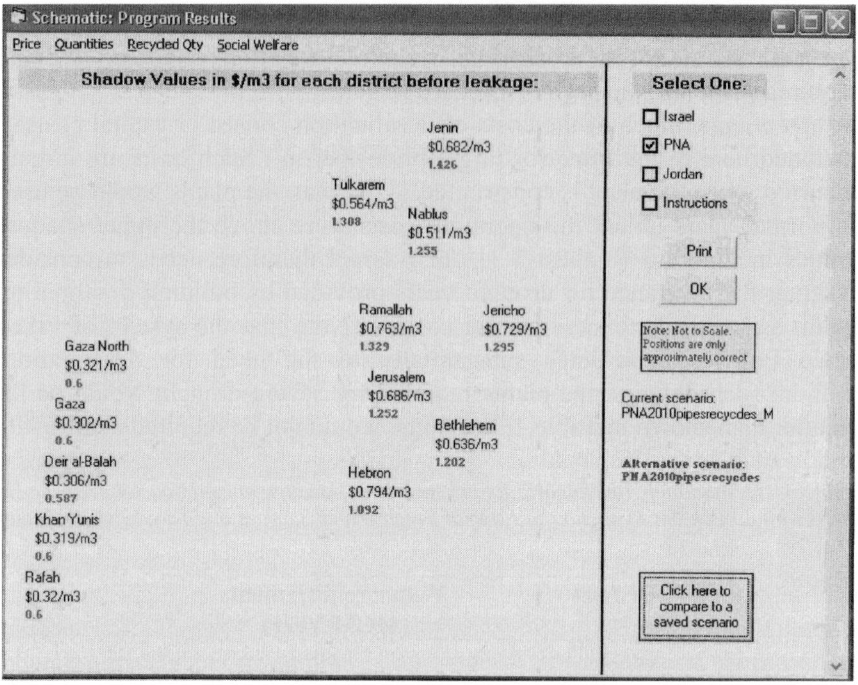

Figure 4: Comparison of Full Infrastructure Scenario in PNA in 2010 with and without Double the Quantity from the Mountain Aquifer.

The lower shadow values are for the case in which Palestine has only its current natural water resources. We see that desalination at $.60/m^3 is efficient in two of the Gazan districts. But the reason for this is not the obvious one of population growth in Gaza. Rather, it is because with its limited water resources on the Southern West Bank, it would actually pay Palestine to desalinate water in Gaza and *pump it uphill* to Hebron! This occurs because, without such pumping, and with Palestine allocated so little water on the West Bank, the difference between the shadow value in Hebron (the value of an additional cubic meter of water there) and that in Gaza exceeds the cost of conveyance. If the Palestinian West Bank water were doubled, and the lower shadow values obtained, desalination would cease to be efficient at prices higher than $.356/m^3. Of course, this result is for a year of normal hydrology and for a middle estimate of Gazan population growth, but the main point remains the same. Without more water or cooperation in water with Israel (see below), Palestine should build one or more desalination plants at Gaza by 2010; but with more water on the West Bank or with cooperation with Israel, that necessity will disappear. Even in 2020, the need for Gazan desalination plants will remain a close question in years of normal hydrology, our results suggesting that such plants would be barely cost-efficient at costs above $.55/m^3. An important implication of these results will appear when we consider cooperation below.

5.3. Jordan and the Interdependence of Infrastructure Decisions

For Jordan (where seawater desalination is currently possible only at Aqaba on the Red Sea), we report results on other issues.

Without action, Jordan faces an increasing water crisis in Amman and nearby districts. Indeed, our results show that if nothing were done, the shadow value of water in Amman would reach roughly $27/m^3 by 2020 (and that too in years of normal hydrology). This is not a tenable situation, and the value of $27/m^3 is not presented as a value that people will pay for water but as an indication of the coming water-scarcity crisis. To alleviate this, Jordan has various options:

1. Jordan has plans to increase the capacity of the conveyance line that takes the Jordan River to Amman from 45 MCM/year to 90 MCM/year no later than 2005. This would reduce the shadow value in Amman in 2020 from $27.23 to $10.56 /m^3. The gain in net benefits in 2010 is approximately $2 million/ year, which, by 2020, reaches almost $500 million/year. (Our evaluation of the other options assumes this conveyance line to be in place.)
2. Jordan could attempt to reduce the large leakage in pipes in Amman and other districts. We find that, by 2020, this would result in an increase in Jordanian water benefits of about $250 million/year, probably making it worth the capital costs involved—not counting the disruption to the population. Nevertheless, this does not satisfactorily alleviate the crisis and only reduces the shadow value in Amman to about $6.43/m^3, which is still unacceptably high.
3. Jordan is considering the construction of a conveyance line from the Disi fossil aquifer to Amman. This will help considerably. If the conveyance line will carry about 100 MCM per year by 2020, then the benefits from its construction will reach more than $300 million per year by that date. The resulting shadow value in Amman would be about $1.44/m^3, which is still high, but not catastrophically so. Adding leakage reduction to this would take the value down to about $1.13, but, of course, such reduction might not be worth the

capital costs involved, with the added benefits as of 2020 falling from \$250 million per year in the absence of the Disi-Amman pipeline to about \$93 million per year in its presence. It should also be noted that, given the expansion of the conveyance line from the Jordan River, the Disi-Amman pipeline would not be used in 2010.

4. There are grand plans for the oft-discussed Israeli-Jordanian construction of a canal to take water from the Red Sea to the Dead Sea, the so-called "Peace Canal." While the canal, if constructed, will largely be built for other reasons, there would be water benefits associated with it. In particular, it is planned to use the downfall of water in the canal to generate electricity, and then to use that electricity to desalinate some of the seawater involved and pump it to Amman. It is estimated that it would cost about 22¢/m^3 to pump such water uphill to Amman. Assuming that the shadow value of water in Amman is at least \$1.13/m^3, as a result of the combination of leakage reduction and the transfer of water from the Disi aquifer, this would be efficient if such desalination would cost less than about \$.91/m^3. This seems guaranteed *if the main capital costs of canal construction and electricity generation are allocated to other uses* and the capital costs of desalination include only the construction of the desalination plant and the laying of the pipeline from the plant to Amman. The energy costs involved in operating costs would surely be lower with hydroelectric generation than with fuel-fired plants.

But note the following. The effects of the Red Sea–Dead Sea project would undoubtedly reduce the shadow value of water in Amman to a figure well below \$1.13/m^3 in 2020. If the shadow value in Amman were at such a level, it would no longer make sense to transport water to Amman from the Disi Aquifer. In such a case, that water could efficiently be used in the Aqaba district, quite possibly forestalling the necessity of a desalination plant there.

This does not mean that it would be a mistake to build the Disi-Amman pipeline. Far from it. First, the Red Sea–Dead Sea Canal may never be built. Second, if it is built, it will be a long time before it is complete. During that period, and after 2010, the Disi-Amman pipeline may very well be highly necessary to avert the Amman water crisis.[12]

---

[12] If the only problem in Jordanian water management were the coming crisis in Amman, then this could be readily solved by a further expansion of the conveyance system bringing water from the Jordan River to the capital. (It is interesting to note that expansion of the conveyance system, *not* additional water *ownership*, is what would be directly involved.)

Note how the benefits of an infrastructure project depend on what other projects have been undertaken. Note further how WAS can be used to investigate such interdependencies.

6. WATER OWNERSHIP AND THE VALUE OF WATER

The view of water as an economic, if special, commodity has important implications for the design of a lasting water arrangement that is to form part of a peaceful agreement among neighbors. WAS can be used to explore resolution of water disputes.

There are two basic questions involved in thinking about water agreements: the question of water ownership and the question of water usage. One must be careful to distinguish between these questions.

All water users are effectively buyers irrespective of whether they own the water themselves or purchase water from another. An entity that owns its water resources and uses them itself incurs an opportunity cost equal to the amount of money it could otherwise have earned through selling the water. An owner will thus use a given amount of owned water if and only if the use of it is valued at least as much as the money to be gained through its sale. The decision of such an owner does not differ from that of an entity that does not own its water and must consider buying needed quantities of water: the nonowner will decide to buy if and only if it values the water at least as much as the money involved in its purchase. *Ownership only determines who receives the money (or the equivalent compensation) that the water represents.*

Water ownership is thus a property right entitling the owner to the economic value of the water. Hence, a dispute over water ownership can be translated into a dispute over the right to monetary compensation for the water involved.

The property rights issue of water ownership and the essential issue of water usage are analytically independent. For example, resolving the question of where water should be efficiently pumped does not depend on who owns the water. While both ownership and usage issues must be properly addressed in an agreement, they can and should be analyzed separately.[13]

However, this would divert the river water from its current principal use in which it is mixed with wastewater and used in agriculture in the Jordan Valley. Jordan could not then continue to subsidize Jordan Valley agriculture. The effects of such an action are not readily captured without an analysis of the social consequences.

[13] This is an application of the well-known Coase Theorem of economics. See Coase 1960.

The fact that water ownership is a matter of money can be explained in a different way. It is common for countries to regard water as essential to their security because water is essential for agriculture and countries wish to be self-sufficient in their food supply. This may or may not be a sensible goal, but the possibility of desalination implies the following:

Every country with a seacoast can have as much water as it wants if it chooses to spend the money to do so. Hence, so far as water is concerned, every country with a seacoast can be self-sufficient in its food supply if it is willing to incur the costs of acquiring the necessary water. Disputes over water among such countries are merely disputes over costs, not over life and death.

The monetization of water disputes may be of some assistance in resolving them. Consider bilateral negotiations between two countries, A and B. Each of the two countries can use its WAS tool to investigate the consequences to it (and, if data permit, to the other) of each proposed water allocation. This should help in deciding on what terms to settle, with a possibility to trade water for other, non-water concessions. Indeed, if, at a particular proposed allocation, A would value additional water more highly than B, then both countries could benefit by having A get more water and B get other things that it values more. (Note that this does not mean that the richer country gets more water. That only happens if it is to the poorer country's benefit to agree.)[14]

Of course, the positions of the parties will be expressed in terms of ownership rights and international law, often using different principles to justify their respective claims. The use of the methods described here in no way limits such positions. Indeed, the point is not that the model can be used to help decide how allocations of property rights should be made. Rather the point is that water can be traded for non-water concessions, with the trade-offs measured by WAS.

Moreover, such trade-offs will frequently not be large. For example, water on the Golan Heights (see Figure 1) is often said to be a major problem in negotiations between Israel and Syria, because the Banias River that rises on the mountains of the Golan is one of the three principal sources of the Jordan River. By running the Israeli WAS model with different amounts of water, we have already evaluated this question.

In 2010, the loss of an amount of water roughly equivalent to the entire flow of the Banias springs (125 million cubic meters annually) would be

---

[14] If trading off ownership rights considered sovereign is unacceptable, the parties can agree to trade short-term permits to use each others' water. See below.

worth no more than $5 million/year to Israel in a year of normal water supply and less than $40 million/year in the event of a reduction of thirty percent in naturally occurring water sources. At worst, water can be replaced through desalination, so that the water in question (which has its own costs) can never be worth more than about $75 million/year. These results take into account Israeli fixed-price policies towards agriculture.

Note that it is *not* suggested that giving up so large an amount of water is an appropriate negotiating outcome, but water is not an issue that should hold up a peace agreement. These are trivial sums compared to the Israeli GDP (gross domestic product) of roughly $100 billion/year or to the cost of fighter planes.

Similarly, a few years ago, Lebanon announced plans to pump water from the Hasbani river—another source of the Jordan. Israel called this a *casus belli* and international efforts to resolve the dispute were undertaken. But whatever one thinks about Lebanon's right to take such an action, it should be understood that our results for the Banias apply equally well to the Hasbani. The effects on Israel would be fairly trivial.[15]

Water is not worth war.

7. COOPERATION: THE GAINS FROM TRADE IN WATER PERMITS

Monetization of water disputes, however, is neither the only nor, perhaps, the most powerful way in which the use of WAS can promote agreement. Indeed, WAS can assist in guiding water cooperation in such a way that all parties gain.

The simple allocation of water quantities, after which each party then uses what it "owns," is not an optimal design for a water agreement. Suppose that property rights issues have been resolved. Since the question of water ownership and the question of water usage are analytically independent, it will generally not be the case that it is optimal for each party simply to use its own water.

Instead, consider a system of trade in water permits—short term licenses to use each other's water. The purchase and sale of such permits would be in quantities and at prices (shadow values) given by an agreed-on version of the WAS model run jointly for the two (or more) countries

[15] Of course, the question naturally arises as to what the effects on Syria and Lebanon, respectively would be in these two situations. Without a WAS model for those two countries, I cannot answer that question. Both countries would surely profit from such a model, but, as of yet, they have not been willing to cooperate in building one.

together. (The fact that such trades would take place at WAS-produced prices would prevent monopolistic exploitation.). There would be mutual advantages from such a system, and the economic gains would be a natural source of funding for water-related infrastructure.

Both parties would gain from such a voluntary trade. The seller would receive money it values more than the water given up (else, it would not agree); the buyer would receive water it values more than the money paid (else, it would not pay it). While one party might gain more than the other, such a trade would not be a zero-sum game but a win-win opportunity.

The WEP has estimated the gains to Israel and Palestine from such cooperation, and finds them to exceed the value of changes in water ownership that reflect reasonable differences in negotiating positions.

To illustrate this, we begin by examining the gains to Israel and Palestine from such cooperation starting from varying assumptions about the ownership of the Mountain Aquifer (see Figure 1). To simplify matters, the case we examine is one in which Israel owns all of the water of the Jordan River. This is to be taken as merely an assumption made for the purposes of generating illustrative examples; it is *not* a political statement as to the desirable outcome of negotiations. We find such gains generally to exceed the value of changes in such ownership that reflect reasonable differences in negotiating positions.

Figures 5A and 5B illustrate such findings and more for years of normal hydrology. In those figures, we have arbitrarily varied the fraction of Mountain Aquifer water owned by each of the parties from 80 percent to 20 percent. (The present division of the water is about 76 percent to Israel and 24 percent to Palestine.[16] Results for that division can be approximated by interpolation, but are, of course, fairly close to those for 80 percent Israeli ownership.)

The two line graphs in Figure 5A show the gains from cooperation in 2010 for Israel and Palestine, respectively, as functions of ownership allocations.[17] Israeli price policies for water ("Fixed Price Policies") are assumed to be the same as in 1995, with large subsidies for agriculture and much higher prices for households and industry.

---

[16] The "Mountain Aquifer" actually consists of several sub-aquifers. It is very difficult to secure accurate information on how the water in each of these is now divided. The 76 percent–24 percent split mentioned in the text is therefore an approximation applying to the total. In the runs reported below, where necessary, we have used that split to represent existing circumstances. Of course, the general conclusions are not affected by this, and the quantitative results cannot be far off.

[17] The results discussed in this section are all for years of normal hydrology. Results for drought years are not qualitatively different, although all numbers are larger.

Starting at the left, we find that Palestine benefits from cooperation by about $170 million per year when it owns only 20 percent of the aquifer.[18] In the same situation, Israel benefits by about $12 million per year. As Palestinian ownership increases (and Israeli ownership correspondingly decreases), the gains from cooperation fall at first and then rise. At the other extreme (80 percent Palestinian ownership), Palestine gains about $84 million per year from cooperation, and Israel gains about $36 million per year. In the middle of the figure, total joint gains are about $84–95 million per year.[19]

It is important to emphasize what these figures mean. As opposed to autarky, each party benefits as a buyer by acquiring cheaper water. Moreover, each party benefits as a seller *over and above* any amounts required to compensate its people for increased water expenses.

Why do the gains first decrease and then increase as Palestinian ownership increases? That is because, at the extremes, there are large gains to be made by transferring water from the large owner to the other party. Palestine has large benefits, seen at the left-hand side of the diagram, because it can obtain badly needed water; it has large gains at the right-hand side because, when it owns most of the Mountain Aquifer water, it can gain by selling relatively little-needed water to Israel (which gains as well). The same phenomenon holds in reverse for Israel—although there the effects are smaller, largely because Israel is assumed to own a great deal of water from the Jordan River.

One might suppose that the gains would be zero at some intermediate point, but that is not the case. The reason for this is as follows:

It is true that a detailed, noncooperative water agreement could temporarily reduce the gains from cooperation to zero. That would require that the agreement exactly match in its water-*ownership* allocations the optimizing water-*use* allocations of the optimizing cooperative solution. That is very unlikely to happen in practice (and, if it did, would only reach an optimal solution that would not last as populations and other factors changed). In our runs, it does not happen for two reasons.

1. We have not attempted to allocate ownership in the Mountain Aquifer in a way so detailed as to match geographic demands. Instead, we have allocated each common pool in the aquifer by the same percentage split.

---

[18] Here and later, the gains with this division are so large as to dominate the scale of the Figure. This must be taken into account in examining the results.

[19] While the qualitative conclusions remain the same, the quantitative results are substantially different from those presented in Fisher, et al. (2002). This is due partly to

2. There are gains from cooperation in these runs that do not depend on the allocation of the Mountain Aquifer. For example, it is always efficient for treated wastewater to be exported from Gaza to the Negev for use in agriculture.

There are further results to be read from Figure 5A. The heights of the various bars in the figure show the value to the parties, *without cooperation*, of a change in ownership of 10 percent of the Mountain Aquifer (about 65 MCM per year or nearly half of the amount of Mountain Aquifer water now taken by the Palestinians). These are shown as functions of ownership positions midway within each 20 percentage point

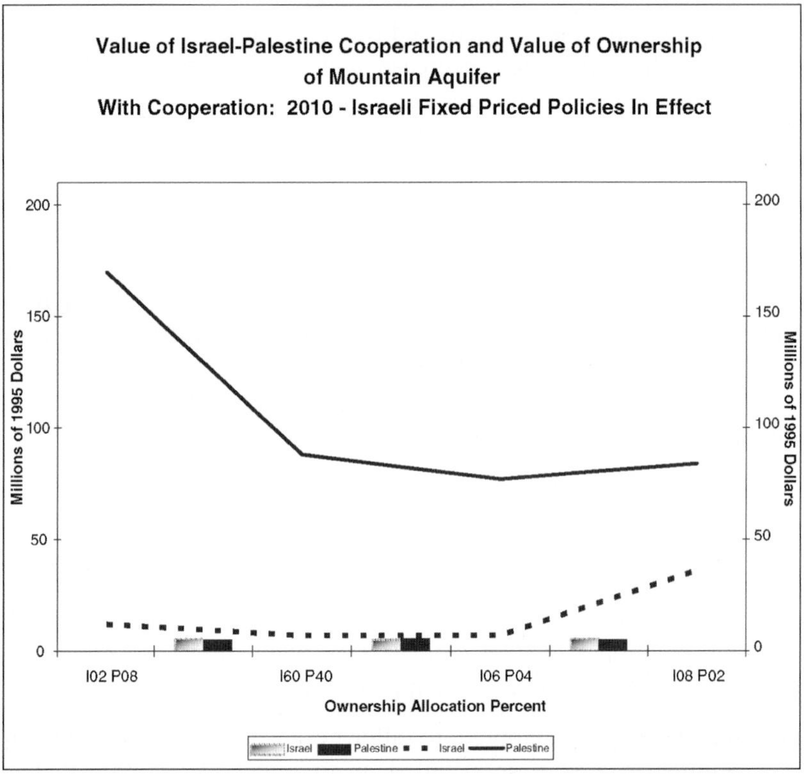

Figure 5A: Value of Israel-Palestine Cooperation and Value of Ownership of Mountain Aquifer: Without Cooperation

---

improved data, but mostly from a more realistic treatment of intra-district leakage in Palestine, which affects the value of water.

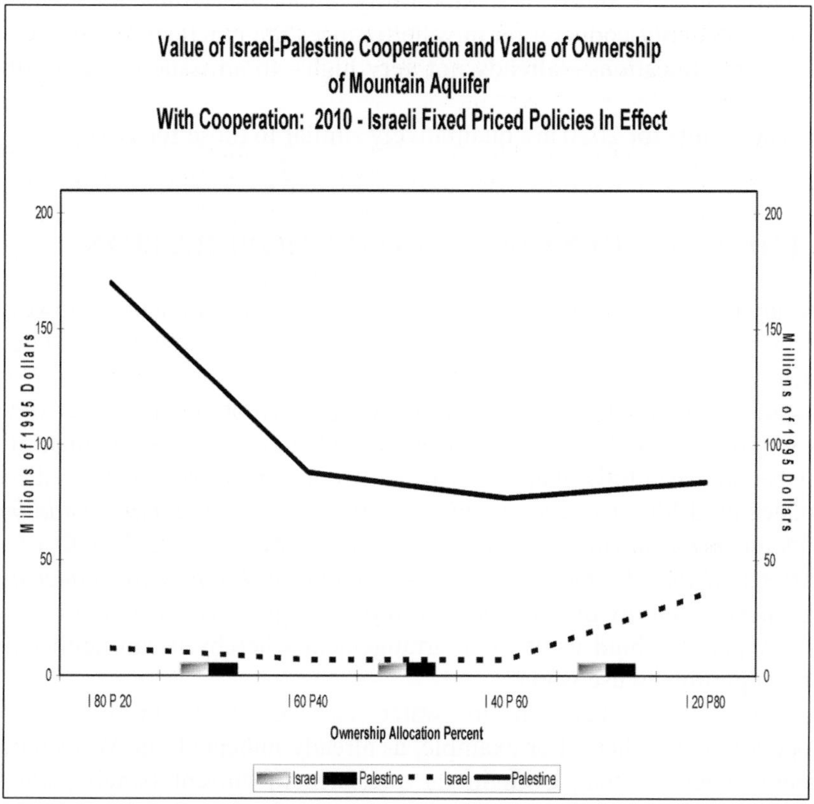

Figure 5B: Value of Israel-Palestine Cooperation and Value of Ownership of Mountain Aquifer: With Cooperation.

interval. For example, the left-hand-most set of bars shows the value to each of the parties of an ownership shift of 10 percent of the Mountain Aquifer starting at an allocation of 70 percent to Israel and 30 percent to Palestine; the next set of bars examines the value of a such a change starting at 50-50. Note that the value of cooperation is generally greater than, or at least comparable to, the value of such ownership changes. This is especially true for Palestine, but holds for Israel as well.

Further, now look at Figure 5B. This differs from Figure 5A only in the height of the ownership-value bars. In Figure 5B, the height of those bars represents the value of shifts of 10 percent aquifer ownership *in the presence of cooperation.* That value is about $8 million/year. The lesson is clear:

Ownership is surely a symbolically important issue, and symbols really matter. But cooperation in water reduces the practical importance of ownership allocations—already not very high—to an issue of very minor proportions.

The results for 2020 are qualitatively similar to those for 2010.

8. THE REAL BENEFITS FROM COOPERATION

The greatest benefits from cooperation may not be monetary, however. Beyond pure economics, the parties to a water agreement would have much to gain from an arrangement of trade in water permits. Water allocations that appear adequate at one time may not be so at other times. As populations and economies grow and change, fixed water quantities can become woefully inappropriate and, if not properly readjusted, can produce hardship. *A system of voluntary trade in water permits would be a mechanism for flexibly adjusting water allocations to the benefit of all parties and thereby for avoiding the potentially destabilizing effect of a fixed water quantity arrangement on a peace agreement.* It is not optimal for any party to bind itself to an arrangement whereby it can neither buy nor sell permits to use water.

Moreover, cooperation in water can assist in bringing about cooperation elsewhere. For example, as already indicated, the WAS model strongly suggests that, even in the presence of current Israeli plans, it would be efficient to have a water treatment plant in Gaza, with treated effluent sold to Israel for agricultural use in the Negev where there is no aquifer to pollute. (Indeed, since this suggestion arose in model results, there has been discussion of this possibility.) Both parties would gain from such an arrangement. *This means that Israel has an economic interest in assisting with the construction of a Gazan treatment plant.* This would be a serious act of cooperation and a confidence-building measure.

9. PROBLEMS AND CONCLUSIONS

Naturally, there are a number of issues that arise when considering such a cooperative arrangement. Chief among them is that of security. What if one of the partners to such a scheme were to withdraw? Of course, such withdrawal would be contrary to the interest of the withdrawing party, but, as we have sadly seen, people and governments do not always act in their own long-run self interest.

The main cost of such a withdrawal would occur if the non-withdrawing party had failed to build infrastructure that would be needed without cooperation but not with it. In the case of Israel and Palestine, it might appear that such risk would be chiefly Palestinian, since they, but *not* Israel, would need desalination plants in the absence of cooperation but not in its presence. Israel, by contrast, already has a highly developed system of water infrastructure and any decision to build desalination plants does not depend on a decision to cooperate or not cooperate with Palestine.

Interestingly, this conclusion may not hold. We saw above that the WAS results show that it will not be cost-effective (at least in years of normal hydrology) for Palestine to build desalination facilities in the Gaza Strip (its only seacoast) simply to supply the growing Gazan population. Rather, with water ownership greatly restricted on the West Bank, it would pay (without cooperation) to build such facilities and expensively pump desalinated water uphill to Hebron. But this result (which holds only with Palestine owning less than 20 percent of the Mountain Aquifer) also implies that a withdrawal by Israel from a cooperative agreement could be met by Palestine temporarily pumping more than permitted by treaty on the West Bank while building a desalination plant at Gaza.[20] This reduces the importance of the security issue under discussion.

Hence, for both parties, cooperation appears to be a superior policy to autarky. In an atmosphere of trust, cooperation would be likely to benefit Palestine even more than Israel, at least in the short run. But, of course, such an atmosphere does not currently exist. Cooperation requires a partner, and, as of late 2004, that did not appear to be immediately likely. Each party is likely to suspect the good faith of the other, even though the proposed arrangement would benefit both.

Despite this, I continue to believe that cooperation is both valuable and possible. As already discussed, water is not worth conflict and can become an area for confidence-building measures. Further, if autarky is truly desired, then one should simply build desalination plants as needed. Autarky in naturally occurring water is a foolish policy except as a money-saving device—and the money it saves is not great. Every country with a seacoast can have as much water as it wants if it chooses to spend the money to do so. Hence, every country with a seacoast can be self-sufficient if it is

---

[20] Of course, there would also have to be a conveyance line. A refusal by Israel to permit this would be a serious matter—and an act whose principal intent would be to harm Palestine. In such an event, Palestinian pumping of the Mountain Aquifer beyond agreed-upon amounts would have to continue, but there would be larger, non-water issues to worry about.

willing to incur the costs of acquiring the necessary water. As a result, disputes over water among such countries should be merely disputes over costs, not over life and death.

10. AFTERWORD: VIEWING TWO OTHER CASES FROM A “WAS” PERSPECTIVE

To show the versatility of the WAS tool, we end by briefly looking at two other chapters in this volume from this perspective

10.1. The Schoengold and Zilberman Chapter

In their review of various problems involved in a sensible water policy—especially of incentive problems—Schoengold and Zilberman (henceforth “S&Z”), not surprisingly, find that such problems abound. They state: “Many regions have a perception of water crisis, because existing water resources are not sufficient to meet growing needs. In most cases, the real problem *is water management crisis*. Incentives for efficient and sociably responsible management of water are lacking. Water projects that cannot be justified economically, and are damaging environmentally, are being built. Users are paying well below the value of the water they use, and are encouraged to consume water. Polluters of water bodies are not penalized. … To achieve sustainable water use, water policies and institutions have to be reformed.”

They list numerous examples of mismanagement, including:

- ignored environmental costs of water projects
- waterborne diseases
- displacement of native populations
- downstream positive externalities produced by upstream canal maintenance
- future costs of potential water projects such as increased salinity and water logging of soils
- depletion of fossil aquifers such as the Oglala Aquifer in the Western United States.

S&Z then go on to consider water trading schemes and their failings. All of these problems need to be handled, and partial analysis will not do so satisfactorily, nor, given the externalities present will private water markets. WAS, on the other hand, will handle at least some of the problems and, in particular, will permit a system-wide view of the effects of different actions. As we have seen, WAS will also generate shadow values for water in different locations; these are the efficiency prices to be charged unless overridden for special social purposes. Further, WAS can be used to guide water trading schemes.

That is not to say that WAS can do everything. Consider environmental issues, for example. Such issues are of two general kinds: environmental concerns over the state of the water system itself (salinity, pollution, the effects of pumping rates on the aquifers, etc.) and the effects of water actions on water-related parts of the environment (species survival, green lands, etc.). The present version of WAS handles both of these in the same way. The user must specify restrictions on water use designed to preserve environmental values. Examples include setting aside water for environmental purposes, preventing more than a certain fraction of irrigation water from being treated wastewater, and restricting water extraction from a source to the annual renewable amount, thus preventing overpumping.

But note that WAS will not tell the user what it is worth to impose such restrictions. Instead, it will provide the costs (both total and marginal) of doing so. By varying the restrictions, the user can then try to decide how tight they should be by explicitly or implicitly estimating the value of their environmental effects.

There would be more help from an expanded WAS, one that is not a single-year but a multi-year model. Such a multi-year WAS would optimize the present value of net benefits over a number of years. In so doing, it could directly include hydrological models of water sources, especially aquifers, and internalize the effects of one year's actions on the state of the water system in later years. It would also assist analysis of the interdependence of infrastructure projects and the order in which they should be built.

Note, in particular, that such a model would readily handle the problem of the rate at which fossil aquifers should be depleted, and other issues of the first kind listed above. It would also allow systematic investigation of the effects on optimal policy of the stochastic nature of the climate.

Such a model has been on the drawing boards of the WEP for some time, and has been constructed since this chapter was presented. Even so, problems will still remain. Not even a multi-year WAS can directly handle environmental issues of the second kind—issues such as the benefits of affecting species survival or preserving green open space. Here the user can, as before, find out the system-wide costs of various actions that can be taken, but the benefits of such actions must be estimated externally.

But, as S&Z show, a system-wide analysis of water policy is crucial, and WAS provides a tool with which this can be done systematically.

10.2. The Güner Chapter

Güner's work, "Evolutionary Explanations of Syrian-Turkish Water Conflict," presents an evolutionary game in which Syria desires water concessions from Turkey, while Turkey seeks Syrian recognition of Turkish sovereignty over a certain territory, Hatay. Each population is divided into hawks and doves, and hawks never concede anything. Hawks and doves meet each other randomly. The "fitness" of one side's hawks increases when a hawk of that side obtains a concession from a dove of the other and a greater "fitness" of hawks than of doves in the same country leads to an expansion of the number of hawks therein. The model evolves into a permanent state of either instability or noncooperation.

This model is not, of course, highly realistic—nor is it intended to be—but its depressing conclusion points to an important issue.

There is an explicit recognition among the doves of the two modeled countries that water can be traded off for other things (here, territorial recognition). Notwithstanding this, such a trade does not take place, because of the existence of large segments of the two populations that will not concede and the disappointing nature of the interactions of the doves of one side with the hawks of the other. This results in the doves becoming discouraged, as it were, and the political power of the hawks growing.

This case illustrates two important points about the potential role that WAS, an economic model designed to optimize benefits for different groups, can play in attenuating water conflicts:

1. While the use of WAS should prevent war from being a *casus belli*, it will not succeed in bringing about a solution to international water conflicts unless the countries involved are interested in arriving at a peaceful arrangement. Otherwise, water can be used harshly as a negotiating lever or even as a weapon.
2. It is not enough for only some people in the disputing countries to understand the message that water cooperation and trading is a win-win situation. Even if water experts understand it, there will be no progress unless political leaders do so. Further, even if political leaders understand it, there may be no success without a full-scale program of public education on the subject. That is devoutly to be wished.

EVOLUTIONARY EXPLANATIONS OF SYRIAN–TURKISH WATER CONFLICT*

Serdar Ş. Güner
Department of International Relations
Bilkent University
Ankara, Turkey
sguner@bilkent.edu.tr

Abstract: The issues of water and territory dominate relations between Syria and Turkey, upstream and downstream riparians in the Euphrates and Tigris basin. This chapter proposes an evolutionary game to explore eventual trajectories of riparian relations. Turkish hawks are defined as those Turkish foreign policies that support no water concessions. Turkish doves can instead support the flow of an increased amount of water to Syria on the basis of an international agreement. Syrian hawks are those Syrian foreign policies that do not recognize Turkish sovereignty over Hatay—also known as the Sandjak of Alexandretta. Syrian doves can in turn accept that the territory belongs now to Turkey. It is found that evolutionary stability does not depend upon the values territory and water represent for the fitness of Syrian and Turkish foreign policies. No evolutionary stability is possible unless doves are cooperative towards hawks. If doves are cooperative towards hawks, the unique evolutionarily stable outcome implies their extinction. Riparian relations will ultimately evolve into mutual intransigence.

Keywords: upstream-downstream water conflicts, game theory, Turkey-Syria water conflicts

1. INTRODUCTION

Two contentious issues underlie Syrian–Turkish relations: water and territory. Syria has requested Turkey, the upper-riparian in the Euphrates river basin, to allow a flow of more than 500 m^3/s (cubic meters per second) of water, making Turkey uncomfortable with Syria's nonrecognition of its sovereignty over Hatay province (the Sandjak of Alexandretta). Given

---

* An earlier version of this chapter was presented at the conference "Mountains: Sources of Water, Sources of Knowledge" held in Sion, Switzerland, October 8–10, 2002. I am grateful to a referee for helpful suggestions and comments on this earlier version.

E. Wiegandt (ed.), Mountains: Sources of Water, Sources of Knowledge, 149–159.

these dimensions of conflict, how will Syrian–Turkish relations evolve? We generate an answer through an evolutionary game where we find that the current status quo characterized by no concessions on both issues is the unique evolutionarily stable outcome. The stability of the status quo is insensitive to different assumptions about behavior traits and to values of water and territory for the riparian countries.

The next section broadly describes water and territory issues in Syrian–Turkish relations. The subsequent section presents the basic assumptions of the evolutionary framework. The implications and interpretations of results follow the presentation of the model. The final section concludes the analysis.

2. CONFLICT DIMENSIONS IN SYRIAN–TURKISH RELATIONS

Turkey and Syria, the upstream and the midstream riparian states in the Euphrates–Tigris basin, have a long history of antagonistic relations around water, territory, and terrorism. Together these contributed to adverse relations and ensuing tension between the two riparian countries until the capture of Abdullah Öcalan, the leader of the PKK (acronym for Kurdish Workers Party), by Turkey in 1999. Now known as KADEK (acronym for the Congress for Freedom and Democracy in Kurdistan), the PKK aims at forming an independent Kurdish state in Turkey. During the 1980s and 1990s, the party received Syria's support with the twin objectives of extracting Turkish concessions on water and territory issues Beschorner (1992), Cohen (1993), Frey (1993), Olson (1992), Robins (1991), Starr (1991).

The issue of territory centers on Hatay which became a part of the French Mandate of Syria after the First World War. In 1936, France signed an agreement with Syria thereby ending its mandate. Turkey requested France to grant Hatay the same status as Syria and Lebanon, as Hatay was not cited in the agreement. Following the French refusal, Turkey and France brought the issue before the League of Nations, which conferred the status of special enclave (entité distincte) on Hatay in 1937. Although the territory was put under joint Turco-French guarantee, Hatay became independent in governing its interior affairs and adopted Turkish as its official language. Following general elections, Hatay proclaimed itself an independent republic in 1938 and became a part of Turkey through a plebiscite in 1939. To this date Syria considers this decision as invalid.

The water issue between Turkey and Syria concerns the Euphrates. Syria has opposed the building of dams and other projects that harness waters of the Euphrates River in upstream Turkey. Joined by Iraq, it also insists that Turkey fix a water quota between 700 m^3/s and 1,000 m^3/s for downstream flow. This exceeds the agreed annual average water flow of 500 m^3/s that was fixed in 1987 by Turkey in return for Syrian cooperation in security matters. According to Bağış (1997), a Turkish water flow lying between 700 m^3/s and 1,000 m^3/s would preclude optimal performance of the GAP (Turkish acronym for the Southeastern Anatolia Project), including several dams and hydroelectric power plants. Syria and Iraq also fear that both the quantity and quality of water will diminish once the GAP fully becomes operational. While there exists a tripartite commission composed of technical personnel from the riparian countries that irregularly meets to discuss these issues, none of the commission's activities or other proposals (such as the Turkish three-stage plan) have thus far resolved conflicting demands in the basin (Lowi 1993; Kolars and Mitchell 1991; Bağış 1991; Naff and Matson 1984).

The Turkish government is uncertain whether Syria has given up its claim over Hatay. In fact, a Turkish diplomat indicated that Turkey would not like Syria to bring the Hatay issue up following an agreement over water (Sarıibrahimoğlu 1995, p. 8). Such linkages seemingly constitute a political norm in all river basins (LeMarquand 1977). For example, Syria reduced the water flow to Iraq in 1974 and 1975 as Iraq favored the closure of all negotiation possibilities with Israel after the Yom Kippur War (Walt 1987, p. 133; Lowi 1993, p. 59).

There are three game-theoretic analyses of the terrorism-water issue-linkage. These studies assume, in general, rationality of players and that the Syrian support to terrorism constitutes a major foreign-policy tool. Güner (1997) models the conflict as a war of attrition, Güner (1998) searches for the implications of Syrian uncertainty about Turkish preferences with respect to the mutual conflict outcome, and Güner (1999) investigates the conditions of various alliance combinations in the basin among Turkey, Syria, and Iraq. The first study reveals that unilateral concessions depend on whether the associated costs and benefits are equal for the riparian states, i.e., costs of continuing the conflict and the benefits from water. Otherwise, if riparian states evaluate future costs and benefits asymmetrically, the propensity to unilaterally concede depends not only on costs of conflict and water benefits but also on discounted future benefits and costs. The second piece of work indicates that Syrian misperceptions about Turkish costs of conflict have no impact upon Turkish water policy. Different Syrian beliefs about how the issue-linkage harms Turkey do not result in a Turkish concession. The third study points towards the possibility of Turkish–Iraqi and Syrian–Iraqi alliances with Iraq acting as a balance to threats in the basin.

A Turkish–Iraqi alliance is likely if Iraq perceives that Turkey would concede provided that Syria ceases its support to the Kurdish separatists. An Iraqi–Syrian alliance is by no means automatic as Turkey is the upstream riparian state in the basin and can give significant advantages to Iraq at the expense of Syria. With terrorism off the agenda in Syrian–Turkish relations, the situation could lead to an alliance between Turkey and Iraq.

3. EVOLUTIONARY FRAMEWORK

There are two major advantages of using an evolutionary game: first, it does not require the restrictive assumption of purely rational players, and, second, it offers the possibility of dynamic interpretations of interactions (Boulding 1991; Kandori, Mailath, and Rob 1993; Maynard Smith 1982; Maynard Smith and Price 1973; Selten 1991; Young 1993). In this game, Syria and Turkey are not assumed to be unitary or rational players; they can make mistakes. They are not assumed to make conscious choices or to be able to compute and anticipate every respective move. The relaxation of the rationality assumption does not imply that strategies bringing low benefits are selected. Success is imitated. Thus, learning still occurs under limited capacities of information processing and of computation of best moves. Simply actions bringing higher rewards are more frequently chosen.

Evolutionary games model interactions within or between large populations. Syrian and Turkish foreign policies with respect to the issue-linkage are assumed to make up the respective populations. Interaction is not between unitary rational players, Syria and Turkey, but between masses of Syrian and Turkish foreign policies. These large populations contain different foreign-policy types or templates. Some Syrian templates may adopt a hostile stand with respect to Turkey but some others can be more cooperative recognizing (directly or indirectly) Hatay as a Turkish territory. Similarly, some Turkish foreign policies may favor a water concession but some others would raise objections against such a move. Either Syrian or Turkish decision makers such as political leaders, diplomats, or experts can adopt a specific template. If a template proves to be successful, more decision makers use it.

An evolutionary framework has two major implications in this case. First, matched foreign policies do not necessarily constitute best replies. Syrian and Turkish foreign policies are randomly matched in pairs resulting in an aggregate behavior rather than conscious and calculated moves. Only successful foreign policies survive and unsuccessful policies ultimately become extinct. Second, foreign policies do not suddenly change directions. Even if a foreign policy is not optimal, its fitness excess is sufficient for it being chosen in the next period (Fudenberg and Levine 1998, p. 71). The evolution of foreign policies thus takes time.

In theoretical terms, hawks and doves respectively depict aggressive and nonaggressive behavioral traits in a population. The recent Syrian signals about a possible acceptance of Hatay as belonging to Turkey and the Syrian–Turkish agreement of 1987 indicate that doves and hawks in fact exist in both populations. We assume that hawks and doves fully describe existing behavioral traits of large populations of Syrian and Turkish foreign policy possibilities. Those hostile and noncooperative foreign policies are labeled hawks and doves those that are in general cooperative. The fitness of each foreign policy is defined as its aggregate or average payoff in its repeated encounters with opposing hawks and doves. Foreign policies that bring greater expected payoffs than average become dominant, otherwise they slowly disappear.

4. EVOLUTIONARY GAME

Territory and water are assumed to be paramount issues representing intrinsic values for Syria and Turkey. To assert that water is a valuable resource around that corner of the world is a truism. As for Hatay, it constitutes an undeniable burden in Syrian–Turkish relations and a substantial value for both countries. Huth (2001, p. 241) cites Hatay as a potential dispute between Syria and Turkey indicating that Syria can still claim it. Syrian–Turkish relations contain a genuine likelihood of war with territory being a central issue (Vasquez 1996, p. 534). Both issues are associated with significant gain and loss prospects. A water concession implies a diminution of Turkish welfare but an increase in Syrian welfare. Similarly, a Syrian recognition of Hatay as a Turkish territory means a loss for Syria but positively affects Turkish welfare. Consequently, water and territory issues shape the fitness of any foreign policy.

We assume that Syrian hawks, denoted by Y, $0 \leq Y \leq 1$, are those foreign policies that consider Hatay as a Syrian territory. In contrast, Syrian doves, denoted by $1 - Y$, are those Syrian foreign policies that regard Hatay as belonging to Turkey when paired with Turkish doves. Similarly, Turkish hawks, denoted by X, $0 \leq X \leq 1$, are defined as those Turkish foreign policies that reject water concessions. Turkish doves, denoted by $1 - X$, instead support an increased amount of water flow to

Syria on the basis of an international agreement. Pairs of foreign policies from respective populations are repeatedly selected to interact.

Syrian doves are cooperative and recognize Turkish sovereignty over Hatay when they are paired with Turkish doves. However, if they are paired with Turkish hawks, they can be either cooperative or not. Let q denote the likelihood that Syrian doves recognize Hatay as a Turkish territory in their encounters with Turkish hawks. Similarly, Turkish doves give a water concession when matched with a Syrian dove. They can become less cooperative facing Syrian hawks. Let p denote Turkish doves' propensity to give a water concession in their encounters with Syrian hawks. Hawks are perpetually noncooperative. Turkish hawks never give a water concession and Syrian hawks never recognize Hatay as a Turkish territory. Hence, unlike hawks of both populations, doves, either Syrian or Turkish, are assumed to behave differently towards hawks and doves of the opposing population. Doves are cooperative towards doves but not so cooperative toward hawks of the opposing population. When two doves from respective populations are matched, the result is mutual cooperation: Syrian dove accepts Hatay as belonging to Turkey and Turkish dove signs an agreement increasing the water quota flown downstream.

No water concession or a territorial recognition is obtained in Syrian hawk–Turkish hawk interactions. This is the current status quo. For convenience, the fitness of Turkish and Syrian hawks is normalized to 0 in this case. A Syrian hawk can obtain a water concession encountering a Turkish dove. The only difference between this outcome and the normalized status quo is the likelihood of a water concession by Turkish doves denoted by p. Therefore, when matched, Syrian hawks and Turkish doves respectively obtain pwS and – pwT, where wT and wS are positive parameters measuring respectively the importance of water for Turkish and Syrian populations. The fitness of Syrian hawks increases and that of Turkish doves decreases.

When matched with a Syrian dove, Turkish hawk's fitness increases if Hatay is recognized as a Turkish territory. The difference between the status quo and this outcome is the likelihood of territorial recognition by Syrian doves. The fitness of Turkish hawks therefore becomes qh and that of Syrian doves $-qh$ where the positive parameter h measures Turkish hawks' fitness increment deriving from Syrian territorial recognition. Syrian doves are assumed to lose what Turkish hawks gain in terms of territorial recognition.

Finally, in interactions among doves, there are concessions over both issues. Hatay is recognized as a Turkish territory and Syria obtains a water concession. The recognition of Hatay lowers the fitness of Syrian doves but boosts that of Turkish doves. Similarly, the water concession boosts the fitness of Syrian doves but lowers the fitness of Turkish doves. The fitness of Turkish and Syrian dove respectively becomes h – wT and wS – h. These assumptions imply the game below:

		Syria	
		Hawk (Y)	***Dove (1 – Y)***
Turkey	***Hawk (X)***	$0, 0$	$qh, -qh$
	Dove (1 – X)	$-pw_T, pw_S$	$h - w_T, w_S - h$

Figure 1: Syrian–Turkish Water/Territory Conflict as an Evolutionary Game.

5. EVOLUTIONARY STABILITY

Assuming that doves, either Syrian or Turkish, concede or do not and that the fitness parameters of water and territory can or cannot be equal, the general game in Figure 1 implies four evolutionary variants. In addition, there are two parameters related to water and territory for each population. For each population these parameters can be equal or unequal with either territory or water weighing more than the other. Hence, nine possible cases exist in each variant with two parameters taking three distinct values each. These nine cases are:

1) wS = h; wT = h,
2) wS = h; wT < h,
3) wS = h; wT > h,
4) wS < h; wT = h,
5) wS < h; wT < h,
6) wS < h; wT > h,
7) wS > h; wT = h,
8) wS > h; wT < h,
9) wS > h; wT > h.

With four variants and nine cases, there are thirty-six evolutionary games in total. The equilibrium analysis is simple. The concept of evolutionary equilibrium, also known as evolutionary stable strategy (ESS), corresponds to strict Nash equilibrium in games where there are two different populations (Syrian and Turkish populations in our case) each possessing two distinct behavioral traits (hawks and doves). ESS and strict

Nash equilibrium concepts are equivalent in these games (Gardner 2003, p. 226). Strict Nash equilibrium implies that all unilateral deviations from equilibrium strategies induce a payoff reduction. The deviator may lose nothing when others stick to their equilibrium strategies in non-strict Nash equilibria. As to the ESS, it indicates the ultimate state that interacting populations will evolve into: both populations will consist of one behavioral trait only. If an ESS implies, for example, dove–dove interactions, this means that respective populations will ultimately contain doves and no hawks.

Variant 1: p = q = 0

Either Syrian or Turkish doves do not concede when matched with hawks of the opposing population. The stage game matrix becomes, as seen in Figure 2, below:

		Syria	
		Hawk (Y)	*Dove (1 − Y)*
Turkey	*Hawk (X)*	$0\,,\,0$	$0,0$
	Dove (1 − X)	$0\,,\,0$	$h - w_T\,,\, w_S - h$

Figure 2: Variant 1.

In none of the nine cases a strict Nash equilibrium and therefore an ESS exists. There is no evolutionary stability under these conditions. No predictions can therefore be made about the evolution of Syrian and Turkish foreign policies. For example, if wS = wT = h, all entries are zero in the game matrix, all outcomes constitute Nash equilibrium and none is strict.

Variant 2: p = q = 1

Either Syrian or Turkish, doves concede when matched with hawks of the opposing population. The stage game reduces to (Figure 3):

		Syria	
		Hawk (Y)	*Dove (1 − Y)*
Turkey	*Hawk (X)*	$0\,,\,0$	$h, -h$
	Dove (1 − X)	$-w_T\,,\, w_S$	$h - w_T\,,\, w_S - h$

Figure 3: Variant 2.

There exists a unique strict Nash equilibrium, and therefore an ESS, in all nine cases: hawk–hawk. Hence, populations will finally contain only hawkish foreign policies.

Variant 3: p = 0, q = 1

Syrian doves concede when matched with Turkish hawks (unlike Turkish doves matched with Syrian hawks). The stage game reduces to (Figure 4):

		Syria	
		Hawk (Y)	*Dove (1 − Y)*
Turkey	*Hawk (X)*	$0, 0$	$h, -h$
	Dove (1 − X)	$0, 0$	$h - w_T, w_S - h$

Figure 4: Variant 3.

There is no strict Nash equilibrium, and therefore no ESS in any of the nine cases under these conditions. No predictions can be made.

Variant 4: p = 1, q = 0

Turkish doves concede when matched with Syrian hawks (unlike Syrian doves matched with Turkish hawks). The stage game becomes, as seen in Figure 5:

		Syria	
		Hawk (Y)	*Dove (1 − Y)*
Turkey	*Hawk (X)*	$0, 0$	$0,0$
	Dove (1 − X)	$-w_T, w_S$	$h - w_T, w_S - h$

Figure 5: Variant 4.

Again, no strict Nash equilibrium and therefore an ESS exists in any of the nine cases. Hence, similar to the previous variant, no prediction can be made regarding the evolution of Syrian–Turkish relations.

6. EVOLUTIONARY IMPLICATIONS AND INTERPRETATIONS

If cooperative Syrian and Turkish foreign policies act like hawks encountering hawkish foreign policies of the opposing population, no stability is reached in Syrian–Turkish relations. Hence, even if populations fully contain hawks, doves would find a fertile environment to proliferate, and, similarly, if they contain only doves and no hawks, hawks could by mistake enter the population and proliferate. Accordingly, there will be a constant change in the relative frequencies of Syrian and Turkish foreign policy traits over time. It is impossible for one foreign-policy type to totally out-compete the other. No state of bilateral relations will be immune to invasion.

Doves and hawks will continually replace each other in respective populations if Turkish doves do not support water concessions matched with Syrian hawks, yet Syrian doves recognize Hatay as a Turkish territory in their encounters with Turkish hawks. In this case, Turkish doves are cooperative only with respect to Syrian doves. Syrian doves are however cooperative, even matched, with hawkish Turkish foreign policies. The indeterminacy in Syrian–Turkish relations still rules when this situation is reversed: Turkish doves give a water concession encountering obstinate Syrian foreign policies in the territory issue and Syrian doves act like hawks matched with intransigent Turkish foreign policies in the water issue.

If doves, either Syrian or Turkish, are cooperative even when they encounter conflictual types of opposing populations, then an evolutionary stability is reached. This is the only instance where both foreign policy populations ever reach stasis: the state of the equilibrium will ultimately consist fully of hawkish foreign policies. Those foreign policies that never concede in either water or territory issues will finally dominate respective populations. Those dovish foreign policies that enter populations by mistake (mutants) will become extinct, i.e., not survive at all. Unconditional cooperation leads to stable conflict. Thus, those doves that behave differently depending upon the type they encounter do not become extinct. Instead of producing mutual cooperation in riparian relations, unconditional cooperation in dovish foreign policies stimulates noncooperative foreign policies in both populations. This type of doves can be sporadically observed in riparian relations, yet they can never dominate foreign-policy populations. Syrian–Turkish relations cannot evolve into a stable peaceful outcome.

Allan (1996) indicates that wheat imports in the region make up "virtual water" as wheat is a water-intensive commodity. Thus, as long as these countries can import food in international markets, imported wheat can be substituted for the scarce water resources at home. Under the condition that Syria can easily make up for water shortages through wheat importations, or, if the Turkish Southeast Anatolian Project (comprising more than twenty dams upstream over the Euphrates and the Tigris rivers) ultimately facilitates cheap agricultural trade in the region by its completion, water may become a lesser concern than territory.

The model implies that none of the above possibilities carry any weight in Syrian–Turkish relations, however. The evolution of foreign policies will not change direction irrespective of whether water-intensive commodities are cheaply produced and play a central role in riparian countries' trade or Hatay suddenly loses its importance for Turkey. What counts is the concessional behavior of dovish foreign policies.

7. CONCLUSIONS

These results do not follow game rules assuming rational, omniscient players who can predict each move but rules of thumb, boundedly rational and aggregate behavior of large populations. Yet, they provide support for earlier game-theoretic analyses and permit alternative explanations as well. Reformulations of fitness functions for hawks and doves in both countries and alternative criteria of evolutionary dynamics can be considered as extensions of the present research. In general, the model indicates that an inherent instability underlies Syrian–Turkish relations. If foreign policies in both countries ever evolve into a stable state, they will only be conflictual.

WATER USE AND RISK: THE USE OF PROSPECT THEORY TO GUIDE PUBLIC POLICY DECISION-MAKING*

Raymond Dacey
College of Business and Economics
University of Idaho, United States
rdacey@uidaho.edu

Abstract: Water is fundamental to human existence and it has no substitutes. As such, policy disputes over water can be expected to be more severe than policy disputes involving almost any other resource. This chapter considers two types of policy disputes over water and water usage. The first type of policy dispute concerns development projects, which are usually very large-scale projects like dams, irrigation systems, and desalination plants, and involve major investments by governments and/or corporations. The second type of policy disputes concerns conservation projects, which typically involve restricting flow rates through existing facilities or otherwise altering existing utilization habits, and thereby involve small measures taken by a large number of individuals. The purpose of this chapter is two-fold—to explain the source of these disputes and to examine the usual attempts at managing the resolution of these disputes. Both the source and the resolution of these problems rest on the psychology of human decision-making.

Keywords: risk analysis, prospect theory, expert, lay person, Safe Minimum Standard model

1. INTRODUCTION

A central goal of the conference "Mountains: Sources of Water, Sources of Knowledge" was "to propose and evaluate water management policies and thereby minimize natural disasters and contribute to sustainnable and fair

---

* Acknowledgements: Prepared for inclusion in the book developed from the conference "Mountains: Sources of Water, Sources of Knowledge," held at the Institut Universitaire Kurt Bösch, Sion, Switzerland, as part of the Alpine Environment and Society Program, October 8–10, 2002. I particularly want to thank the two referees for their remarkably helpful comments.

E. Wiegandt (ed.), Mountains: Sources of Water, Sources of Knowledge, 161–178.

water use." This chapter attempts to contribute to this goal by presenting an analysis of the difficulties involved in making policy decisions of the kind envisioned here.

Water is fundamental to human existence and it has no substitutes. As such, policy disputes over water can be expected to be more severe than policy disputes involving almost any other resource. The problems that underlie these disputes are well known (Postal 1999 and 1992; Shiva 2002; Stauffer 1999) and graphically documented (Mitchel 2002; Montaigne 2002). Furthermore, water-related disputes engender conflicts that range from fairly low key regional disagreements to interstate wars (Shiva 2002, 69–82).

This chapter considers two types of policy disputes over water and water usage. The first type of policy dispute concerns development projects, which are usually very large-scale projects like dams, irrigation systems, and desalination plants, and involve major investments by governments and/or corporations. The second type of policy disputes concern conservation projects, which typically involve restricting flow rates through existing facilities or otherwise altering existing utilization habits, and thereby involve small measures taken by a large number of individuals.

Both types of policies often engender disputes between experts and lay people. Development projects and conservation projects typically enjoy the support of the experts but are opposed by the lay people. The purpose of this chapter is two-fold—to explain the source of these disputes and to examine the usual attempts at managing the resolution of these disputes. Both the source and the resolution of these problems rest on the psychology of human decision making.

2. THE BASIC POLICY DECISION PROBLEM

Margolis (1996) considers conflicts in policy decision making that involve clashes between experts and lay people. Examples include nuclear power generation, biologically engineered foods, industrial and agricultural subsidies, and sanctions. The clashes arise from different resolutions of policy decision problems. In its generic form, the policy decision problem involves two choices, acting and not acting, and two states of nature, success and failure. If the decision maker acts and is successful, then there is a gain, in amount G, and if unsuccessful, there is a loss, in amount L. The choice of acting results in a gain with probability p. If the decision maker does not act, then there is neither a gain nor a loss. The payoffs for the basic decision problem are given in Table 1.

Table 1: The Basic Policy Decision Problem.

	ACT	NOT ACT	Probability
SUCCESS	G	0	P
FAILURE	−L	0	1−p

The values of G, L, and p, for a particular policy decision problem, can be determined or estimated in various ways. For example, the chapter by Manfreda (this volume) presents a statistical model, which can be used to determine or estimate the values of the decision variables. The chapter by Fisher et al. (this volume) presents an extensive economic model that incorporates both optimization and simulation. The remarkable photograph of the high mountain hydroelectric dam at Verzasca presented in Romerio (this volume) provides a graphic visualization of the possible scale of L. Simply picture a rupture of such a dam and the resulting flood in the valley below.

Following Margolis, I will presume that the expert and the lay person agree on the objective details of the problem. "Subjectively they see different situations, though both agree that objectively it is the same situation" (Margolis 1996, p. 79). Thus, the payoffs for the decision problem are commonly understood to be those given in Table 1.

The expert and the lay person, however, have different ways of choosing between ACT and NOT ACT. Margolis makes this point by noting that "expert intuition is focused narrowly on statistical expectations ..." and that "what is most important about the contrast [between expert and lay intuition] is that the experience of experts makes them focus narrowly on what they have been trained to attend to" (Margolis 1996, p. 80). Lay intuition, on the other hand, is "not so fluent at responding to subtle information about statistical risks" (Margolis 1996, p. 80). I take these comments to mean that the expert makes decisions as would an actuary, an engineer, or a scientist, and the lay person makes decisions as would a "normal" or "naive" person. Thus, the expert makes decisions in accord with the maximum expected (i.e., actuarial) value rule. I will presume that the lay person behaves as a normal person and "sees" the problem in the terms characterized by prospect theory (Kahneman and Tversky 1979, 1992). The justification for this presumption is straightforward. Prospect theory was inferred from experimental investigations involving many individuals from many different cultures and pursuits (i.e., "normal" people). Thus, prospect theory captures the behavior of a normal person. Prospect theory was presented in two main papers, Kahneman and Tversky (1979) and Tversky and Kahneman (1992), and is well discussed in Camerer (1995) and Rabin (1998). The key papers of the literature are collected in Kahneman and Tversky (2000).

Margolis considers two types of policy disputes. He notes that "The most stubborn cases have been those where the public is more worried than the experts, but it is important to notice that there are also many converse cases, in which the public is (for some substantial time at least) unresponsive to expert warnings, as for cigarette smokers and drivers who do not bother with seat belts" (Margolis 1996, p. 7). Development projects are among the "most stubborn cases," and conservation projects are among the "converse cases." As Margolis notes, the former cases are far more "stubborn" and as such, more interesting: "I will have something to say about converse cases. But my prime focus will be on cases where high passion is aroused about what, to experts, appear to be very modest dangers" (Margolis 1996, p. 8).

In what follows we will see how disputes over development and conservation projects arise, how they can be managed, and why the former are stubborn and the latter are not. Before we can examine these two classes of problems, we require an analysis of expert and lay decision making.

3. HUMAN DECISION-MAKING

Experts make decisions on the basis of the maximum expected value rule, and lay people make decisions on the basis of prospect theory. The former involves the very familiar value function that is linear in payoffs and linear in probabilities. Thus, an expert chooses between ACT and NOT ACT on the basis of the expected values of the two options, as follows:

$$\mathrm{EV(ACT)} = \mathrm{pG} + (1-\mathrm{p})(-\mathrm{L})$$

and

$$\mathrm{EV(NOT\ ACT)} = \mathrm{p}(0) + (1-\mathrm{p})(0) = 0$$

The expert chooses the act with the greater expected value. Thus, the expert prefers ACT to NOT ACT if and only if

$$\mathrm{pG} + (1-\mathrm{p})(-\mathrm{L}) > 0,$$

i.e., if and only if

$$\frac{\mathrm{p}}{1-\mathrm{p}} > \frac{\mathrm{L}}{\mathrm{G}}.$$

Lay people, i.e., untrained or "normal" people, make decisions in accordance with prospect theory. Prospect theory is a descriptive alternative to the normative expected value theory of Pascal (Arnauld and Nicole 1662) and the normative expected utility theory of von Neumann and Morgenstern (1944). Prospect theory was developed to explain observed behavior that could not be explained by the theories of Pascal and von Neumann and Morgenstern. In particular, individuals are regularly observed purchasing both insurance and lottery tickets.

There are two key components to prospect theory. First, prospect theory posits a valuation function that represents individuals' risk attitudes. Specifically, the valuation function, here denoted by v, is a reference dependent function defined over changes in wealth that represents risk aversion over gains and risk preference over losses, and is more steeply sloped over losses than over gains. While the valuation function of prospect theory is a piecewise defined von Neumann–Morgenstern utility function (Fishburn and Kochenberger 1979), the presence of both risk aversion and risk preference distinguishes prospect theory from expected

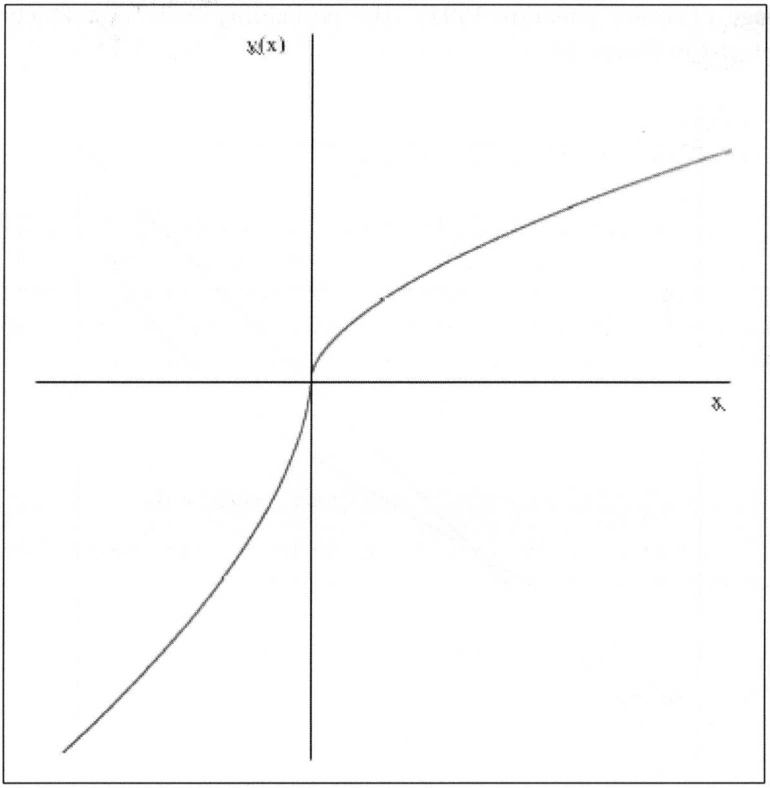

Figure 1: The Valuation Function.

utility theory. Applications of expected utility theory have almost uniformly employed utility functions that represent either risk aversion or risk preference, but not combinations of the two. The notable exception is Friedman and Savage (1948). Kahneman and Tversky, via their many experiments, established that normal individuals behave in accordance with a valuation function that represents both risk attitudes (Kahneman and Tversky 1992). The valuation function is displayed in Figure 1.

Second, prospect theory posits a probability weighting function that represents an individual's mis-valuation of probabilities. Specifically, the probability weighting function, here denoted by w, over-values low probabilities and under-values medium and high probabilities (Tversky and Kahneman 1992: 298). The probability weighting function extends prospect theory beyond expected utility theory, and is required to explain observed decision behavior that cannot be explained solely on the basis of the S-shape of the valuation function of prospect theory or adaptations of the utility function of expected utility theory as proposed by Friedman and Savage (Tversky and Fox 1995). The probability weighting function is displayed in Figure 2.

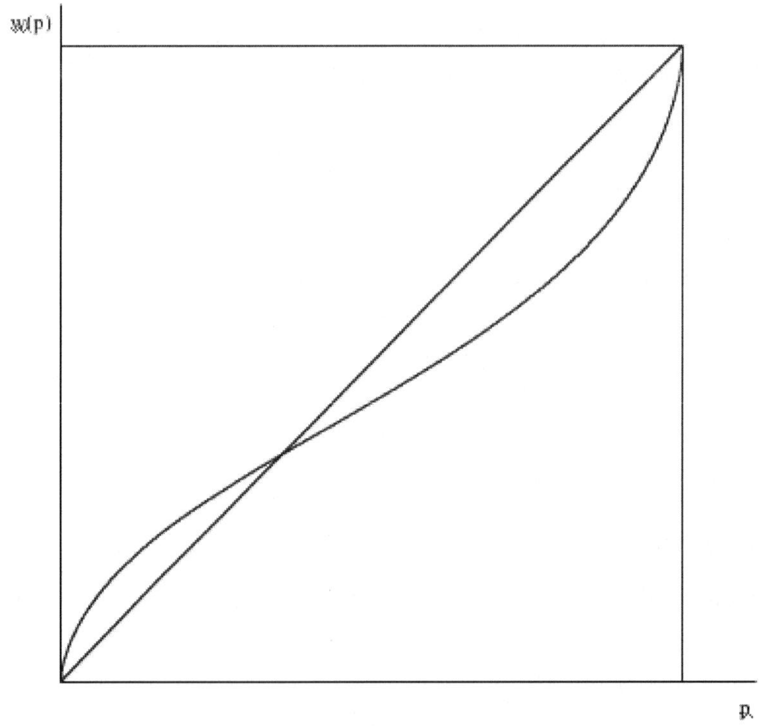

Figure 2: The Probability Weighting Function.

A lay person chooses between ACT and NOT ACT on the basis of the values

$$ev(ACT) = w(p)v(G) + w(1-p)v(-L)$$

and

$$ev(NOT\ ACT) = w(p)v(0) + w(1-p)v(0)$$

For convenience we will take $v(0) = 0$, so that ev(NOT ACT) = 0. The symbol "e" is used in lieu of the traditional "E" because the function w is sub-additive (i.e., $w(p) + w(1-p) < 1$), so that ev is not the expected value of v.

The lay person chooses the act with the greater ev-value. Thus, the lay person prefers NOT ACT over ACT if and only if

$$w(p)v(G) + w(1-p)v(-L) < 0$$

i.e., if and only if

$$\frac{w(p)}{w(1-p)} < \frac{-v(-L)}{v(G)}.$$

We now have all of the tools we need to examine the source and the resolution of policy disputes involving development projects.

4. DEVELOPMENT PROJECTS

Development projects are instances of the basic policy decision problem wherein L is larger than G and p is high. Since L is larger than G, and since the valuation function v is more steeply sloped over losses than gains, we know that $\frac{-v(-L)}{v(G)} > \frac{L}{G} > 1$. Since the probability weighting function is reverse S-shaped, we know that for $0 \leq p < 0.5$, $p > w(p)$ and $1-p < w(1-p)$. Thus, for $0 \leq p < 0.5$, we have $\frac{p}{1-p} < \frac{w(p)}{w(1-p)}$. Similarly, we know that for $0.5 < p \leq 1$, $p < w(p)$ and $1-p > w(1-p)$. Thus, for $0.5 < p \leq 1$, we have $\frac{p}{1-p} > \frac{w(p)}{w(1-p)}$. Clearly, at $p = 0.5$, $p = 1-p = 0.5$ and $w(p) = w(1-p) = w(0.5)$, so that $\frac{p}{1-p} = \frac{w(p)}{w(1-p)} = 1$.

Now define the critical risk probabilities p_{EX} and p_{LP} as follows:

$$\frac{p_{EX}}{1-p_{EX}} = \frac{L}{G}$$

and

$$\frac{w(p_{LP})}{w(1-p_{LP})} = \frac{-v(-L)}{v(G)}.$$

Since $\frac{-v(-L)}{v(G)} > \frac{L}{G}$ and $\frac{p}{1-p} > \frac{w(p)}{w(1-p)}$ for all values of $p > 0.5$, and since $\frac{p}{1-p}$ and $\frac{w(p)}{w(1-p)}$ are each monotonically increasing in p, clearly $p_{EX} < p_{LP}$. Therefore, for any p between p_{EX} and p_{LP}, there exists a dispute wherein the expert chooses ACT and the lay person chooses NOT ACT. The diagram for the development project is given in Figure 3.

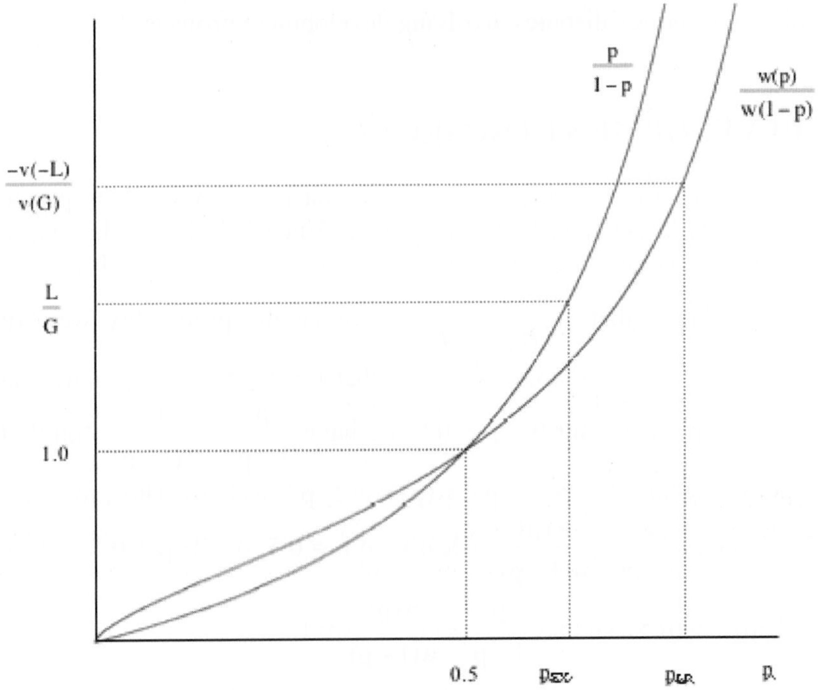

Figure 3: The Development Project.

The explanation for the existence of a dispute rests on the probability of success. If $p < p_{EX}$ or if $p > p_{LP}$, then there is no dispute. In the former case both the expert and the lay person agree on NOT ACT, and in the latter case both agree on ACT. Thus, not all development projects involve disputes. However, if $p_{LP} < p < p_{EX}$, then there is a dispute.

4.1. Management of the Dispute

A dispute over a development project is resolved when both the expert and the lay person agree on an action. The resolution of a dispute is managed by changing one or more of the parameters (G, L, and p) so as to induce a resolution. Those individuals who prefer ACT will attempt to drive G and p upward and L downward; those who prefer NOT ACT will attempt to drive G and p downward and L upward. For example, each side can conduct studies that are more refined and precise than extant studies of the situation at hand, and then make public the results of the studies. In an ideal situation, all studies would be refined and precise, and all information would be made public. In reality, it is often the case that studies are unrefined and/or imprecise, and that unfavorable information is suppressed. An account of this particular political game goes beyond the analysis at hand. In what follows, we presume that the available information comes from refined and precise studies, and that all available information has been made public.

The simplest case involves the experts providing information that drives p upward (leaving G and L unchanged), and the lay people providing no information. Let us consider this seemingly simple case. If the experts provide information that drives the probability of success upward so that $p > p_{LP}$, then agreement is reached and both sides select ACT. In some disputes this is quite easy to achieve. However, in general, it is not easy. Note that for $p > 0.5$, $w(p) < p$, $\frac{w(p)}{w(1-p)} < \frac{p}{1-p}$, and $\frac{w(p)}{w(1-p)}$ increases more slowly than $\frac{p}{1-p}$ as p increases. Thus, what appears to an expert to be a decisive value of p (i.e., a value of p greater than p_{EX}) or a sizeable increase in p (i.e., a sizable increase over p_{EX}) need not be a decisive value or a decisive increase to a lay person. Thus, the expert must be prepared to provide information that drives p to a very high level.

4.2. Stubbornness of the Dispute

It remains to explain why disputes over development projects are "stubborn". The key to the explanation rests on the combined effects of the probability weighting function w and the down-side slope of the valuation function v. This point is treated in greater detail in Dacey (2002a). The basic point to note is that in the case of a development project, the spread between $\frac{L}{G}$ and $\frac{-v(-L)}{v(G)}$ can be very large, thereby making the spread between p_{EX} and p_{LP} relatively large. For example, if we adopt the valuation function advanced in Kahneman and Tversky (1992), then $v(x) = x^{\beta}$ for $x \geq 0$ and $v(x) = -\alpha(-x)^{\beta}$ for $x \leq 0$. Kahneman and Tversky (1992) find α is between 2.0 to 2.5 and that β is between 0.8 and 1.0. Thus, $\frac{-v(-L)}{v(G)}$ is greater than $\frac{L}{G}$ and less than $2.5\frac{L}{G}$. Since $L > G$, the spread between $\frac{L}{G}$ and $\frac{-v(-L)}{v(G)}$ can be very large, as claimed. However, note that the spread between p_{EX} and p_{LP}, while relatively large, is a spread between two probabilities and therefore can only be a fraction. Furthermore, note that when we are dealing with probabilities near unity, the functions $\frac{p}{1-p}$ and $\frac{w(p)}{w(1-p)}$ are rising rapidly and the latter is approaching the former from below. Thus, the spread between p_{EX} and p_{LP} can only be relatively large and must be diminishing as the probabilities increase.

A brief account of the spread between p_{EX} and p_{LP} can be given by example. Suppose that the valuation function and the weighting function are modest simplifications of those estimated by Kahneman and Tversky (1992), so that $v(x) = x^{.88}$ for $x \geq 0$ and $v(x) = -2.25(-x)^{.88}$ for $x \leq 0$, and $w(p) = \frac{p^{.65}}{\left(p^{.65} + (1-p)^{.65}\right)^{\frac{1}{.65}}}$. Then the values of p_{EX} and p_{LP} for various values of $\frac{L}{G}$ are given in Table 2.

Note that the spread between p_{EX} and p_{LP} decreases as the ratio $\frac{L}{G}$ increases, and the spread goes to zero as $\frac{L}{G}$ increases without bound. As

Table 2: Critical Probability Values.

$\frac{L}{G}$	p_{EX}	p_{LP}
1	0.50000000	0.77688152
2	0.66666667	0.89898521
5	0.83333333	0.96852304
10	0.90909091	0.98744412
50	0.98039216	0.99856315
100	0.99009901	0.99943734

noted, this is because $\frac{p}{1-p}$ and $\frac{w(p)}{w(1-p)}$ are rising rapidly for high values of p and the latter ratio is approaching the former from below. Thus, the spread between p_{EX} and p_{LP} can only be relatively large, as claimed. Indeed, at $\frac{L}{G} = 1$, relatively large amounts to a difference of about 0.28, whereas at $\frac{L}{G} = 100$, relatively large amounts to a difference of about 0.009. It is interesting to note that the spread between p_{EX} and p_{LP} is greatest when G and L are approximately equal.

There is a dispute if p is between p_{EX} and p_{LP}. The dispute is resolved in favor of ACT if p is increased beyond p_{LP}. For example, if L is twice as large as G, and p is between 0.6666… and 0.8989…, then there is a dispute, and the dispute is resolved in favor of ACT only if p is increased beyond 0.8989. Increasing p from some value just slightly larger than 0.6666 to a value greater than 0.8989… can be very difficult.

Note that if L is remarkably larger than G, then a dispute exists only if p is already very large. For example, if L is 100 times larger than G, then p must be greater than 0.9900… in order to have the expert choose ACT. Thus, as expected, if L is remarkably larger than G, then it is not likely that a dispute will exist. In general, we should expect disputes over development projects to exist when L is larger, but not remarkably larger, than G, and we should expect such disputes to be "stubborn" if p is close to p_{EX} since the gap between p_{EX} and p_{LP} will be relatively large.

Since disputes over development projects are stubborn given only a change in p, it would seem reasonable to suggest that the management of such disputes will involve changing G and L along with p. The analyses of these strategies are not presented here.

5. CONSERVATION PROJECTS

Conservation projects are instances of the basic policy decision problem where G is larger than L. Conservation projects typically involve a situation wherein the lay people have chosen ACT and the experts attempt to move the lay people to NOT ACT. Here ACT represents actions that are not sustainable. Such acts are very likely to produce long-term damages to the environment, but are such that they produce sizable short-term gains, at low probability, to the individual. Continuing with ACT is collectively wasteful but not individually costly. Thus, the experts seek to get the lay people to switch to NOT ACT, which amounts to adopting more conservative ways.

For example, an extreme form of ACT might be the activity of (irrationally) consuming household water as though it were a free good. A less extreme form of ACT might be the rational use of water where water is priced below its replacement cost. In the former case the expert must get the lay person to "see" the decision problem as it is; in the latter case the expert must get the lay person to agree to change the pricing of water. In such cases, NOT ACT involves not acting in the old way, thereby bringing the system at least to a break-even level. As such, conservation projects are instances of the basic policy decision problem wherein G is larger than L and p is low. As such, to choose ACT is to choose a lottery, and to switch to NOT ACT is to abandon the lottery.

Disputes over conservation projects are of the form where the lay people have chosen ACT, and the experts choose NOT ACT. Thus, such a dispute exists if and only if

$$\frac{p}{1-p} < \frac{L}{G}$$

and

$$\frac{w(p)}{w(1-p)} < \frac{-v(-L)}{v(G)}.$$

Since G is larger than L, $\frac{L}{G} < 1$. Since p is small, $p < 0.5$. The ratio $\frac{-v(-L)}{v(G)}$ can take on any value greater than 0. Define p_{EX} and p_{LP} as before:

$$\frac{p_{EX}}{1-p_{EX}} = \frac{L}{G}$$

and

$$\frac{w(p_{LP})}{w(1-p_{LP})} = \frac{-v(-L)}{v(G)}.$$

If $\frac{-v(-L)}{v(G)} > \frac{w(p_{EX})}{w(1-p_{EX})}$, then no disagreement can exist. Thus, we are interested in the case where $\frac{-v(-L)}{v(G)} < \frac{w(p_{EX})}{w(1-p_{EX})}$. (Note that this kind of condition was not needed in the development project case.) If $\frac{-v(-L)}{v(G)} < \frac{w(p_{EX})}{w(1-p_{EX})}$ and p is between p_{LP} and p_{EX}, then a dispute exists. The diagram for the conservation project is given in Figure 4.

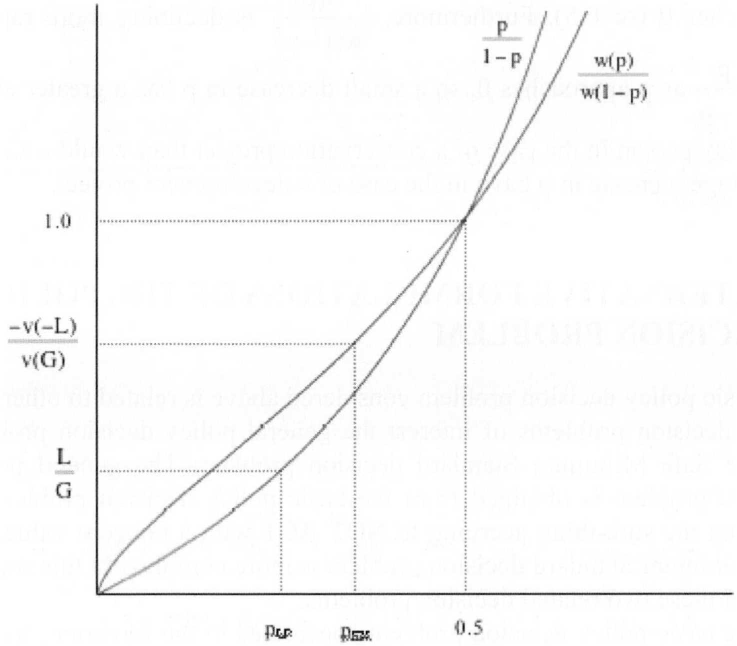

Figure 4: The Conservation Project.

5.1. Management of the Dispute

A dispute over a conservation project is resolved in favor of NOT ACT if the expert can provide information that reduces p below p_{LP}. Disputes over conservation projects are reasonably easy to resolve. Typically, the resolution is achieved by driving p close to 0, as is the case in a drought. This is usually fairly easy to achieve, particularly for something as indispensable as water. For example, it is reasonably easy to get a large number of people to adopt water conservation measures during a period of water shortages, as caused by a drought or a mechanical failure.

5.2. Stubbornness of the Dispute

Disputes over conservation projects are typically not stubborn. The reason is that the functions $\frac{p}{1-p}$ and $\frac{w(p)}{w(1-p)}$ are equal at both p = 0 (and p = 0.5), and therefore the gap between $\frac{w(p)}{w(1-p)}$ and $\frac{p}{1-p}$ is shrinking as p approaches 0 (or 0.5). Furthermore, $\frac{w(p)}{w(1-p)}$ is declining more rapidly than $\frac{p}{1-p}$ as p approaches 0, so a small decrease in p has a greater effect on the lay person in the case of a conservation project than would a similar percentage increase in p have in the case of a development project.

6. ALTERNATIVE FORMULATIONS OF THE POLICY DECISION PROBLEM

The basic policy decision problem considered above is related to other two policy decision problems of interest the general policy decision problem and the Safe Minimum Standard decision problem. The general policy decision problem is obtained from the basic policy decision problem by replacing the sure-thing accruing to NOT ACT with a nonzero value; the Safe Minimum Standard decision problem is more complex. In this section we treat these two related decision problems.

The basic policy decision problem considered in the foregoing section of this chapter is a short run decision problem. A long run version of the decision problem is generated by including a nonzero payoff to NOT ACT.

This decision problem is more general than the basic policy decision problem considered here, but it is also more algebraically complex, and the algebraic complexity obscures the analyses of the interesting public policy issues. Nonetheless, the policy-specific results advanced in this chapter remain unchanged for the general policy decision problem. The payoffs for the general policy decision problem are given in Table 3.

Table 3: The General Policy Decision Problem.

	ACT	NOT ACT	Probability
SUCCESS	G	SQ	p
FAILURE	−L	SQ	1−p

SQ represents the status quo, and, to avoid trivialities we assume that $G > SQ > -L$. Note that SQ can be positive, negative, or zero. The basic policy decision problem is the special case where SQ = 0. To generate the analysis of the general decision problem from the analysis of the basic decision problem, the ratios $\frac{L}{G}$ and $\frac{-v(-L)}{v(G)}$ must be replaced. The ratio $\frac{L}{G}$ is replaced with $\frac{SQ-(-L)}{G-SQ}$ everywhere the former appears in the analysis of the expert. However, things are not so simple for the analysis of the lay person. The ratio $\frac{-v(-L)}{v(G)}$ is replaced with:

$$\frac{v(SQ)-v(-L)}{v(G)-v(SQ)}+\frac{1-w(p)-w(1-p)}{w(1-p)}\frac{v(SQ)}{v(G)-v(SQ)}$$

where the former appears in the analysis of the lay person. The latter structure greatly increases the algebraic complexity of the analysis of lay person decision making. A variation of this problem is analyzed in detail in Carlson and Dacey (2006).

However, the introduction of the nonzero status quo also allows the examination of two very interesting issues. First, as raised by a referee, the assumption that SQ = 0 is justifiable in the short run but not in the long run and certainly not in the very long run. The discounted present value of the future derived from present inaction (i.e., NOT ACT), and measured by SQ, can be significant. For example, continuing extraction of water from an aquifer that is not recharging at or above the drawdown rate will result in increasingly costly water or no water at all. Thus, the introduction of a nonzero SQ is required for an analysis of the long or very long.

Second, also as raised by the same referee, the rate used to discount the future provides another mechanism to draw distinctions between and among experts, lay people, and politicians in the arbitration of policy disputes related to costly projects. Thus, the introduction of SQ as the discounted present value of a future value is required if we wish to examine the effects of differing discount rates.

A very specific version of the policy decision problem is examined in the Safe Minimum Standard literature (e.g., Ciriacy-Wantrup 1952; Berrens 2001; Berrens et al. 1998; Bishop 1978; Castle and Berrens 1993). The decision problem treated in the Safe Minimum Standard (SMS) literature combines development and conservation in one decision problem. Consider an economic development project that puts an endangered species at risk. The decision problem of interest involves two acts and two states of nature. The two acts are to intervene or not intervene, by adopting or not adopting a plan put forward by the relevant (governmental) agency, to protect the endangered species. Following the SMS literature, the two states of nature involve the economic value of the species in question (Bishop 1978, pp. 12–13). Let S denote the (net present) economic value of the species, let P denote the (net present) value of the project, and let p denote the probability that S is large.

We denote the choice of intervention by ACT, the choice of nonintervention by NOT ACT, the state of high species value by $S \gg 0$, and the state of negligible species value by $S \approx 0$. The payoffs for the SMS decision problem are given in Table 4.

Table 4: The Safe Minimum Standard Decision Problem.

	ACT	NOT ACT	Probability
$S \gg 0$	$S-P$	$-S$	p
$S \approx 0$	$-P$	0	$1-p$

This table is a direct translation of the SMS loss-based game matrix presented in Bishop (1978, 12). As Bishop notes, $S-P$ can be a loss (i.e., $S-P < 0$) or a gain (i.e., $S-P > 0$). If $S-P < 0$ and $S-P < -S$ (i.e., if $P > S > 2P$), then NOT ACT is a dominant act (Bishop 1978, 13). Thus, if $S-P < 0$, then we must restrict the model by assuming that $S-P > -S$ to avoid trivialities. If $S-P > 0$, then no such restriction is needed.

The analysis of the SMS decision problem is straightforward and parallels the analysis presented in the body of this chapter. For the simpler case where S−P > 0, $\frac{L}{G}$ is replaced with $\frac{2P}{2S-P}$, everywhere the former appears in the analysis of the expert. While things are not so simple for the analysis of the lay person, they are not nearly as messy as in the general case of the policy decision problem (i.e., where SQ ≠ 0). Again assuming S−P > 0, $\frac{-v(-L)}{v(G)}$ is replaced with $\frac{-v(-P)}{v(S-P)-v(-S)}$ everywhere the former appears in the analysis of the lay person. The latter does not greatly increase the algebraic complexity of the analysis of lay person decision making. However, the need to address the case where 0 > S−P > −S greatly complicates the overall analysis.

The SMS model is quite odd in that the model confounds realized costs (i.e., −S) with opportunity costs (i.e., −P) and net opportunity benefits (i.e., if S−P > 0) or net opportunity losses (i.e., if S−P < 0). This is because the foregoing decision problem attempts to combine development and conservation in one problem, and because the decision problem was developed as a two-person game between society and nature (Bishop 1978, 12−13). When reformulated as a set of decisions made under risk, the SMS model provides an interesting, though remarkably tedious, account of decision making involving experts, lay people, and politicians (Dacey 2002c).

7. CONCLUSIONS

We have provided accounts of development and conservation projects, shown how disputes over these kinds of projects arise, how the disputes can be managed, and explained why disputes over development projects are stubborn and those over conservation projects are not. The two kinds of projects are shown to be special cases of the basic policy decision problem. Disputes arise because experts and lay people employ different decision resolution procedures. Specifically, experts, by training, behave so as to maximize expected (i.e., actuarial) value. Lay people, on the other hand, behave so as to maximize the objective function posited by prospect theory. Note that we have not claimed that the experts are rational and the lay people are not. The two share a common decision problem, objectively agreed to. However, the lay person displays various "biases," which are represented by the probability weighting function w and the S-shaped valuation function v.

The disputes can be managed by introducing information that moves the probability p to a level where agreement is achieved. Depending on the kind of dispute, this may or may not be relatively easy. If it is not relatively easy, then the dispute is stubborn. In particular, disputes over development projects are stubborn exactly because the curvatures in the functions w and v work against achieving agreement by increasing p, whereas disputes over conservation projects are not stubborn exactly because the curvatures of the two functions work toward achieving agreement as p decreases.

INDIGENOUS KNOWLEDGE; TECHNICAL SOLUTIONS

DISASTERS, DEVELOPMENT, AND GLACIAL LAKE CONTROL IN TWENTIETH-CENTURY PERU*

Mark Carey
Department of History
Washington and Lee University, United States
careym@wlu.edu

Abstract: During the past 65 years, glacier melting in Peru's Cordillera Blanca mountain range has caused some of the world's most deadly glacial lake outburst floods and glacier avalanches. Since the onset of these catastrophes in 1941, various groups have understood glacier hazards in distinct ways. Scientists and engineers saw them as technical problems. Economic developers and government officials believed glacier hazards threatened vital hydroelectric, irrigation, and tourism projects. And local residents feared glaciers and glacial lakes, though they ranked natural disasters among other social, political, and economic risks. Despite these marked differences in defining glacier hazards, local residents, authorities, developers, and scientific experts generally sought the same solution to Cordillera Blanca glacier disasters: draining glacial lakes to avoid outburst floods. Thus, risk perception varied, but each group proposed similar strategies to prevent glacier disasters. This chapter also suggests that development interests can help reduce the risk of natural disasters for local people and that local, marginalized populations can influence their degree of vulnerability to natural disasters.

Keywords: glacier retreat, natural disasters, history, risk, Cordillera Blanca

---

* I would like to thank Ellen Wiegandt, Ben Orlove, Benjamín Morales, Chuck Walker, Susan Carey, and anonymous reviewers for comments that improved this paper. It was originally prepared for the conference "Mountains: Sources of Water, Sources of Knowledge," held at the Institut Universitaire Kurt Bösch, Sion, Switzerland, October 8–10, 2002. I gratefully acknowledge the S.V. Ciriacy-Wantrup Postdoctoral Fellowship at the University of California, Berkeley and the Mabelle McLeod Lewis Memorial Fund for supporting writing and revisions. Field research in Peru was funded by the American Meteorological Society, the International Dissertation Field Research Fellowship Program of the Social Science Research Council with funds provided by the Andrew W. Mellon Foundation, the Pacific Rim Research Program, and the Agricultural History Center at the University of California, Davis.

E. Wiegandt (ed.), Mountains: Sources of Water, Sources of Knowledge, 181–196.

1. INTRODUCTION

During the past 65 years, glacier melting in Peru's Cordillera Blanca mountain range has caused some of the world's most deadly glacier disasters. In 1941, a glacial lake outburst flood claimed 5,000 lives and demolished a third of the Ancash Department capital city of Huaraz. A 1945 outburst flood in Chavín killed approximately 500 people, devastated much of the town, and destroyed ruins and artifacts from one of Peru's oldest organized societies. In 1950, the Los Cedros outburst flood left 200 people dead and destroyed the nearly completed Cañón del Pato hydroelectric station. Beyond the floods, glacier retreat also triggered deadly avalanches in 1962 (4,000 deaths in Ranrahirca) and 1970 (15,000 deaths in Yungay). No society in the world has experienced such devastation as a result of melting glaciers (Carey 2005).

Since the onset of these catastrophes in 1941, various groups have understood glacier hazards in distinct ways. First, experts (scientists and engineers) saw a technical problem that required scientific studies, glacier monitoring, and engineering projects that used technology to contain unstable glacial lakes. Second, economic developers and government officials believed disaster mitigation was necessary to protect vital infrastructure; but they also recognized that disaster mitigation programs created an opportunity to bolster modernizing projects in hydroelectricity, irrigation, and tourism. Third, local residents' fear of additional catastrophes—and their concerns with a series of additional risks only tangenttially related to natural disasters—led them to demand immediate draining of dangerous Cordillera Blanca glacial lakes.

Scholars have recognized discrepancies among groups in both perceiving risks and proposing solutions to natural hazards (see Dacey, this volume). These differences have become increasingly clear in recent years as analysts now understand that natural disasters often result from social factors that force marginalized segments of the population into those areas or situations most vulnerable to natural disasters (Cannon 1994; Hewitt 1997; Oliver-Smith and Hoffman 1999; Wisner et al. 2004). To understand these distinct forces influencing vulnerability, researchers often divide local and expert perceptions of risk into two paradigms (Maskrey 1994). On the one hand, experts comprehend natural hazards through a single focus on an environmental problem—a view that generally neglects the social dimensions of natural disaster. On the other hand, local people face a host of risks, and they rank potential natural disasters far below other, more immediate risks to their survival. Not simply differences, these variations in risk perception are embedded in power structures whereby dominant social groups, governments, developers, and experts exert significantly greater power over most marginalized, vulnerable populations (Davis 1998; Steinberg 2000). Nevertheless, while recognizing that poor, marginalized people do suffer disproportionately from natural disasters, this view can both victimize local residents and assume that local people's demands for disaster mitigation differ from those of experts, authorities, and developers. However, by recognizing the historical agency of vulnerable populations and understanding their perceptions and demands, this tendency to victimize vulnerable populations can be avoided.

This essay suggests that while distinct groups may perceive risk differently, they can agree about the solutions to natural hazards. Local residents living beneath unstable Cordillera Blanca glacial lakes faced a host of risks that transcended the threat of natural disaster. Yet the most vulnerable population—the urban and wealthiest people in the Cordillera Blanca region—believed that the best way to protect themselves was to drain glacial lakes. They thus chose a scientific-engineering solution that failed to account for their vulnerability to natural disasters (their habitation of flood zones). Despite different understandings and worldviews, the experts, developers, policymakers, and local residents thus agreed that the best strategy for disaster mitigation was the draining and damming of Cordillera Blanca glacial lakes.

2. THE ONSET OF GLACIAL LAKE DISASTERS

The Cordillera Blanca mountain range runs approximately 180 km north-south through the Department of Ancash (see Figure 1). Approximately 600 glaciers cover the range and account for about one quarter of the world's tropical glaciers (Georges 2004). Large valleys—called the Callejón de Huaylas along the west side and the Callejón de Conchucos to the east—run parallel to the Cordillera Blanca. Varying between 2,000 and 3,500 m above sea level, these two valleys are densely populated with nearly a half million people; the majority inhabit the Callejón de Huaylas. The largest towns in the Callejón de Huaylas—such as Huaraz, Carhuaz, Yungay, and Caraz—cling to the banks of the Santa River, which flows through the bottom of the valley. The river descends from above 4,000 m to the Pacific Ocean, carrying 70 percent of Cordillera Blanca runoff through the Callejón de Huaylas and then through the steep, narrow gorge at Cañón del Pato, where a 260 megawatt hydroelectric station exists today.

People living around the Cordillera Blanca—like Peruvians throughout the country—have historically divided their society into geo-racial binaries

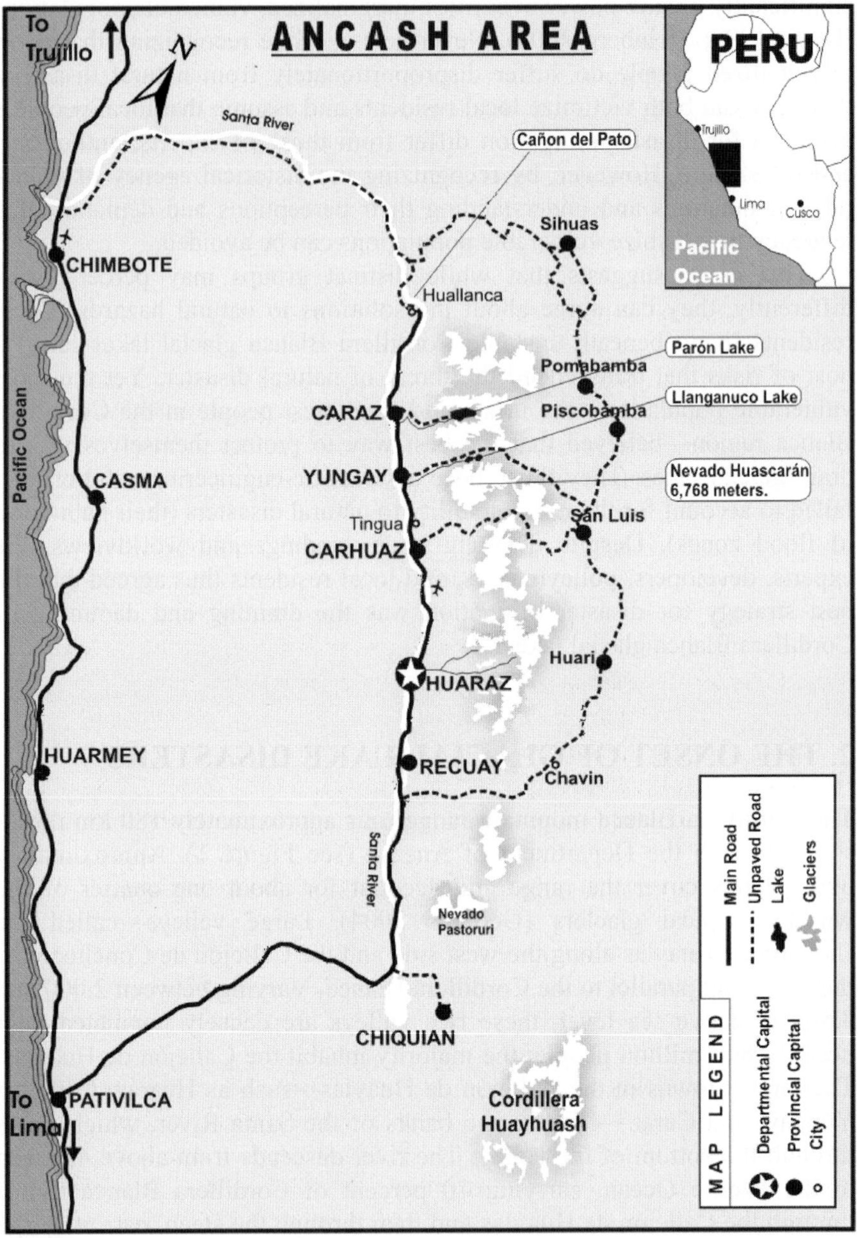

Figure 1: Ancash Area and Location in Peru (map drawn by Tito Olaza).

that separated indigenous people from *mestizos* (of mixed Spanish-indigenous descent). Although literacy, education, language, and dress have always distinguished these groups, most Peruvians have since the nineteenth century equated indigenous people with rural and highland areas; on the other hand, they associated mestizos with urban and lowland areas. Peruvians have generally identified a third group, called *cholos*, of indigenous people who exhibit mestizo customs and behavior, such as formal education or residence in urban areas (Mangin 1955; Stein 1974; Oliver-Smith 1977). A continuum has thus existed in Peru: moving inland from the coast to the highland Andes, society becomes more rural and thus more indigenous, regardless of phenotype (Orlove 1993).

Within the highland Andes, people also maintain that populations on valley floors are more urban and mestizo than those inhabiting upland slopes. As one Callejón de Huaylas resident explained in the early 1950s, local urban residents believed that the "indigenous race" lives in "rural hamlets, ranches, and high-elevation plains (*punas*); mestizos, on the other hand, generally inhabit district capitals and the occasional hamlet of importance" (Carrillo Ramírez 1953). More than just racial divisions, the mestizo-indigenous distinction also characterized significant power divisions: lowland urban residents based their domination of highland rural populations on the racial hierarchy that gave mestizos higher social status than indigenous people.

Yet, strong as these geo-racial binaries were in people's minds, they always remained fictional in reality. Centuries of miscegenation, increased rural-to-urban migration during the twentieth century, constant movement of labor and commerce, and fluid connections between highland and lowland, between town and country, and even between indigenous and mestizo reveal the imprecision of these culturally constructed racial categories (de la Cadena 2000; Larson 2004). Nevertheless, the terms remain useful for describing the twentieth-century Callejón de Huaylas because they represent the ways in which the historical actors themselves viewed their world.

In the mountains above these human populations, Cordillera Blanca glaciers have been melting since the late nineteenth century (Kaser and Osmaston 2002; Georges 2004). Glacier retreat became deadly because, as glaciers melted, glacial lakes formed on top of glacial ice or immediately below glaciers. The dynamic process of glacier retreat in the Cordillera Blanca has led to a massive increase in the number of these glacial lakes: from 223 in 1950, to 314 in 1983, to 374 in 1997 (Fernández Concha and Hoempler 1953; Zamora Cobos 1983; Electroperu 1997). In many cases, terminal moraines dammed these swelling lakes. Heightened pressure on unstable moraine dams or wave action caused by icefalls into the lake could cause dams to rupture, thereby producing a glacial lake outburst flood (GLOF). As lakes discharged in a GLOF event, often in a matter of minutes, the water mixed with rock, soil, glacial ice, and other debris to produce slow moving but powerful outburst floods with a consistency of wet cement (Trask 1953; Fernández Concha 1957; Lliboutry et al. 1977).

On the morning of December 13, 1941, residents of the Ancash Department capital city of Huaraz awoke to one of these outburst floods descending violently from the Cordillera Blanca. The flood's thunderous howl sent the city's inhabitants running from their beds into the street. Moments later, a wall of water, mud, glacial ice, and rocks slammed into Huaraz, fanning out across one third of the city while overtaking bridges, houses, schools, hotels, streets, and families. Some residents managed to escape or avoid the outburst flood. Others, however, could not escape, like Olguita Ríos who stood confused at her second-story window when the flooding mass engulfed her, or Zoila Gonzáles de Huertas, who sprinted into her house from the street to rescue her children just when the flood swamped them (Zegarra 1941; Anonymous 1941b; Coral Miranda 1962). Eyewitness Reynoldo Coral Miranda (1962) later described that terrifying morning: "On December 13, 1941, the blinding force of nature unleashed its power, making the mountains shake; the avalanche came, killing and swallowing, destroying and demolishing everything in its path. It cut short the lives of thousands of innocent people who just happened to live or be located in this picturesque and beautiful land. [The flood] converted the city and its precious lands into a heaping pile of inert material." The flood destroyed approximately one-third of Huaraz and killed an estimated 5,000 people.

The 1941 Huaraz outburst flood originated from Lake Palcacocha, also known as Lake Cojup because it lies at the head of the Cojup Canyon. Palcacocha burst when an icefall from the retreating glacier above the lake crashed into the lake and triggered giant waves, which quickly eroded and ruptured the moraine dam. The quantity of flood water increased again when it flowed into Lake Jircacocha farther down Cojup Canyon and caused that lake to burst as well. By the time it reached Huaraz, 23 km below Lake Palcacocha, the flood contained 8 million m^3 of water and debris (Anonymous 1941a; Giesecke and Lowther 1941; Zegarra 1941). Not the world's largest outburst flood by volume, the 1941 Huaraz flood was nonetheless one of the most deadly (Carey 2004).

The Huaraz flood was neither the first nor the last Cordillera Blanca outburst flood (Ames Marquez and Francou 1995; Zapata Luyo 2002).

Several other glacial lakes had ruptured since the end of the Little Ice Age, though none caused major destruction or loss of life. In 1945, an outburst flood on the eastern slope of the Cordillera Blanca razed part of the town of Chavín, killed 500 people, and destroyed irreplaceable pre-Columbian indigenous artifacts at one of Peru's most important archeological sites (Trask 1953; Tello 1960). A 1950 GLOF in the Los Cedros Valley killed approximately 200 people and destroyed nearly a decade of construction progress at the Cañón del Pato hydroelectric station (Ghiglino 1950; Spann and Concha 1950; Trask 1953). Cordillera Blanca glacial lakes remain precarious today. A small outburst flood from glacial Lake Safuna destroyed agricultural lands, damaged roads and bridges, and killed livestock in 2002, while Lake Palcacocha (source of the 1941 flood) produced a mini-outburst flood in 2003 that partially destroyed one of the lake's security dams.

In addition to numerous GLOFs during the 1900s, Cordillera Blanca residents also experienced two catastrophic glacier avalanches (Zapata Luyo 2002). In 1962, Glacier 511 on the north peak of Mt. Huascarán (6,768 m—Peru's tallest mountain) unleashed a glacier slide that killed 4,000 people and buried the town of Ranrahirca. The same glacier triggered another avalanche in 1970—the most deadly in world history—that destroyed several towns and killed an estimated 15,000 people. Both of these glacier slides resulted from glacier retreat: the ice had thinned and fractured after decades of melting.

While glacier disasters triggered the most deadly natural disasters in Peru's Cordillera Blanca, GLOFs were neither a new phenomenon nor specific to the Peruvian Andes. In fact, Pleistocene glacial lake ruptures generated some of the largest floods known (Rudoy 2002; Snorrason et al. 2002; Clarke et al. 2003). During the Little Ice Age (~1350–1850), glacial lake outburst floods caused significant destruction and death in the European Alps (Le Roy Ladurie 1971; Grove 1987; Grove 1988; Fagan 2000). And in recent decades, glacier retreat in the Himalayas has led to the formation of precarious glacial lakes that have triggered nearly a dozen outburst floods since 1935 (Kattelmann 2003; Richardson and Reynolds 2000b; Mool 2001; Richardson and Reynolds 2000a). A recent inventory of 4,989 glacial lakes in Nepal and Bhutan revealed 26 potentially dangerous lakes in Nepal and 24 in Bhutan. Efforts are now underway to lower glacial lake water levels to protect inhabitants from outburst floods. The history of glacial lake control in Peru offers important considerations for those concerned with GLOFs in the Himalayas and elsewhere (Clague and Evans 2000).

3. GLOFs AS TECHNICAL PROBLEMS

Programs to study, monitor, and contain Cordillera Blanca glacial lakes began immediately following the 1941 Huaraz outburst flood. Scientists from the Peruvian Institute of Geology conducted several glacial lake studies during the 1940s, while the Peruvian Division of Water and Irrigation partially drained six glacial lakes between 1942 and 1950. Authorities also tried to prevent Huaraz residents from rebuilding in the flood zone, an effort that ultimately failed. Despite such cataclysmic destruction in Huaraz and Chavín, neither the 1940s projects at glacial lakes nor the attempts at urban hazard zoning represented a systematic or comprehensive program to reduce glacial lake hazards. Destruction of the Cañón del Pato hydroelectric facility by the 1950 Los Cedros outburst flood, however, shifted the national government's approach to disaster mitigation: after the Los Cedros catastrophe, the sluggish state responses of the 1940s gave way abruptly to glacial lake control.

On February 20, 1951, the Peruvian government created the Control Commission of Cordillera Blanca Lakes (CCLCB). Its enabling legislation (*Resolución Suprema* No. 70) charged the agency with carrying out studies and conducting disaster mitigation projects to avoid "repetition of those outburst floods that can damage the current and ongoing works of the Peruvian Corporation of the Santa, as well as those of the Chimbote-Huallanca Railroad avoiding at the same time damage to the populations of the Callejón de Huaylas." Although future CCLCB funding hinged on political circumstances and its objectives were justified in large part by the protection of infrastructure rather than humanitarian goals, the technical achievements of the agency (and its successors) have for more than a half century helped protect the region from glacial lake outburst floods.

In its first decade alone, the CCLCB "contained" more than a dozen lakes through its "lakes security projects," which involved draining millions of cubic meters of water from Cordillera Blanca lakes and constructing cement dams to hold back these precarious lakes in the future. By the early 1950s, CCLCB researchers had established an effective system to categorize glacial lakes and their likelihood of producing GLOFs. They analyzed glacial lake dam types (rock, debris, or moraine), dam slopes, and location of glacier tongues in regard to lakes (tongues in contact with lakes were regarded as the most dangerous). They also learned—and explained to the world—what caused glacial lake outburst floods and how waves from icefalls eroded moraine dams (Fernández Concha 1957). Since the CCLCB laid the foundations for Cordillera Blanca glacial lake control during the 1950s, the government has contained thirty-five glacial lakes by lowering lake water levels and constructing erosion-resistant security dams at lake outlets. Moreover, CCLCB historical records of glaciers and glacial lakes provide some of the most complete data for tropical glaciology and glacial lake management.

After 1967, when the Peruvian Corporation of the Santa appointed Benjamín Morales Arnao as director of a new Glaciology and Lakes Security Division, Peruvian scientists and engineers devoted attention to glacier research as well as disaster mitigation. By the 1980s, the agency had completed a national inventory of glaciers, a project important both for Peru and for the World Glacier Inventory (Ames 1988). The long-term glacier research has allowed scientists to understand effects of climate change on mass balance (e.g., Kaser 1990; Hastenrath and Ames 1995; Kaser and Georges 1997), glacier tongue behavior (e.g., Ames 1998), and glacial lake dynamics (Fernández Concha 1957; Morales 1969; Lliboutry et al. 1977; Kaser and Osmaston 2002). As scientists increasingly rely on glaciers for historical climate records, and as other mountain societies struggle against GLOFs, the Cordillera Blanca scientific and engineering advances have become increasingly important, not only for Peruvians but for people worldwide.

4. THE POLITICAL ECONOMY OF GLACIAL LAKE CONTROL

Establishment of the CCLCB marked a decided shift in the state's disaster prevention agenda—and this transformation stemmed largely from development interests that sought protection of hydroelectric infrastructure and expansion of irrigation, tourism, conservation, and transportation. In particular, water developers pursued disaster mitigation as a way both to protect hydroelectricity and irrigation infrastructure and to bolster their development projects. Cordillera Blanca glaciers have always enriched the Department of Ancash with one of its most vital natural resources: water. Glacier runoff on the western slope of the Cordillera Blanca flows into the Santa River, one of Peru's largest rivers flowing into the Pacific Ocean. On its path to the ocean, the Santa River makes a precipitous fall through a narrow gorge at Cañón del Pato. Since 1913, when Santiago Antúnez Mayolo first outlined plans for generating hydroelectricity at Cañón del Pato, engineers and developers have dreamed of exploiting Cordillera Blanca water to help industrialize Peru (Antúnez de Mayolo 1941). It was not, however, until 1943 that the national government created the Peruvian Corporation of the Santa (CPS) to construct and manage the Cañón del Pato hydroelectric station (CPS 1944). The hydroelectric station was 80 percent completed in 1950, when the Los Cedros outburst flood devastated the facility. After this dramatic setback, Cañón del Pato did not begin its first stage of operation (generating 50 megawatts) until 1958. The CPS brought another 50 megawatts on line in 1967 and again increased capacity by 50 more megawatts in 1982 (Electroperu 1989). In 1998, Duke Energy Perú (2002), which bought the facility in 1996, expanded Cañón del Pato to its current capacity of 260 megawatts. While coastal residents and industries in Chimbote utilized the bulk of this energy, Callejón de Huaylas communities have also relied on Cañón del Pato electricity generated from glacier water (Reynolds 1993).

Hydroelectric interests—which were not affected by the 1945 Chavín flood and were not yet established when the 1941 Huaraz flood passed through Cañón del Pato—motivated creation of the CCLCB in 1951. Interestingly, the Huaraz outburst flood had previously stimulated scientists and engineers to analyze the importance of glacial lakes for hydroelectric power generation. Rather than believing that glacial lakes needed to be controlled to protect hydroelectric facilities, however, experts in the early 1940s believed disaster prevention (damming lakes, regulating water flow, etc.) could be combined with increased hydroelectric generation. In 1942, for example, when geologists examined glacial lakes above Huaraz for security purposes, they also noted hydrological resources, prospective dam sites for water storage, and, according to Jorge Broggi (1942), methods to "help regulate the flow of the Santa River, which the State plans to exploit exhaustively for energy and irrigation, and to sustain agriculture along the banks for the river."

An explicit link between water use (hydroelectricity and irrigation) and disaster mitigation did not emerge until after the 1950 Los Cedros outburst flood. The late start for systematic glacial lake monitoring, research, and engineering projects suggests that the destruction of a major hydroelectric facility proved more compelling to the national government than the deaths of more than 5,000 Huaraz and Chavín residents. Through time, hydroelectric and irrigation interests have increasingly guided the state agency charged with disaster mitigation at Cordillera Blanca glacial lakes. Hydroelectric developers benefited from glaciological and hydrological research conducted under the auspices of glacial lakes security, and they utilized access roads and dams that were constructed originally for GLOF prevention. By the 1970s, the office explicitly "oriented its studies toward an evaluation of the hydrological potential of glacial watersheds for use in energy and agriculture" (Electroperu 1975; INGEMMET 1979). Hydroelectric power generation and irrigation became principal objectives of glaciological research and lakes control projects after the 1980s. Increasingly, the office studied glacial lakes for their potential to store water and boost flow of the Santa River during the dry season (UGH 1990). In the 1990s, the office proposed damming sixteen glacial lakes, not for security alone, but for exploitation of the water (Electroperu 1995). Glacial lakes Parón, Cullicocha, Aguascocha, and Rajucolta are now utilized for water storage and streamflow regulation (Duke Energy Perú 2002). Although hydroelectric interests sometimes eclipsed disaster prevention programs, the CCLCB and its successors nonetheless contributed enormously to the safety of the regional population, the protection of infrastructure, the development of CPS initiatives, the expansion of economic activity in Ancash, and the extension of public services into people's homes.

Tourism has also been part of policymakers' and developers' understanding of glacial lake control projects. Since the early twentieth century, the Cordillera Blanca has attracted mountaineers to its majestic peaks, especially Mt. Huascarán (Morales Arnao 2001). By the 1950s, advocates of tourism also promoted Lake Llanganuco, one of Peru's most famous lakes, as a tourist center and, thus, an economic resource for local communities to exploit (Carrión Vergara 1959). As Isaías Izaguirre (1954) noted, the beauty of Lake Llanganuco "has always instilled spiritual grandeur in all people. With Llanganuco, we can convert our towns into paradises of a thousand successes." Llanganuco had by 1954 become the centerpiece of the Automobile Circuit of the Callejón de Huaylas. And in 1964, the Automobile Association of Peru featured Lake Llanganuco on its national tourist guidebook, noting the lake's "unforgettable views" and opportunities for boating and trout fishing (Asociación Automotriz del Perú 1963).

Conservation efforts linked to tourism also helped justify both the control of glacial lakes and the CCLCB's construction of labor camps at glacial lakes and access trails to remote parts of the Cordillera Blanca. Conservation initiatives in the Cordillera Blanca paralleled not only the glacial lake control projects, but also the earliest national park movements in Peru. A year before Cutervo National Park became the country's first national park in 1961, Ancash Senator Augusto Guzmán Robles presented a bill to the national congress to create Huascarán National Park in the Cordillera Blanca. Interest in this park continued and in 1967 two United States Peace Corps volunteers also made formal proposals to create Huascarán National Park. Describing the tourist potential as well as conservation issues, they identified Lakes Llanganuco and Parón as ideal cites for Cordillera Blanca tourism (Slaymaker and Albrecht 1967). The national government officially created Huascarán National Park in 1975, and it became a World Natural Heritage Site in 1985. Glaciers and glacial lakes have become increasingly popular destinations for tourists who now flock to the Cordillera Blanca in the tens of thousands each year (Barker 1980; Bartle 1985; Byers 2000).

Tourism and conservation have been part of glacial lake control projects since disaster mitigation initiatives began in the Cordillera Blanca. As early as 1942, scientists argued that glacial lakes security projects could bolster tourism: access roads and trails would open areas of the Cordillera Blanca to tourists, while the labor camps built for disaster mitigation project workers would later serve as climbing huts for mountaineers (Broggi 1942). When César Morales Arnao founded the nationally and internationally recognized Cordillera Blanca Mountaineering Club in 1952, many of the original thirteen members were scientists who had worked on glacier studies and disaster mitigation at glacial lakes (Grupo Andinista Cordillera Blanca 1952). Through the following decades, the disaster mitigation managers continually noted tourist advantages that stemmed from their projects. As one official explained in 1975, "It is important to note that, indirectly through access routes to these [glacial lakes security] projects, we are providing the initial steps for the region's tourist infrastructure, the fundamental pillar for its future development. We have constructed ten labor camps at our lakes, and these will serve as bases for future high mountain tourist lodges" (Electroperu 1975). Today tourists do not use most of these labor camps, but the trails and roads remain vital for tourism, one of the region's principal industries.

To protect and expand modernizing projects and infrastructure, the Peruvian state developed a specific approach to disaster mitigation: draining and damming glacial lakes rather than reducing people's vulnerability to floods through hazard zoning. On first glance, this could suggest neglect of the local people, cooperation with developers, an over-reliance on science and technology to resolve social issues (vulnerability), and pursuit of political goals rather than the safety of the population (Cannon 1994). Analysis of local people's perceptions of risks and proposed solutions to glacial lake hazards, however, indicates agreement and cooperation instead of government abandonment or deceit.

5. GLOFs AND THE SOCIAL CONSTRUCTION OF RISK

In the wake of the glacier disasters that occurred periodically after 1941, local residents developed a profound awareness of Cordillera Blanca natural hazards. The most vulnerable populations—those in towns and cities along the rivers that descended from the Cordillera Blanca—demanded state action to prevent additional outburst floods. Like populations worldwide, however, Callejón de Huaylas inhabitants ranked various risks that influenced their lives. Fear of glacial lake outburst floods existed alongside a number of other concerns: adequate food and shelter, education, access to public services, crime, social control, and other issues related to people's everyday livelihoods. Safety from outburst floods was not their only, or even their most pressing, concern. Nevertheless, urban residents were traumatized by the glacial lakes and they persistently advocated for disaster mitigation programs.

Although natural disaster scholars argue that the poorest, most marginalized populations are usually the most vulnerable to natural disasters, in the Cordillera Blanca the situation was the reverse. Spanish colonizers' decisions to construct towns along the Santa River and its tributaries led to the establishment of urban areas in flood zones. Indigenous settlements before the sixteenth-century Spanish conquest had been more dispersed and at higher (and thus safer) elevations (Oliver-Smith 1999). Further, Peru's geo-racial binary put the wealthiest people and the dominant socio-racial class in urban areas on the valley floor. Consequently, the dominant classes of urban residents suffered the vast majority of death and destruction from glacial lake outburst floods. Most indigenous residents in the mid-1900s lived on upland slopes, though urbanization, jobs, and fluid urban-rural connections had brought some rural residents to flood zones within regional towns. It was thus the urban population—the wealthiest segment of the population—that was most vulnerable to glacial lake hazards and most concerned about disaster mitigation projects.

The constant threat of catastrophic GLOFs caused these urban residents to vilify glacial lakes and fed their anxieties about impending disaster. Songs, books, poems, literature, and stories depicted glaciers and glacial lakes as both beautiful and villainous. In 1954, a local resident called Lake Palcacocha, source of the 1941 Huaraz flood, a traitor and an assassin that took on a new course of punishing local people (Neogodo 1954). Some residents considered glacial lakes treacherous and secretive, capable of destroying the valleys. Others believed the glacial lakes could suck people in if they got too close (Yauri Montero 1961; Bode 1990). Despite such negative perceptions of the glacial lakes, most Cordillera Blanca inhabitants continued to see the lakes and mountains as defining characteristics of the region. This landscape, according to many residents, makes the region more beautiful—and thus better—than other areas of Peru (Ibérico 1954; Villón 1959; Yauri Montero 2000). This dual view that simultaneously celebrated and vilified glacial lakes became common after the 1940s.

Glacial lake instability has also made people live in fear of impending disaster. Worried that more GLOFs would inundate Cordillera Blanca communities after the 1941 Huaraz and 1945 Chavín outburst floods, a poet (Montes 1945) penned his frustrations with recently formed glacial Lake Cuchillacocha:

> . . . look at the cruel Andean lakes
> that have tormented us
> be careful Cuchillacocha, don't kill me
> like the traitor [Lake] Cojup
> that killed our brave brothers [in Huaraz]. . . .

In 1955 a local newspaper reporter, fearful of heavy rains and rising water levels in lakes, complained that "these hydrological phenomenon that happen much too frequently provoke an anxious restlessness among the Huaraz populations" (Anonymous 1955). Anxieties over glacial lake outburst floods remained high enough that amidst the violent earthquake of May 31, 1970 (7.7 on the Richter Scale), some people feared a GLOF more than the trembling earth itself. Residents of Huaraz, for example, stared intently at the Cojup Canyon, watching and listening for an outburst flood. People in Caraz below Lake Parón, the Cordillera Blanca's largest glacial lake, ran directly for the hills screaming "Here comes Parón!" (Pajuelo Prieto 2002). Fear of glacial lakes has continued in recent years. In 1998 the Quillcay River in Huaraz rose dramatically during a rainstorm, creating a panic in the city that sent many residents running to the hills for safety. In March 2003, Huaraz residents again became alarmed when they learned that Lake Palcacocha had overflowed due to a landslide into the lake.

The mounting social anxieties that accompanied this history of relentless glacier disasters led people to demand the Peruvian government to protect them from glacial lake hazards. After the 1945 Chavín flood, for example, residents criticized the government for wasting time on scientific studies while not acting to eliminate societal threats from glacial lakes: "What we want is that [the government] listens to the voice of the people, which is the voice of God, and drains the lakes without waiting until the rainy season passes and without doing more studies" (Anonymous 1945c). Ten years later, a Huaraz newspaper revealed public fear of floods and frustration with the CCLCB, especially for not preventing the 1950 Los Cedros GLOF and for not investing more in the reconstruction of Huaraz after 1941. Additionally, the newspaper urged the agency to publish its studies so that Huaraz residents would know whether "future catastrophes were possible or impossible" (Ayllón Lozano 1955; Anonymous 1955; Anonymous 1954). Local authorities also demanded that the national government do more to protect their communities from GLOFs. In Caraz, for example, municipal leaders in 1942 requested immediate disaster mitigation work at Lake Parón (Subprefecto Lucar 1942). Although work had begun by the early 1950s, Caraz authorities still complained about slow government response and the continued threat of hydrological catastrophes from Parón (CCLCB 1951–1954). When glacier disasters occurred, many local people blamed the government for withholding glaciological information and thus hiding the true dangers to Cordillera Blanca towns (Oliver-Smith 1986).

Huaraz residents also rejected hazard zoning that would have helped reduce their vulnerability to outburst floods. Instead of complying with building codes that sought long-term disaster prevention, many people re-colonized the floodplain to recover their lost property (Anonymous 1945b, 1956c). During the 1950s, Huaraz residents actively fought against the building restriction and defied the law by constructing homes and buildings on the floodplain, in what many considered the true heart of Huaraz. Residents wrote newspaper editorials that complained about the economic losses from hazard zoning. Others asserted their right to rebuild Huaraz and the "need" to rebuild the city in its previous location. And some simply did not care about restrictions or dangers because the floodplain offered vacant land (Anonymous 1951; Irving 1952; Anonymous 1956b). In 1956, urban residents even formed the Association of Flood Zone Property Owners to "restore the urbanization of this important sector of the city". They argued that recuperating their land and re-urbanizing such an "important sector of the city" was vital not only to their own interests but also to the general recovery from the 1941 disaster (Anonymous 1956a).

Reconstruction of the city was critical to Huaraz residents because it symbolized modernity, civilization, and the social standing of mestizos over indigenous people (Fernández 1942; Anonymous 1945a; Coral Miranda 1962). In a society where geo-racial binaries dictated social status, urban residents of the Callejón de Huaylas sought to preserve the status quo of the class–race hierarchy. Maintaining their cities in the lowest possible sites, which were hazard zones, ranked as a more important risk than the possibility of another glacial lake outburst flood. This was particularly true during the second half of the twentieth century, when rural to urban migration rose steadily in Peru and eroded rigidly defined categories of race, class, and social standing. As Huaraz and other Callejón de Huaylas residents refused to move to less vulnerable locations, they clung to their towns along the Santa River floodplains and demanded that the government drain and dam glacial lakes (e.g., Anonymous 1959a, b; Anonymous 1970; Salazar Bondy 1970). Authorities abandoned the

Huaraz hazard zoning soon after mandating it, thus indicating their preference for glacial lake control rather than the reduction of human vulnerability.

6. CONCLUSION

The most vulnerable Callejón de Huaylas inhabitants believed—as did government officials, developers, and experts—that science, engineering, and technology could effectively protect them from the melting glaciers that periodically unleashed avalanches and outburst floods. Of course, the Peruvian government did respond much more energetically and systematically to economic losses at Cañón del Pato in 1950 than to catastrophic loss of human life in 1941 Huaraz, thereby suggesting the political-economic nature of disaster mitigation rather than a primarily humanitarian endeavor. Unlike some development interests that put people in danger (Steinberg 2000), however, the CCLCB and its successors engaged in significant (and successful) disaster mitigation projects for more than a half century. Thus, it was *because of* development projects—not in spite of them—that disaster mitigation blossomed during and after the 1950s.

Local complaints of government projects have focused on the *speed* of the scientific studies and lake control projects, not on the government's general approach to disaster mitigation. The only disaster mitigation plans that locals rejected were those expert and government initiatives that sought to reduce local residents' vulnerability to future natural disasters. Given their overall agreements—and locals' willingness to reconstruct their destroyed city in a vulnerable hazard zone—it is clear that local people's continued vulnerability to glacier hazards stemmed from their social constructions of risk. Perceptions of risk and understandings of glacier hazards thus differed among distinct groups. But local urban residents, authorities, developers, and experts generally sought the same solution to natural hazards: draining glacial lakes to avoid outburst floods.

WETLANDS AND INDIGENOUS KNOWLEDGE IN THE HIGHLANDS OF WESTERN ETHIOPIA

Alan Dixon
Department of Applied Sciences, Geography and Archaeology
University of Worcester,
Worcester, UK
a.dixon@worc.ac.uk

Abstract: Wetlands in the highlands of western Ethiopia are important natural resources that provide a range of goods and services to local communities. A perceived increase in the drainage and cultivation of these wetlands in the mid 1990s, however, prompted concerns of widespread wetland degradation and unsustainable levels of utilisation, with consequences for food and water security. Drawing upon participatory field research carried out in Illubabor zone, Ethiopia, this chapter discusses the contribution of indigenous knowledge (IK) to wetland management in the area, and assesses its implications for the sustainability of the wetland environment. The results of the research suggest that continuous wetland drainage and cultivation in Illubabor has been ongoing for at least 30 years and that indigenous management practices based on IK have evolved over time, through farmers' experience of the wetland environment. Contrary to initial concerns, these management practices appear to form the basis of sustainable wetland use strategies. It is suggested, however, that recent government initiatives that are not sensitive to indigenous wetland management practices may threaten the sustainability of the wetland system.

Keywords: indigenous knowledge, wetland management, sustainability, Ethiopia

1. INTRODUCTION

In many parts of the developing world, wetlands represent important natural resources in terms of their biodiversity, and the range of functions and products they provide for human populations (Dugan 1990; Hollis 1990; Denny 1994; Roggeri 1998). In arid and semi-arid environments, their capacity to act as natural reservoirs of moisture means they often play a critical role in ensuring food and livelihood security (Scoones 1990; Barbier 1993; Adams 1993; Silvius et al. 2000). This is particularly evident in the highlands of western Ethiopia, where almost every household

E. Wiegandt (ed.), Mountains: Sources of Water, Sources of Knowledge, 197–210.

relies upon wetlands for various products and services (Wood et al. 2002), but particularly for crop production. Increasing reliance on wetlands for food production in recent years has prompted concerns regarding the environmental sustainability of wetland use, and the ability of existing wetland management systems to continue to provide a range of functions and benefits in the long-term (Kebede Tato 1993; Wood 1996).

Drawing upon field research undertaken in western Ethiopia between 1996 and 1998, this chapter argues that these concerns are, for the most part, unfounded and that wetland use in the area is based on locally developed sustainable wetland management practices, rooted in an evolving indigenous knowledge of the wetland environment.

2. INDIGENOUS KNOWLEDGE

The last two decades have seen increasing recognition of the importance of community based natural resource management strategies in developing countries. Although previously blamed for destructive, inappropriate and opportunistic practices, a vast body of literature has emerged that suggests local people in developing countries possess the knowledge and skills to manage their natural resources and environment in a sustainable manner (Brokensha et al. 1980; Chambers 1983; Richards 1985; Warren 1991). The "local" or "indigenous" knowledge (IK) held by communities is regarded as a key component of sustainable natural resource management, since it continuously adapts and evolves over time in a specific location, culture or environment, in response to changing circumstances (Haverkort and Hiemstra 1999). Much research has drawn attention to the role of farmers as experimenters, a process regarded important as a means of generating new knowledge and contributing to adaptation and sustainability (Richards 1985; Millar 1993; Rhoades and Bebbington 1995; McCorkle and McClure 1995). Such is the perceived importance of IK among the international development community that "participation" with local communities and sensitivity to IK has now become a fundamental component of most NGO and state-based development initiatives.

There is, however, a risk of regarding IK as a panacea for rural development in developing countries. Although everyone possesses knowledge of their environment and livelihood system, knowledge is not distributed equally amongst a community. Differences inevitably occur as a result of gender, age, experience and profession (Swift 1979; Mundy and Compton 1995), and IK may simply be flawed or based on inaccurate observations. Moreover, possessing knowledge does not equate to actually applying it, and in many natural resource management strategies the application of knowledge often depends on more mundane issues such as access to credit and tools, and the availability of labour. Unless IK is able to adapt and overcome such issues, and evolve in response to any socioeconomic and environmental change, its presence alone does not guarantee a sustainable natural resource management system.

3. ILLUBABOR AND ITS WETLAND RESOURCES

Illubabor zone (Figure 1) remains one of Ethiopia's most fertile and least exploited regions. It has a total area of approximately 16,555 km^2 and lies between longitudes 33°47′ and 36°52′ east and latitudes 7°05′ and 8°45′ north. The zone has a warm temperate climate, atypical of conditions in the rest of the country, with a mean average temperature of 20.7°C and rainfall in excess of 1,800 mm per annum (Solomon Abate 1994). The undulating topography of the landscape, which ranges between 1,400 and 2,000 m above sea level, combined with the climatic conditions produces an environment characterized by steep-sided river valleys and flat, waterlogged valley bottoms. The accumulation of runoff, poor drainage and a high groundwater table in these valley bottoms promote the formation of permanent and seasonal swamp-like wetlands, ranging in size from less than 10 to more than 300 ha. Government reports suggest wetlands account for between 1.6–5 percent of the total land area of the zone (Afework Hailu et al. 2000).

The most recent government census, carried out in 1994, estimated the population of Illubabor to be 847,048 persons (CSA, 1997). Estimates for 1998 put this figure at 960,431 (of which 91 percent live in rural areas), making Illubabor one of the most sparsely populated areas in highland Ethiopia, with an average of 58 pers/km^2 (Government of Ethiopia 1998). The dominant ethnic groups, who account for 90 percent of the population within the zone, are the Oromo, who settled in the area after their invasion and expansion during the eighteenth century. The remaining 10 percent is composed of Amhara, Tigrayan, Gurage, Mocha and Keffa peoples. The majority of the population relies on agriculture as a means of subsistence and approximately 38 percent of Illubabor's land area is under cultivation. Maize constitutes the major cereal crop and staple food (38.3 percent of total food production), whilst smaller quantities of tef (*Eragrostis tef*), barley and sorghum are also cultivated.

Illubabor's wetlands are important natural resources for local communities. They represent a vital source of water throughout the year in an area which receives half of its annual rainfall between June and August and only 5 percent during the dry season months of December, January and February. The abundance of water in the wetlands supports the growth of dense sedge vegetation known locally as *cheffe* (*Cyperus latifolius*), which is traditionally harvested by local communities for use as a roofing and craft material. Medicinal plants such as *balawarante* (*Hygrophila auriculata*), a treatment for skin diseases, are also collected (Zerihun Woldu 2000). Although it is difficult to trace the origins of wetland cultivation in the western highlands, the small-scale cultivation of wetland margins has traditionally been practiced by the Oromo population, in response to food shortages on the uplands (McCann 1995; Tafesse Asres 1996). At some stage during the last hundred years, wetland cultivation has extended dramatically to include the intensive cultivation of maize in whole wetlands.

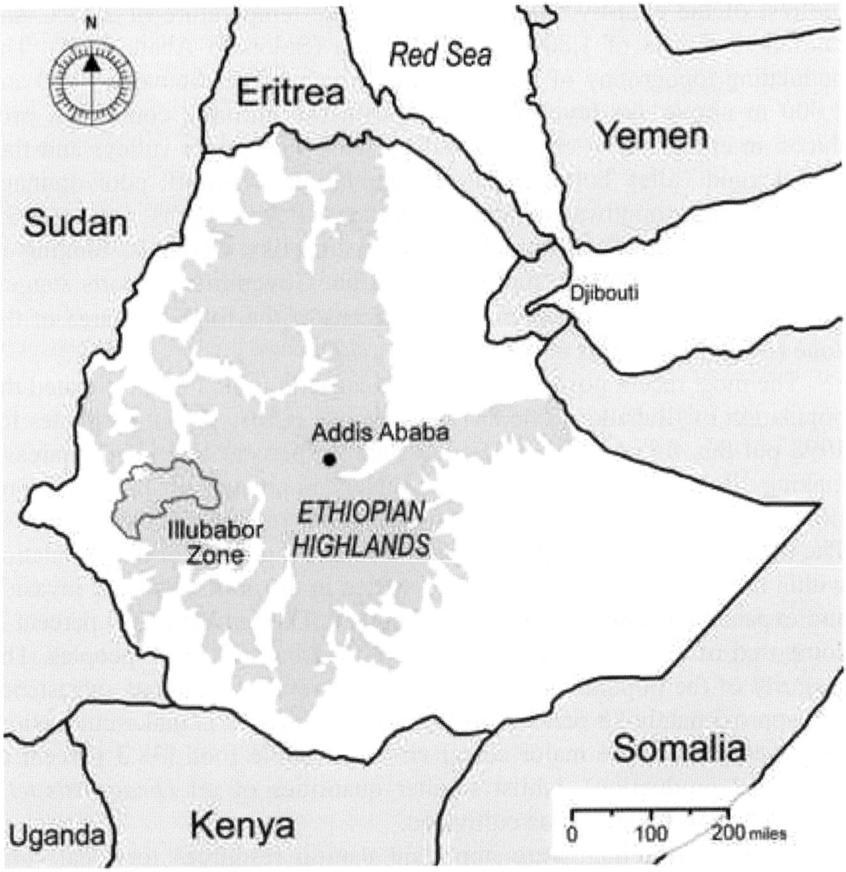

Figure 1: The Location of Illubabor Zone in Ethiopia.

Recurring food shortages on the uplands have been the principal reason for the spread of wetland agriculture throughout Illubabor. In the Menelik and Haile Selassie eras (1913–1974) Ilubabor's feudal landlords distributed wetland plots among peasant farmers and instructed them to cultivate maize in years of famine and drought. Following the overthrow of the Haile Selassie government by the Derg[1] (1974–1991), wetland cultivation was intensified in order to meet regional targets of food self-sufficiency and farmers who failed to cultivate their wetland plots risked the reallocation of their land to the landless or those who were willing to expand into wetland cultivation (Afework Hailu 1998). The expansion of coffee farming on the uplands, growth of the market economy and an influx of settlers from the north during the Derg era, also increased the pressure to produce food supplies from wetlands. In the post-Derg era (1991–present) the Oromiya Bureau of Agriculture has generally encouraged the cultivation of wetlands to meet food production targets.

The increase in wetland agriculture in Illubabor has prompted concerns regarding the environmental sustainability of the wetland system. Reports from Illubabor during the mid 1990s suggested that wetlands were beginning to exhibit signs of over-use and degradation, characterized by falling water tables, soil erosion and the disappearance of *cheffe* vegetation (Kebede Tato 1993; Butcher and Wood 1995; Wood 1996). Under such conditions, wetlands are unable to provide their full range of functions and benefits, and this has implications for food and water security for local communities. In addressing such concerns, the Ethiopian Wetlands Research Programme (EWRP), between 1996 and 2000, embarked upon an extensive program of interdisciplinary research in Illubabor Zone, exploring the functioning of wetlands, and the motivating forces and management practices contributing to sustainable and unsustainable use.

4. RESEARCH METHODS

One key area of EWRP's research focused on the role and significance of local communities and their IK in the design and implementation of wetland management practices. Investigations of farmers' wetland knowledge and the application of this knowledge in wetland management strategies were regarded as critical to a wider understanding of the dynamics of

---

[1] Derg: The government of Ethiopia between 1974 and 1991 in which power was initially shared by a military committee and later centralised into the hands of President Mengistu Haile Mariam.

wetland use, and ultimately wetland sustainability throughout the area. This research, undertaken between 1996 and 1998, focused on five wetlands typical of those found within Illubabor in terms of hydrological regime and geomorphological characteristics. All were less than 20 ha in size and four were undergoing maize cultivation at the time of the study (Table 1).

Table 1: Characteristics of the Study Wetlands.

	Bake chora	Tulube	Dizi	Anger	Supe
Altitude (m)	1,700	1,680	1,560	1,640	1,720
Size (ha)	8	8	4	16	10
Hydrological classification	small headwater	small headwater	small mid-valley	small mid-valley	small headwater
Drainage	artificial drainage	Inter-mittent channel	natural stream & artificial drainage	natural stream & artificial drainage	artificial drainage
Main water source	springs/ runoff	springs/ runoff	Inflow stream	inflow stream	springs/ runoff
Hydrological conditions	low water table throughout year	high water table throughout year	low water table throughout year	spatially variable water table	spatially variable water table
Land use	maize and tef cultivation/ abandoned plots/*cheffe* collection	abandoned wetland/ grazing	maize cultivation	maize cultivation/ abandoned plots/ grazing/ *cheffe* collection	maize & sugar cane cultivation/ *cheffe* collection
Periods of cultivation	c1900 to present	1960s–1989	1950s–1976, 1991 to present	1949–present	1930s to present

A program of meetings with farmers from each of these wetlands was established, in which a range of Participatory Rural Appraisal (PRA) tools were employed in a total of 26 field sessions. Tools, such as resource mapping, seasonal diagrams, transect walks and Venn diagrams (Chambers 1992; IIRR 1996; Grenier 1998) were utilized at each site, facilitating a high level of interaction between the research team and the wetland users. These sessions facilitated the exchange of information on topics ranging from seasonal hydrological and vegetation changes in each wetland, to the past and present sources of wetland management knowledge and techniques. Concurrently, a hydrological monitoring program was undertaken to establish the seasonal hydrological characteristics of the wetlands, enabling comparisons of wetland knowledge and hydrological reality to be made (Dixon 2003).

5. WETLAND KNOWLEDGE AND WETLAND MANAGEMENT PRACTICES

The results from the PRA sessions suggest that wetland management in Illubabor zone is indigenous in nature, having developed through a process of observation, trial and error, and the intergenerational accumulation of information. Those involved in wetland management (mostly farmers) demonstrated a detailed understanding of the wetland environment and the impact of human intervention in the wetlands. Their management practices, which attempt to balance community needs with environmental sustainability, are based on this body of IK.

Critically, farmers were found to possess extensive knowledge of the wetland hydrological system, particularly spatial and temporal variability of the wetland water table. Water is almost universally recognized by farmers as fundamental to the functioning and survival of wetlands, and some suggested that excessive drainage rendered wetlands "lifeless" and "just like bleeding a person to death". Using seasonal diagrams, farmers demonstrated an in-depth knowledge of the seasonal patterns of rainfall around their wetland, the seasonal variations in water table elevation, and the relationship between both. When compared to the data from hydrological investigations (Dixon 2002), farmers' knowledge was found to be remarkably accurate. For example, the comparison between actual rainfall and perceived rainfall outlined in Figure 2, indicates that whilst there is some discrepancy between the perceived quantity of rainfall and actual recorded levels, farmers' knowledge of the trend in rainfall during a typical year remains accurate. Similarly whilst farmers' perceptions of the wetland water table height during the year appear higher than those actually measured the seasonal trends of both are similar (Figure 3). The anomalous weather conditions (an extended dry season) during the 1997–1998 season, arguably accounts for the difference between farmers' knowledge (which was indicative of average conditions) and the water table data for that year.

Farmers were also able to elaborate on the consequences of any deviations from the normal hydrological regime, particularly the effects of flooding or drought. As one farmer from Supe wetland recalled:

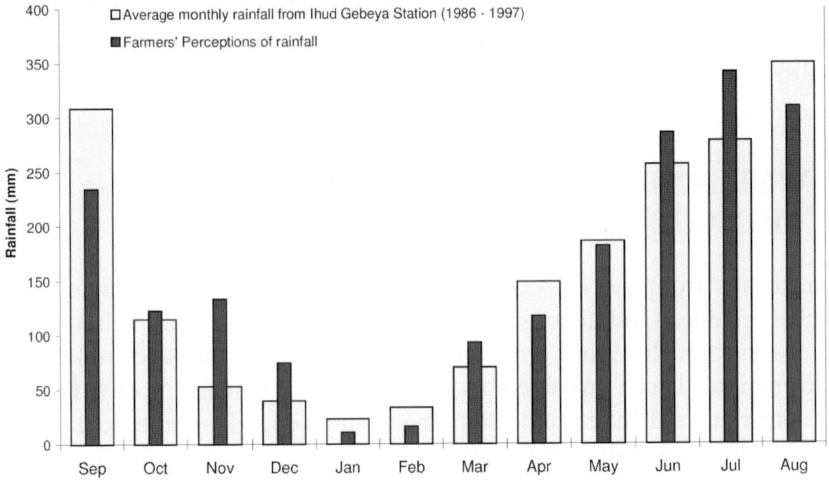

Figure 2: Farmers' Perceptions of Rainfall in Bake Chora Wetland Compared to Rainfall Data.

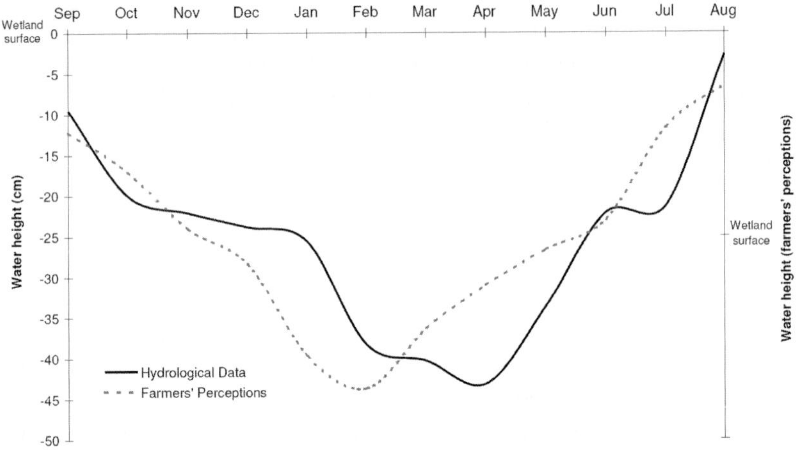

Figure 3: Farmers' Perceptions of the Water Table Elevation in Bake Chora Wetland Compared to Hydrological Data.

"Last year the rainfall in October and November was much heavier than usual. As a result, the upland crops suffered and the coffee was washed away. The wetland was also flooded but it didn't really matter because the harvest was over by then."

Supe farmer (April 15, 1998)

Similarly, knowledge of the spatial variability of hydrological conditions within wetlands was well understood by most farmers, for example:

> *"The lower part of the wetland is always wetter. When we try to drain this area we can't dig it properly. When there's excess rainfall, the bottom becomes wet first and the moisture spreads upwards towards the head of the wetland."*
>
> Supe farmer (April 15, 1998)

Farmers demonstrated an understanding of soil moisture conditions and the need to maintain it at a specific level that facilitates growth, whilst also avoiding waterlogging at the crop root zone. Using their knowledge of the hydrological regime and its spatial and temporal variability, they design and excavate drainage networks that are specifically adapted to hydrological conditions. For example, where excessive waterlogging occurs, permanent drainage ditches up to 1 m in depth and width, are excavated in close proximity to one another. In intermittently flooded areas, shallower ditches are excavated, often when and where necessary. It is also common practice for drainage ditches to be cleared of weeds during the growing season, in order to maintain optimum drainage conditions. Such practices were found to be common to all the sites studied, although the farmers of Bake Chora wetland have, in addition, developed a system of ditch blocking in order to regulate soil moisture conditions in the wetland. The blocking of ditches using soil or crop residue (Figure 4) is employed during the rainy season (June to September), prior to the sowing of maize, as a means of re-flooding drained wetland areas so that the soil fertility and soil structure are more conducive to crop growth. Conversely, where waterlogged conditions prevail, either at the beginning of the growing season (November to December) or as a result of heavy rain, farmers in Dizi and Bake Chora recalled how they remove invasive vegetation or sediment that accumulates in the drains, thereby improving drainage conditions.

The management of the hydrology in each wetland is intrinsically linked to knowledge of soil and vegetation changes in the wetlands. Farmers in each of the study wetlands were found to make decisions on whether to drain and cultivate based on the color and depth of soils. Shallow soils are not considered suitable for crop cultivation and are associated with poor fertility and moisture retention, hence they are left uncultivated. Soils are also classified as dark (*beyo guracha*) or grey (*beyo daleti*). The former is considered to be a more fertile soil with a greater moisture holding capacity than the latter, which is associated with over-cultivated soils. Where shallow *beyo guracha* soils overlay *beyo daleti* subsoils, however, farmers avoid cultivation and instead reserve such areas for the regeneration of *cheffe* vegetation. Although farmers acknowledge that repeated drainage and cultivation can cause the conversion of *beyo guracha* to *beyo daleti*, resulting in lower crop yields, they are also aware that the fertility of the soils is constantly being improved by the input of sediment from the catchment via runoff and flooding. The system of ditch blocking outlined above is indicative of the application of this knowledge.

The grazing of cattle in the wetlands is also regarded by farmers as an important means of providing a nutrient input, although farmers are also aware that intensive grazing leads to soil compaction, erosion and destruction of the natural vegetation. In most cases, farmers prohibit grazing in cultivated wetlands because of the threat of degradation. In Bake Chora wetland in particular, where wetland agriculture has been sustained for over 80 years with little sign of degradation, cattle are not given access to the wetland even after the harvesting of crops.

The relationship between *cheffe* and the wetland water table is also well understood by farmers, who regard it as both a hindrance to effective drainage and cultivation, but at the same time the key to wetland regeneration. The growth of *cheffe* is associated with waterlogged conditions, hence its presence is used by farmers as an indicator of the return of natural wetland characteristics, e.g., a high water table and increased soil fertility, following drainage. Similarly, farmers regard the growth of wetland plants such as *inchinne* (*Triumfetta pilosa*) and *tuffo guracha* (*Asteraceae*) as indicators of recovering soil fertility, whilst *Kemete* (*leersia hexandra*) indicates declining fertility.

From their knowledge of soil, water and vegetation in the wetlands, farmers have effectively developed "indicators of sustainability", that range from changes in soil moisture and color, to the colonization of specific vegetation. Based on these indicators, adaptive management mechanisms that prevent wetland degradation from occurring appear to be in place, such as, for example, when farmers block ditches or restrict cattle access to restore soil moisture and fertility. Indeed, one of the key mechanisms that prevents degradation and promotes the environmental wetland sustainability is the periodic abandonment of wetland plots, which encourages the regeneration of *cheffe*, the wetland water table and soil fertility. Farmers understand that this contributes to the maintenance of water table levels throughout each wetland as a whole, and in Bake Chora in particular, an area of *cheffe* is always reserved at the head of the wetland as a means of regulating water supply but also as a source of roofing material.

Figure 4: Ditch Blocking as a Means of Regulating Soil Moisture in the Wetlands.

6. AN EVOLVING KNOWLEDGE SYSTEM

The study has confirmed that wetland drainage and cultivation practiced in these wetlands is not undertaken haphazardly. Wetland farmers actively acquire knowledge of wetland processes and functions over time, they apply this knowledge, and demonstrate a capacity to modify their practices in response to environmental change, albeit on a small-scale. Although the study did not identify any examples of farmers engaging in formal experimentation, the farmers themselves were keen to point out that wetland use has undergone a gradual process of refinement, and drainage and cultivation practices have evolved over time. The case below of one farmer at Dizi wetland typifies this process:

> *"Here half of the maize is yellow and half is healthy. Before two years there was no difference between them, but last year I lost the whole yield. This year I have a drain down the middle because I thought there was too much water in the soil, but the result has been half good and half bad. I dug the drain through the middle because another farmer told me that the source of water for that area of the wetland was under an avocado tree on the valley side. Next year I will drain along the left side where the valley sides meet the wetland and this should solve the problem. Maybe I will also block the middle ditch which I made this year. The ideal depth of these ditches should be about knee height".*
>
> Dizi farmer (April 26, 1998)

The extent of the evolution of management practices does, however, vary among wetlands. In the *kebele* (community) of Bake Chora, farmers identified a local religious teacher as the main source of knowledge and instigator of a more intensive form of wetland drainage and cultivation than traditionally practiced. Hence, wetland farming knowledge and practices were already well developed when first applied. Elsewhere, farmers report that it was upon instructions from landlords during the Haile Selassie era that they initially held meetings amongst themselves to discuss the requirements of drainage and cultivation. In effect, their first attempts at complete wetland drainage and cultivation simply represented a transfer of upland farming practices into a new environment. In both scenarios, however, there is little doubt that the knowledge and practices of wetland drainage and cultivation has evolved through the acquisition of new knowledge and experience.

7. INDIGENOUS KNOWLEDGE AND SUSTAINABLE WETLAND MANAGEMENT

By demonstrating extensive knowledge of wetland processes, adopting indicators of sustainability, and actively modifying knowledge and practices in response to change, farmers arguably possess important prerequisites for the sustainable management of these wetlands. Whilst it is spurious to base an assessment of sustainability solely on farmers' oral histories and observations, the research did ascertain that all study wetlands have undergone drainage and cultivation for at least 30 years, and, despite farmer reports of degradation during the first seasons of cultivation, crops have subsequently been produced annually. Any further degradation appears to have been mitigated by wetland management practices. Hence it can be suggested that, contrary to initial concerns of widespread wetland loss that precipitated the EWRP study, farmers' wetland management strategies have largely been sustainable.

One key problem highlighted by farmers, however, is that despite crop production continuing year after year, and the wetland hydrology continuing to support production, the level of crop production does not regularly meet the aspirations of farmers. Whilst most possess knowledge of practices and technologies that could sustain crop production at higher than current levels, they recognize there is a problem of operationalizing and applying their knowledge to fulfill their management aims. The application of wetland knowledge is, it seems, restricted by a range of spatially and temporally variable constraints such as unpredictable weather, geological barriers to drainage, the need to retain areas of *cheffe* in the wetlands for construction material, and individual socioeconomic status. The latter is perhaps the most influential factor, since it determines the extent to which wetlands are utilized (Solomon Mulugeta 2004). Wealthy farmers can afford to hire labor, and buy cattle and farming equipment, whilst poorer farmers struggle to manage drainage and cultivation in their wetland plots. Ironically, it is the lack of resources available to poorer farmers that may ultimately lead to the abandonment of wetland plots, the subsequent retention of water and *cheffe* in the wetlands, and overall conditions of environmental sustainability.

Wetland farmers, therefore, regard access to resources rather than lack of knowledge as the limiting factor in their wetland management system. This raises a further question, not unlike the one prompted by the original research: could wetland management remain sustainable if these resource constraints were removed and each farmer was able to apply his knowledge? In fact, these very scenarios have become increasingly common since the establishment of the government Wetlands Task Force in Illubabor in 1999 in response to drought and growing food security problems (UNDP 1999). In a policy echoing those of the Derg era, farmers have been instructed to bring all wetlands into cultivation year after year, and in some instances abandoned wetland plots have been redistributed to those either willing, or having the capacity to cultivate. Critically, the importance of preserving *cheffe* vegetation in wetlands has not been recognized by the Wetlands Task Force, and although most farmers are also keen to see an increase in crop production and the redistribution of abandoned land to those who can cultivate effectively, they are in conflict with the government over this issue.

8. CONCLUSION

In addressing initial concerns that wetland drainage and cultivation in Illubabor is unsustainable, leading to widespread degradation, the research outlined here draws attention to the sustainability of a locally developed system of wetland management. Those farmers involved in wetland management have, over a relatively short time, developed extensive and accurate knowledge of the wetland environment and its dynamics. Experience and the acquisition and evolution of knowledge have produced a repertoire of wetland management practices which, in several cases, has facilitated annual crop production in the wetlands for over 50 years, with little apparent environmental degradation. Critical to the success of these management practices has been farmers' understanding and adoption of plant, soil and hydrological indicators of sustainability, which periodically determine either the implementation of regenerative practices or the resumption of drainage and cultivation.

From the farmers' point of view, the major limitation to this wetland management system is the lack of capacity to apply their knowledge and practices. Whilst food production from wetlands could increase if farmers had access to more resources, there is some doubt whether environmental sustainability could be maintained, since it is the abandonment and preservation of natural vegetation that keeps degradation in check. Whether farmers would continue to incorporate such restorative practices under scenarios of intensification, remains to be seen. Given more recent pressure to intensify wetland use emanating from the Wetlands Task Force, this would appear unlikely, unless the government recognizes the value of indigenous practices, or if intensification in wetland use stimulates the indigenous development of new regulatory mechanisms that ensure sustainability.

A NEW ANCIENT WATER MILL: REMEMBERING FORMER TECHNIQUES

Michel Dubas
Haute Ecole Valaisanne
Sion, Switzerland
michel.dubas@hevs.ch

Abstract: Waterwheels are one of the most ancient and most common machines. They appeared at the end of the second century B.C. and remained the most important source of mechanical energy beside that of humans and animals till the Industrial Revolution, driving mills, saws, pumps, bellows or hammers. Through a description of their design found in old texts, it has been possible to trace the conception of technology, beginning with the craft industry and ending with mathematical physics. In order to revive this extremely important technique, a water mill was constructed near Sion, using historical methods and workmanship. It includes a wooden waterwheel as well as all gears necessary to allow two millstones to rotate at a specified speed. Moreover, measurements were made on a scale model of the waterwheel with its ancient geometry, which has shown an efficiency of nearly 80 percent.

Keywords: waterwheels, energy, ancient techniques

1. INTRODUCTION

During the twentieth century, many waterwheels ceased their operation and were replaced by more powerful machines. However, for more than 2000 years they were employed as one of the most important sources of energy throughout Europe and Asia where water was abundant enough to power them. The development of water-related technologies was especially important in the difficult terrain of mountainous regions. Not only did people's requirements for potable and irrigation water need to be met, their lives, houses and cultures had to be protected from the damage water could cause when it was overabundant. Further water also had to be channeled and harnessed to power the mills for grinding cereals to provide food.

To understand the technology that developed to meet these complex and at times contradictory needs, we shall first look briefly at the history of waterwheels and see how they were treated in ancient miller's manuals.

E. Wiegandt (ed.), Mountains: Sources of Water, Sources of Knowledge, 211–220.

We then describe the mechanical parts of the new water mill in Nendaz near Sion, which was constructed using the historical methods and workmanship available in the past, and we finally present some results of efficiency measurements made on a model wheel.

2. A BRIEF HISTORY OF THE WATERWHEELS

According to current knowledge (see, e.g., Brentjes 1978; Capocaccia 1973; Daumas 1962; Singer 1954), waterwheels appeared independently in China and in the Mediterranean area at the end of the second century B.C. It seems that the horizontal wheel, also called Greek wheel because the first evidence of it was found in Athens, was a contemporary of the vertical or Roman wheel, of which examples could be found throughout the entire Roman Empire (Figure 1). Although known, these machines were, however, not in common use, either in Europe or in China until much later. This may have been due to the availability of large amounts of manpower, in the form of either handworkers or slaves, and to the will of the ruling classes to preserve employment and livelihood for their population despite the economic advantages that a more powerful technique might have brought. The contempt of technique and arts (both designated by the same word τεχνη in Greek) held by Greek and Roman philosophers and politicians could also have inhibited the broader development of waterwheels, although the craft industry was highly considered in these civilizations. However, the most probable cause for the failure to develop the use of these machines was the simple lack of the technical knowledge and skills amongst the population of the period to build, operate, and maintain them.

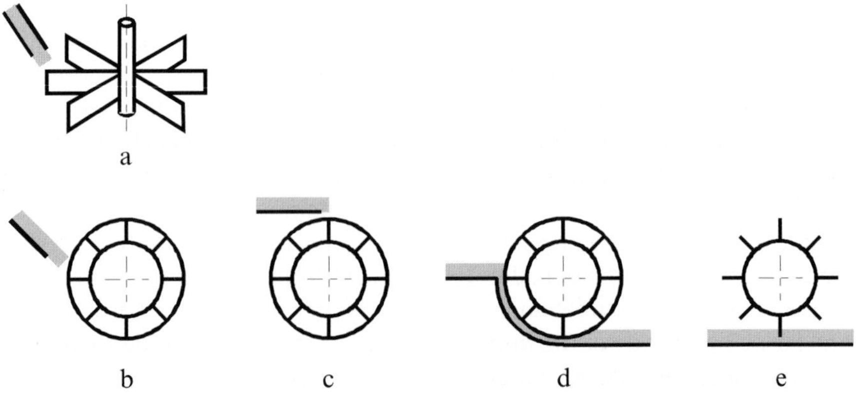

Figure 1: Main Types of Waterwheels. Top: a: Horizontal Wheel. Bottom: (Vertical Wheels). b: Impact Wheel; c: Overshot Wheel; d: Breast Wheel; e: Undershot Wheel.

A first industrial revolution occurred in Europe during the eleventh and twelfth centuries through the impetus of the Cistercian monks. In the context of a political environment in which wars were relatively infrequent, there was a significant expansion of the population, development of commercial trade, the creation of new markets, and an extension of agriculture requiring the clearing of large wooded areas to make arable land. At the same time, new technical advances were introduced such as the collar harness or shoes for horses, the rudder for ships, and the wind mill with a horizontal axis. The use of hydraulic wheels spread considerably, greatly increasing the amount of energy at the disposal of the population. As a consequence, the quantities of flour, oil and other goods produced by these machines increased steadily, leading to the development of the first real industries.

However, the Black Death that occurred in Europe during the fourteenth century reduced the population by almost one third, bringing with it a fundamental change in both society and its supporting structures. With fewer people there were fewer demands on the use of the soil for basic agricultural food crops, and more luxurious cultures began to develop. In the Alps, cow herding gradually replaced much of the rye, wheat, barley and oats fields, leading to new needs for water, now required to irrigate the extensive meadows. In Valais, water canals (called *bisses* in French and *Suonen* in German) were built in many places to meet this demand (see Giovanola 1999; Reynard 2002). These water courses, just like the water mills, were installed and run as common goods by whole villages, a practice which itself considerably influenced the development of social customs and policies in these communities. In mountainous regions, where the supply of water for drinking, irrigation, and for turning mills was scarce, the collective management of this precious resource required an organization with a high degree of order, solidarity and interdependence, where a joint sense of shared responsibility could lead to a coordination of labor (on this subject, see the contribution by E. Wiegandt to this book).

Techniques evolved slowly bu interactions between merchants, craftsmen, or soldiers were able to efficiently spread advances in knowledge over nearly the entire world (see the article of D. Crook in this volume and Pelet 1988, 1991, 1998), including the less accessible valleys in the mountains even though the transportation and communication systems were not as quick as those of today. As a result, each village became the owner of one water mill with a power of roughly 1 kW, which can be compared to the power represented by the equipment of a modern kitchen or of an automobile motor. This situation remained largely unchanged until the Industrial Revolution of the nineteenth century, when steam engines, or later combustion engines, as well as hydraulic turbines became new power sources. In remote regions such as in the Alps, however, some water mills continued to run into the 1980s; P.-L. Pelet reports for example (Pelet 1988) that in 1956 the Swiss Federal Wheat and Grain Administration granted a credit to renovate a water mill in Taesch near Zermatt with the aim of guaranteeing the flour supply in case of an international crisis.

Thus, for about twenty centuries waterwheels were essential as the most common prime movers for all sorts of machines for practically the entire northern hemisphere: they primarily drove mills, but also pumps, saws, hammers, bellows, etc. Their basic design remained largely unchanged although undoubtedly many parts were improved or also introduced, such as the crankshaft that appeared during the late Middle Ages, or the rounding of vanes that became possible when iron replaced wood during the Industrial Revolution.

3. WHAT OLD TREATISES SAY ABOUT WATER MILLS

As very common machines present in every day life, waterwheels appeared in the arts: poems, songs, or paintings. They were depicted in so-called "Theatres of machines", although with more of an artistic rather than a technical aim. They were also described in numerous miller's manuals and, later, in books about mechanical engineering. Thanks to these works, the development of techniques through time can be followed quite accurately. Building a mill, and especially a waterwheel, was long considered a craft rather than a technology as we understand it today. The methods used were essentially traditional, based on experience and not on theoretical knowledge, with the rules for their design and construction passed along orally and kept secret among the persons belonging to the concerned guilds.

Therefore, in our eyes, the descriptions of waterwheels appearing in historical texts suffered both in their quality as well as their completeness. As Sturm puts it in the preamble to his work (Sturm 1815) published at the beginning of the eighteenth century (translation from the German), "the views in such books are mostly perspective and not drawn according to geometric rules, rather sketched only by eyesight and freehand, so that no dimension or subdivision can be gained from them. The second fault is that few feasible schemes are presented, and many have been put together with bits and pieces, and are not suitable to practical use. Thirdly, when good inventions are present, the authors have intentionally distorted them, as if they had intended to show the connoisseurs that they knew the secrets but have concealed them with application so that those who ignore them cannot imitate them. Fourth, they have given no distinct explanation, nor knack, nor reason, nor calculation, which yet matters." Illustrations were a problem indeed, being difficult to draw and difficult to print. It is precisely for this reason that Sturm's work, with its fifty engraved plates, enjoyed such a great success, with no fewer than six editions appearing between 1718 and 1819. He did his best to show in each case a horizontal view and an elevation, and to provide comments on his drawings (Figure 2). Nevertheless, several of the constructions that he presents are inventions of his own and not proven achievements which can be recommended for practical purposes.

The text part of the Encyclopedia by Diderot and d'Alembert, published between 1751 and 1775, tends to be more precise, containing figures under the article entitled "Mills" describing the dimensions of the millstones, the number of teeth of the gears and the rotational velocities, but the wheel itself does not seem to be worthy of a detailed description. In the illustration part (Diderot 1762), several undershot waterwheels are shown, in all cases as part of a complete factory: corn-mill, sawmill, tannery, smithy, etc. (Figure 2).

In the middle of the eighteenth century, a decisive step forward was taken after mathematics, especially differential and integral calculus had made important advances. As soon as the theoretical works of Johann and Daniel Bernoulli and of Leonhard Euler on hydraulics and hydraulic machines appeared, the design of waterwheels began to be influenced by science, opening the way to the introduction of hydraulic turbines that a number of factories started to manufacture in about 1830. As soon as the physical laws and mathematical tools became better understood, they were used to describe and predict how machines work, as can be seen for example in the treatise of Bélidor published from 1737 to 1753 (Bélidor 1737). More and more complete technical descriptions were available in what can be considered as real textbooks on the mechanical design of mills, such as in the work of Benoît (Benoît 1836) in 1836 which is already devoted to industrial machinery. Until about 1950, the subject of waterwheels also played an important role in lectures on mechanical engineering.

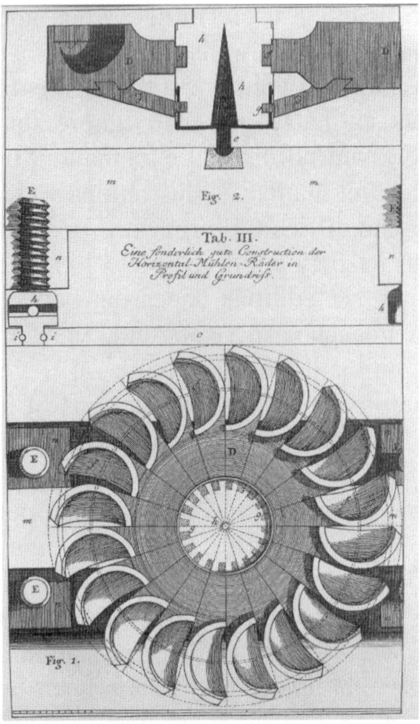

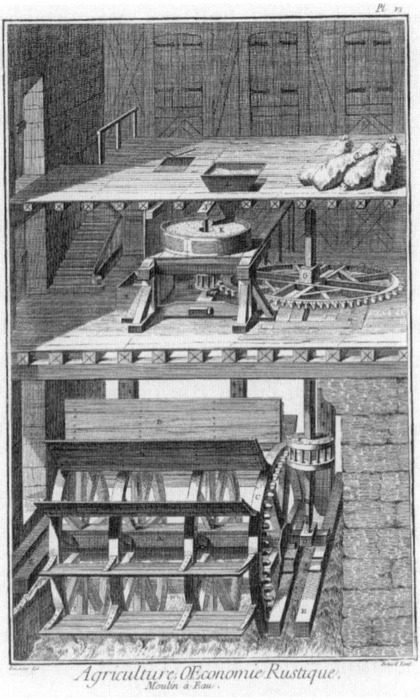

Figure 2: Waterwheels in Sturm's Manual (Sturm 1815, left) and in the Encyclopedia (Diderot 1762, right).

4. THE NEW MILL OF LE TSÂBLO AT NENDAZ

As we already said, a number of water mills continued to work as recently as 1950, particularly in mountainous regions where industrialization developed late. This explains why Valais, for example, still has a great number of these machines, some abandoned and in poor condition, others renovated and running on certain days for demonstration purposes. In the village of Nendaz above Sion, the last mill was demolished in 1984 to make room for a road; only some of the millstones and parts of wooden gear wheels were preserved. In the mid 1990s, the decision was made to build a new mill on the same stream as the old one but in another location. In order for the current generation of locals and visitors to see what the past looked like, the new construction adopted the ancient techniques formerly in use in the region, Although carpenters are nowadays still capable of copying or repairing existing ancient wheels or gears, the knowledge of determining their dimensions and their geometry has been lost. The Haute Ecole Valaisanne was therefore asked to design an overshot waterwheel as well as the gearing necessary to drive two pairs of millstones. Because the information given in the old treatises just described is so imprecise, additional data had to be collected from still existing installations.

Figure 3: Overshot Waterwheel of Le Tsâblo (at that time still without water channel).

As is typical for a mountainous area, while steep slopes are available, the quantity of water is small. Under these circumstances, the outer diameter of the waterwheel was fixed at 3 m (Figure 3), to avoid a construction that would be too large, and, in addition, water is diverted from a canal to achieve sufficient discharge to let the mill run during demonstrations. On the basis of a rough estimate of the friction between two stones and by comparisons with other mills, the power required for driving two mills was calculated to be about 1 kW (which is about 1/4 of what Benoît indicated in 1836 for industrial machines). Assuming a conservative value of 70 percent for the efficiency, this led to a flow rate of 44 l/s. Following Müller's advice (Müller 1939), the circumferential velocity of the waterwheel was chosen as 1.3 m/s, which implied a rotational speed of 8.3 rpm. The number and the geometry of the buckets were also designed according to the recommendations of this author, which are precise on this point. As for the millstones, their rotational speed was fixed at 60 rpm, coinciding with what was usual in the region and also to the value given in Diderot's Encyclopedia (Benoît's millstones rotate about 1.5 times quicker).

Figure 4: Gear Transmission of Le Tsâblo. Left: Gear on the First Shaft, which carries the Waterwheel. Right: Second Horizontal Shaft with One of the Two Wheels with Lateral Teeth.

To connect the waterwheel and the millstones, a gear ratio of about 7 was necessary. In order to use gear wheels of the same dimensions as the preserved ones, two stages were planned: a first horizontal shaft carries the waterwheel as well as a big gear wheel, and on a second horizontal shaft are mounted a pinion and two wheels with lateral teeth which drive the lantern pinions of the two vertical shafts carrying the rotating millstones (Figure 4). These lantern pinions have only one disk with inserted cylindrical teeth and they can glide on their shafts, thus forming clutches that allow one millstone or the other to be stopped. All parts, except the shafts carrying the rotating stones, are made of wood (larch for the wheel; beech for the teeth) and are calculated to be able to bear the maximal possible loads that can occur when the mechanism is jammed and the wheel is full with water.

5. MODEL TESTING

In order to confirm the values of the efficiency of overshot waterwheels such as those described in Benoît or Müller for instance, tests were carried out on a model at a scale of 1:4.16, as is commonly done for hydraulic turbines. A test rig was built that allowed the measurement, on the one hand, of the hydraulic power at the disposal of the waterwheel, i.e., the flow rate and the net head, and, on the other hand, of the delivered mechanical power, i.e., the torque and the rotational speed. During these measurements, due attention was paid to the similarity laws, using dimensionless coefficients defined in a similar way as for Pelton turbines (more details can be found in Dubas 2005).

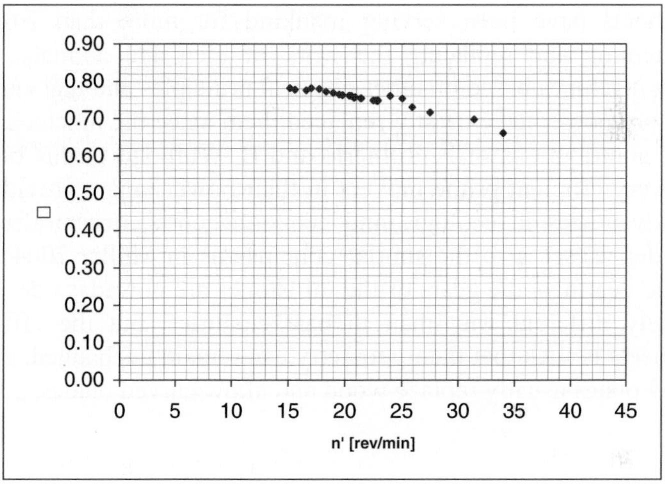

Figure 5. Efficiency of the Model Waterwheel in Function of the Rotational Speed. Channel Position and Flow Rate (1.2 l/s) Are Kept Constant.

These tests delivered the whole hillchart diagram of the efficiency. The top of the hill has its summit at 79 percent and is very flat, which means that the efficiency diminishes only slightly when parameters such as the flow, the rotational speed or the slope of the water channel are varied (see the section of the hillchart shown in Figure 5 with dimensional quantities). Moreover, when the wheel rotates more slowly, the buckets fill more, so that the torque increases. Conversely, when the wheel accelerates, less water enters the buckets and the torque decreases. This type of motor is thus particularly well suited to driving machines with varying loads, for instance saws. It was noted that energy was lost if the buckets contain too much water, in that water is poured off from the buckets before the lowest point of the wheel. This, along with friction problems leading to a possible locking of the wheel, imposes a lower limit on the rotational speed. Conversely, a too high wheel velocity is not recommended because water is ejected by centrifugal force. As this is the case for Pelton turbines, it can be assumed that the efficiency and the behavior of the model is very precisely the same as that of the prototype.

6. CONCLUSION

Waterwheels have been serving mankind for more than 2,000 years, supplementing and replacing the work of men and animals. Although different machines based on other physical principles and that can generate much more power have mostly replaced them since the nineteenth century (see the contributions of F. Romério and E. Wuilloud to this book), they remain very efficient prime movers in their power range, providing significant advantages in ecologic and economic costs, decentralization and independence (see also the similar conclusions in Müller 2004). Furthermore, the bearings, the gears or the millstones are nowadays designed in a completely different way than in past centuries; but the efficiency of waterwheels is such that their geometry can remain unchanged, except that thin steel plates usually replace wood and allow curved blades.

WATER-RELATED NATURAL DISASTERS: STRATEGIES TO DEAL WITH DEBRIS FLOWS: THE CASE OF TSCHENGLS, ITALY*

Walter Gostner
Patscheider & Partner Engineers, Ltd.
Mals, Italy
W.Gostner@ipp.bz.it

Gian Reto Bezzola
Federal Office for the Environment FOEN
Hazard Prevention Division
Bern, Switzerland

Markus Schatzmann
Basler & Hoffmann Consulting Engineers AG
Zürich, Switzerland

Hans-Erwin Minor
Laboratory of Hydraulics Hydrology and Glaciology
Swiss Federal Institute of Technology
Zürich, Switzerland

Abstract: The chapter reports on a case study of how people and public administrations dealt in the past and currently deal with the danger of debris flow. After a brief description of main debris flow features, the time series of debris flow events and the history of training works at the Tschengls torrent are reported. Finally, a modern approach based on a theoretical background of debris flow research is described. The integral analysis allows us to assess the debris flow activity and intensity in satisfactory detail and therefore allows us to derive recommendations for structural and nonstructural measures.

Keywords: debris flow, mitigation of debris flow hazards, time series analysis

* The study was financed by the Public Department for Hydraulic Structures of the Province of South Tyrol, Italy. A special thanks for the support goes to the director, Rudolf Pollinger.

E. Wiegandt (ed.), Mountains: Sources of Water, Sources of Knowledge, 221–241.

1. INTRODUCTION

Water resources play a key role for humanity. Without water, no life is possible; however, at the same time water, can represent a major threat for mankind. Floods cause huge damages and loss of human life.

Through the centuries, people in mountainous regions have had to face different features of floods. Clear water floods do not occur only in the wide plains closed to the outflows of the rivers into the ocean, but also in the main mountain valleys where large rivers flow. Flooding becomes hazardous when great depths of water or strong currents occur in the flooded region. Manfreda, this volume, describes examples of practical research and explains how the scientific community and public administrations deal with the problem.

In addition to clear water floods, mountainous regions, where slopes are steep and loose material potential are pronounced, are affected also by the phenomenon of debris flows. Compared to ordinary floods, debris flows may mobilize and transport significantly larger volumes of solids, thus causing severe damage mainly in the deposition areas, i.e., on the fans of the torrents.

A close observation of debris flow events seems to indicate that there are three regions where debris flows are more prevalent: (1) semi-arid regions, (2) alpine regions and (3) volcanic regions (Mainali and Rajaratnam 1991). Thus, the province of Tyrol, situated in the heart of the Alps, is subjected to this phenomenon.

In earlier periods, because large rivers were not yet managed and floodplains were often marshy and inhospitable, our ancestors tended to settle in elevated locations on the alluvial fans. By constructing buildings and infrastructure on these fans, a large potential for damage was created over the course of time.

The village of Tschengls, situated in the province of South Tyrol in Italy, represents a typical example of the struggle between people and nature. The village is situated on the fan of the Tschengls torrent and has been affected several times by debris flow events. Since then, the inhabitants of the village have faced the torrent with great respect. Downstream of the village, for instance, there is an almost untouched alluvial forest which has been reserved for debris flow deposition. The existence of this forest in the middle of cultivated land is a testament to wise land use planning.

This chapter first describes the general characteristics of debris flows. After presenting the watershed of the Tschengls torrent, the time series of debris flow events is illustrated. Then, several strategies that people have developed to deal with debris flows in the Tschengls torrent are described.

Finally, an integrated approach, based on the essential processes of debris flow formation, propagation and deposition, is presented. The example shows how increased understanding of the processes of debris flow contributes to optimize mitigation measures.

2. CHARACTERISTICS OF DEBRIS FLOW

2.1. General Aspects

A typical debris flow occurs in one or several pulses. The flow consists of a small percentage of water mixed with a much larger percentage of solids (clay, silt, sand, gravel, boulders, and wood). Typically three distinct zones (Figure 1) may be distinguished along the path of a debris flow (1) The initiation zone where the debris flow starts, (2) the transition zone where erosion and/or deposition may occur, but often balance each other, (3) the deposition zone where the debris flow stops and the material is deposited.

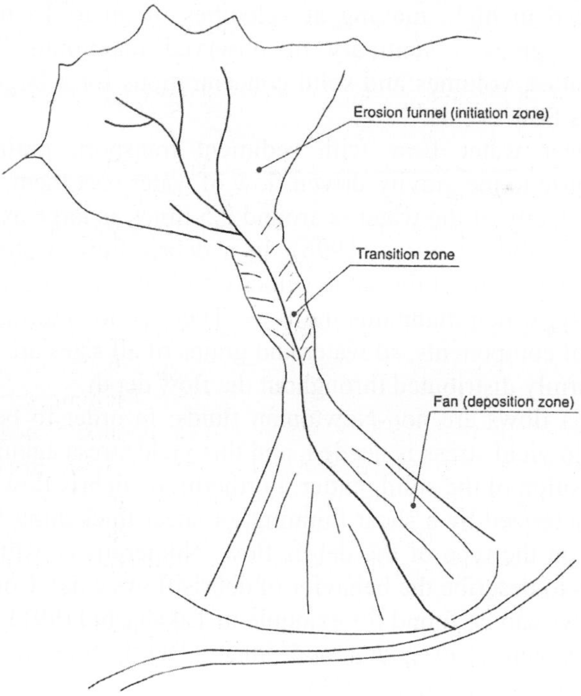

Figure 1: Definition of Erosion Zone, Transition Zone, and Deposition Zone (Bezzola 2000).

Debris flows can be distinguished from clear water flows by the following characteristics:

1. Debris flows generally mobilize much more solid material than clear water floods with bed load transport. Due to its erosion capacity, a single debris flow event can bring about larger changes in the landscape than can longer periods of continuous geomorphologic activity, particularly in the erosion and deposition zones.
2. The solid concentration of debris flows is impressively high. According to O'Brien (2001) the solid concentration of a debris flow surge can reach 40 – 50 percent, whereas Coussot (1996) reports solid concentrations larger than 80 percent. In clear water floods, solid concentrations usually do not exceed 10 percent.
3. The maximum discharge from debris flows far exceeds the maximum discharge of clear water floods in watersheds of the same size. Therefore, maximum flow depths and velocities are also generally higher. Furthermore, the density of debris flows can reach values of up to 2.200 kg/m^3. The high velocity and high density of debris flows are thus responsible for the enormous destructive potential of such flows. Debris flow waves have been reported to be up to 6 m high, moving at velocities of up to 15 m/s. Coussot (1996) gives a summary of observed maximum flow depths, velocities, volumes and solid concentrations for a large number of debris flow events.
4. In clear water flow with sediment transport, grains move in response to the gravity-driven flow of water past them. In this case, the velocity of the water is around 2.5 times as large as the velocity of the solids (Coussot 1996). In a debris flow, water and solids move with almost the same velocity and all components are subject gravity, which maintains the flow. There is no significant segregation of components, so water and grains of all sizes are more or less uniformly distributed throughout the flow depth.
5. Debris flows are non-Newtonian fluids: in order to be initiated, a certain yield stress is needed, and this yield stress again leads to the deposition of the solid matter. Furthermore, debris flows usually are characterized by a shear thinning, or shear thickening flow, depending on the type of the debris flow. Numerous constitutive approaches to describe the behavior of debris flows exist. Comprehensive reviews can be found for example in Takahashi (1991), Mainali and Rajaratnam (1991), Jan and Shen (1993), Iverson (1997), and Iverson and Vallance (2001).

2.2. Flow Profile of Debris Flows

Within a typical debris flow surge, three parts can be defined (Figure 2). The first part is the front where usually the largest flow depths can be observed. Occasionally the front is almost dry or only partially saturated, since at the front water often infiltrates in the underground (Tognacca 1999) and large boulders accumulate at the front.

A clear distinction between the front and the debris flow body, which follows the front, often is not possible, as there is a continuous exchange of material between the front part and the body of a debris flow.

In the debris flow body, the concentration of solids usually diminishes. However, concentration and grain size distribution may vary according to the specific local conditions.

The third part is the more fluid tail, where discharge and sediment concentration usually are considerably lower. Larger grains concentrate in the vicinity of the bed and the debris flow changes into a hyperconcentrated flow.

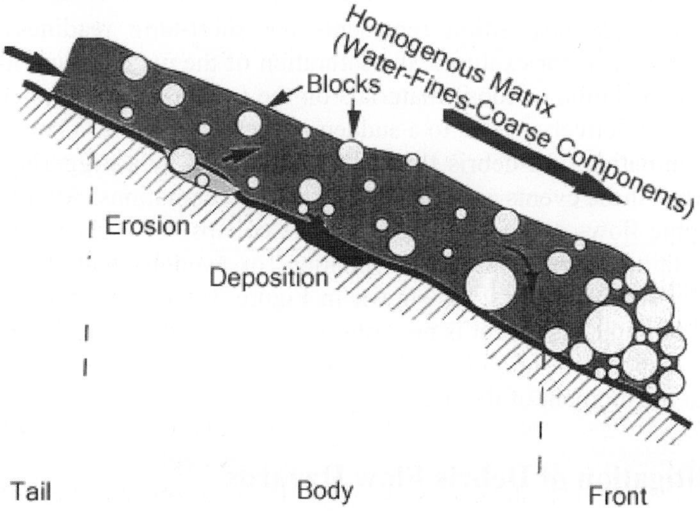

Figure 2: Conceptual Scheme of the Flow Profile of a Debris Flow Pulse (Coussot 1996).

2.3. Conditions for the Occurrence of Debris Flow

For the occurrence of debris flows in a watershed, two conditions must exist: debris sources and a minimum torrent slope. According to Kienholz (1995), the presence of these conditions represents the basic disposition (long-term tendency) of a torrent to show debris flow activity.

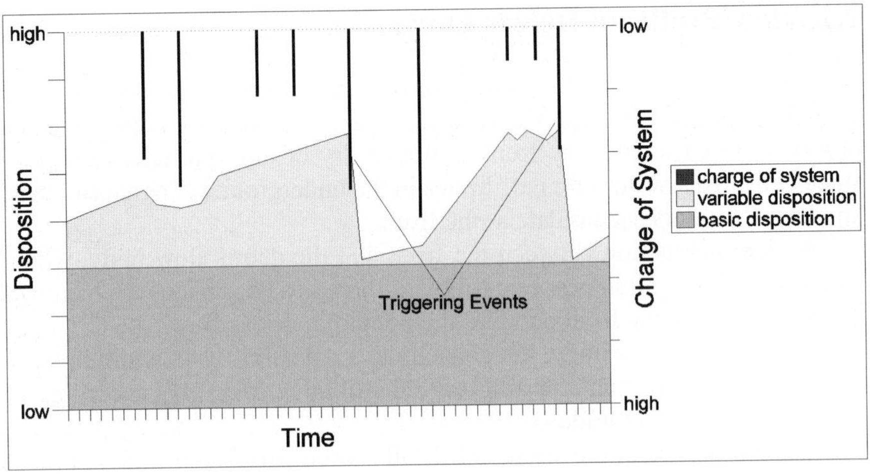

Figure 3: Link between Basic Disposition, Variable Disposition and Triggering Events (Kienholz 1995).

The variable disposition represents the short-term readiness for a debris flow event, for example, the saturation of the underground material or the accumulation of loose materials on the torrent bed (due to continuous sediment delivery or due to a sudden event, for example, a landslide).

The initiation of a debris flow finally is caused by a triggering event. In the Alps, these events are primarily intense precipitations. Alternatively, catastrophic flows can be originated by the collapse of a glacial moraine, by dam failures, by strong snow melting, or even by soil liquefaction caused by seismic events. As shown in Figure 3, the triggering intensity required to initiate an event is not constant in time and is directly related to the total disposition, i.e., the sum of basic and variable disposition, in the torrent at the moment of the triggering event.

2.4. Mitigation of Debris Flow Hazards

When dealing with mitigation of debris flow hazards, generally three types of measures can be distinguished: (1) structural measures, (2) nonstructural measures, (3) emergency planning.

Structural measures can be undertaken in the initiation zone, in the transition zone and in the deposition zone of the torrent. The most important structural measures are listed in Figure 4.

Measure	Purpose
Initiation zone	
Reforestation / controlled harvest	Reduce debris production due to logging or natural loss of forest cover
Forest road construction control	Eliminate / prevent unstable cuts and fills that could represent debris sources or initiation areas
Stabilisation of debris sources (channel linings / check dams)	Stabilise channel bed and side slopes in potential source areas
Transportation zone	
Training by chutes, channels, deflecting walls, dykes	Ensure the passage of surges down a predetermined path without blockage or overflowing
Channel diversion	Diversion of debris flow away from a potentially endangered area
"Sacrificial" bridges, mobile bridges, fords	Prevent channel blockage due to obstruction by bridges with inadequate clearance
Bypass tunnel beneath torrent bed / protection gallery	Protect transportation route without/with modification of the torrent channel
Deposition zone	
Open debris deposition basins, dykes or walls	Control the extent of a natural deposition area
Closed / permeable retention barriers and basins	Create a controlled deposition space
Debris flow breaker screen	Stop smaller debris flows at a given point
Structures designed for debris flow impact and for burial	Prevent damage to structures located in potential deposition zones
Debris sheds (galleries) or cut-and-cover tunnels	Place transportation route beneath potential deposition zone

Figure 4: List of the Most Important Structural Measures against Debris Flows (Bezzola 2000).

Nonstructural measures for the most part comprise hazard and risk zone assessment and consequent spatial planning. By identifying and classifying the concerned areas, suitable land use can be accordingly planned in order to manage the further development of these areas with respect of the existing natural hazards.

Emergency planning includes the installation of warning devices, the implementation of evacuation plans and the organization of measures such as the temporary closure of openings with stop-logs or sandbags. It should be noted that emergency planning must be done properly in advance —a difficult task as warning time generally is very short. In the case of the Tschengls torrent, for example, the span between the initiation of a debris flow and its arrival in the village is around 30 minutes.

3. THE CASE STUDY OF TSCHENGLS

3.1. The Tschengls Torrent Watershed

The Tschengls torrent has a catchment area of 10.6 km^2 and is a tributary of the Etsch, the river with the second largest catchment area of Italy. The highest point of the watershed is at an altitude of 3,375 m a.s.l., the lowest point is at 875 m a.s.l. The main reach of the torrent has a length of 7.1 km with a mean slope of 34 percent.

Geologically, the Tschengls watershed is located in the schist zone of the Vinschgau valley with presence mainly of mica schist and gneiss. In the upper area of the watershed, amphibolites can be found.

The Tschengls torrent is a typical alpine mountain torrent. The main debris sources are in the upper basin area and have a total surface area of around 0.75 km^2. Several secondary creeks run through the debris sources and, after having passed a rocky portion with cascades, they join and form the main reach of the Tschengls torrent. Both the steep slope and the large volume of available debris in the upper basin area contribute to a high likelihood of the occurrence of debris flows in the Tschengls torrent.

The upper basin area of the Tschengls torrent is surrounded by very steep rocky walls that provoke thunderstorm cells to burst, resulting in intensive precipitation, which is an important triggering condition for the initiation of debris flows (Figure 5). Therefore the upper basin area is characterized by a high debris flow activity.

Downstream from the cascades, the transition zone begins. Observations of past events show that the material balance in this reach is almost neutral: i.e., the amount of erosion and deposition during a single debris flow event, as well as over longer periods of time, balance each other, as the reach has been trained with a series of check dams. In the lower part of the transition zone, there are two debris retention basins. Deposition of solid material in these two basins reduces the volume of debris flows. The basins are regularly cleared.

At the end of the transition zone, the Tschengls torrent enters the main valley of the Etsch where the village of Tschengls is located and where deposition processes occur. Since the last ice age, the Tschengls torrent has formed an impressive debris fan with a volume of around $11 \cdot 10^6$ m^3.

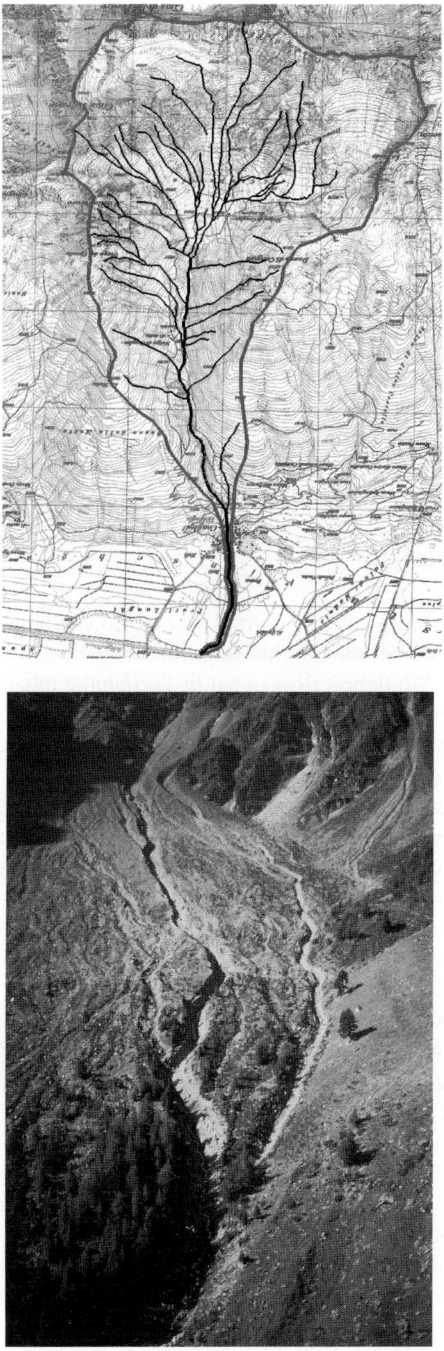

Figure 5: The Watershed and the Upper Basin Area of the Tschengls Torrent.

3.2. Time Series of Debris Flows in the Tschengls Torrent

The debris fan is a clear and evident index for the debris flow activity of the Tschengls torrent. Physical evidence is corroborated by written sources recording debris flows that date back to the eighteenth century.

Table 1 reports the time series of recorded debris flow events. The following sources have been used: (1) records of the village chronologist (Raffeiner 1990), (2) documentation of catastrophic natural events in Tyrol

Table 1: History of Debris Flows in the Tschengls Torrent.

Year	Description of event	Source
1719	Flood with debris flow in Tschengls	1,2
1768	Flood with debris flow in Tschengls	1,2
1784	Flood with debris flow in Tschengls	1,2
1850	June 18th floods in whole valley, debris flow event in Tschengls	2
1865	April 10th debris flow event in Tschengls and neighboring villages	2
1868	July 23rd after 5 days of rainfall, debris flow event in Tschengls and in the neighboring torrents	2
1887	September 8th debris flow event in Tschengls; inhabitants blame authorities for slow execution of structural measures at the torrent. The longitudinal protection walls in the village are destroyed over a length of 80 m	1,2
1889	Debris flow with partial destruction of the protection works	1
1902	Debris flow with partial destruction of the protection works	1
1911	Debris flow with large damage in meadows downstream of village. Structural protection measures withstand the event	1
1929	Debris flow with destruction of the two bridges across the river in the village	1
1931	Debris flow event in Tschengls	1
1933	Debris flow with new destruction of bridges	1
1948	Debris flow again with destruction of bridges	1
1956	On August 21st flood (supposed debris flow event) with damage of the protection walls in the village	3
1971	August 28th debris flow event with obstruction of the first bridge in the village. Subsequently the debris flow leaves the torrent bed and floods village on both sides of the torrent	1,3,4
1999	August 16th debris flow event. Maximum discharge almost reaches the structure of the bridges. In the alluvial forest downstream from the village, the debris flow deposits over an area with a surface of 13 ha.	4
1999	September 20th a smaller debris flow event occurs, exactly one day after the debris retention basins had been cleaned	4

(Fliri 1998), (3) technical reports of the Department for Hydraulic Structures of South Tyrol, (4) eyewitnesses.

Assuming that the time series is complete, in the past 280 years 18 debris flows reaching the village of Tschengls occurred, 15 of them during the last 150 years. This activity corresponds to an average occurrence of debris flow events reaching the village of Tschengls every 10–15 years.

However, as shown in Table 1 and Figure 6, debris flow activity on the Tschengls torrent is rather erratic. Frequently after big debris flow events, a series of smaller events occurs (for example after the events in the years 1865, 1887, 1929, and 1999). This observation leads to the hypothesis that the zone where debris flows develop is destabilized by the first event, encouraging the development of subsequent smaller debris flows.

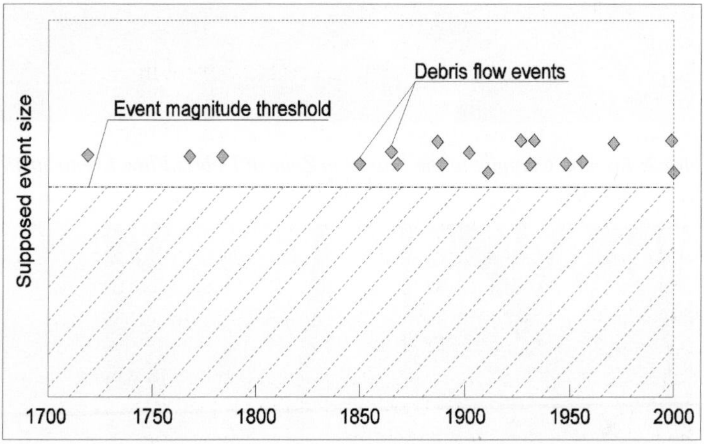

Figure 6: Time Series of Debris Flow Events at the Tschengls Torrent.[1]

Figures 7 and 8 show the initiation zone and deposition zone respectively of the debris flow events in 1999, whose main characteristics could be reconstructed quite well from field evidence, a video recording and eyewitnesses accounts.

The total volume of the two debris flows was estimated by comparing the volume of channel erosion with the volume of the deposits. Both debris flows together mobilized a volume of 70,000 to 100,000 m^3. Based on their duration, velocity, and depth, as observed by eyewitnesses, the volume of the first event was estimated to be twice the volume of the second event.

[1] The graph shows the debris flow that exceeded the threshold of event magnitude which is necessary to reach the village of Tschengls (according to Ouarda et al. 2002). Beneath the threshold other events that did not reach the village might have occurred.

Figure 7: Erosion Channel in the Initiation Zone of Debris Flow Events in 1999.

Figure 8: Deposition Zone of Debris Flow Events in 1999.

The maximum discharge was observed during the first event and was seen to decrease along the flow path. An estimation of the maximum discharge was possible as the traces of the event, corresponding to the maximum flow depths were still visible on the torrent embankments. In the upper basin area, the discharge was around 160 m^3/s, in the transition zone around 100 m^3/s and in the urbanized area about 70 m^3/s.

3.3. History of Hazard Management in the Tschengls Torrent

After a debris flow event, public administrations and affected populations generally intensify their efforts to undertake protection measures. This being said, planning and execution of structural measures are closely related to public funds, which are not always available. The history of hazard management in the Tschengls torrent must be seen in this light.

Written sources (Raffeiner 1990) reveal that in 1747 a municipality law required the inhabitants of Tschengls to come to the channel during floods or debris flows in order to prevent an outburst. If a citizen protected only his own property, he could expect punishment.

After the debris flow event of 1768, the population of the village envisaged construction of protective structures. As the village could not raise the money for it, the works were not undertaken.

The systematic training of the Tschengls torrent began in the nineteenth century. Longitudinal protection walls with a total length of 380 m on both sides of the torrent bed were constructed through the village. The year of completion has been found engraved on the left wall.

Figure 9: Year of Construction, Engraved on the Protection Wall (Raffeiner 1990).

Table 2: History of Training Works at the Tschengls Torrent.

Year	Description of work	Position of work	Source
1870	Longitudinal protection walls in village (total length 380 m, maximum height 3 m)	Lower reach	1
1882–1883	Extension of protection walls at the downstream side	Lower reach	1
1882–1883	Construction of 130 small check dams (main channel and secondary creeks)	Middle and Upper reach	1
1904	Construction of a debris retention dam ("Kohlstattl")	Middle reach	2, 3
1908–1909	Protection works	Lower and Middle reach	1
1934–1935	Construction of 4 check dams	Lower reach	1, 3
1935	Paving of discharge section in the village	Lower reach	1
1938–1939	Construction of 6 check dams	Middle reach	3
1951	Construction of 3 check dams	Middle reach	3
1957	Rehabilitation of 2 check dams and protection walls	Lower reach	3
1966	Rehabilitation of 3 check dams	Middle reach	3
1972	Construction of a check dam series (9 dams) and rehabilitation of 3 check dams	Middle reach	3
1973	Construction of 4 check dams in main channel and 5 metal gabions in lateral creeks	Middle reach	3
1973	Construction of a debris retention dam (maximum height 14.4 m)	Lower reach	3
1974	Construction of 1 check dam and of a paved trapezoidal profile in the village (length 260 m), reconstruction of 2 bridges	Lower reach	3
1975	Continuation of the paved trapezoidal profile at the downstream end (length 250 m)	Lower reach	3
1982	Rehabilitation of check dam and heightening of debris retention dam "Kohlstattl"	Middle reach	3
1984	Construction of 1 check dam	Middle reach	3
1990	Construction of protection wall (length 30 m)	Lower reach	3

The first intensive period of structural measures to contain the Tschengls torrent coincided with a rash of other similar efforts undertaken in South Tyrol between 1883 and 1893. After some works had been carried out in 1908 and 1909, activities stopped during World War I. They were reinitiated following the events of 1929, 1931, and 1933. During World War II, work again ceased, to be reinitiated only after 1950. The period between 1972 and 1975 was characterized by intense training works following the catastrophic event of 1971. After the completion of these works, public opinion was satisfied that the protection objectives had been fulfilled and the residual risk lowered to an acceptable level. Nevertheless, the debris flow events of 1999 raised the question of whether the protection strategy should be revised and additional measures should be planned.

Table 2 presents the most important training works at the Tschengls torrent. The sources are: (1) Raffeiner (1990), (2) Stacul (1979) and (3) technical reports of the Public Department for Hydraulic Structures of South Tyrol. The Figures 10–12 show the most important structural measures at the Tschengls torrent.

Figure 10: Check Dam Series in the Transition Zone of the Torrent (construction: 1972–1973).

Figure 11: Debris Retention Basin Immediately Upstream of the Village (construction: 1973)

Figure 12: Trapezoidal Channel through the Village (construction: 1974–1975).

3.4. A Modern Approach for Debris Flow Assessment

3.4.1. Methodology

The 1999 events raised the question of whether the protection strategy should be revised and which additional measures should be planned. In order to answer these questions, the Public Authority for Hydraulic Structures of South Tyrol financed a study in which an integrated approach has been proposed (Gostner et al. 2003).

The methodology used for the integrated analysis of debris flow hazards at the Tschengls torrent is based on the theoretical background of debris flow research and assesses the applicability of recent laboratory

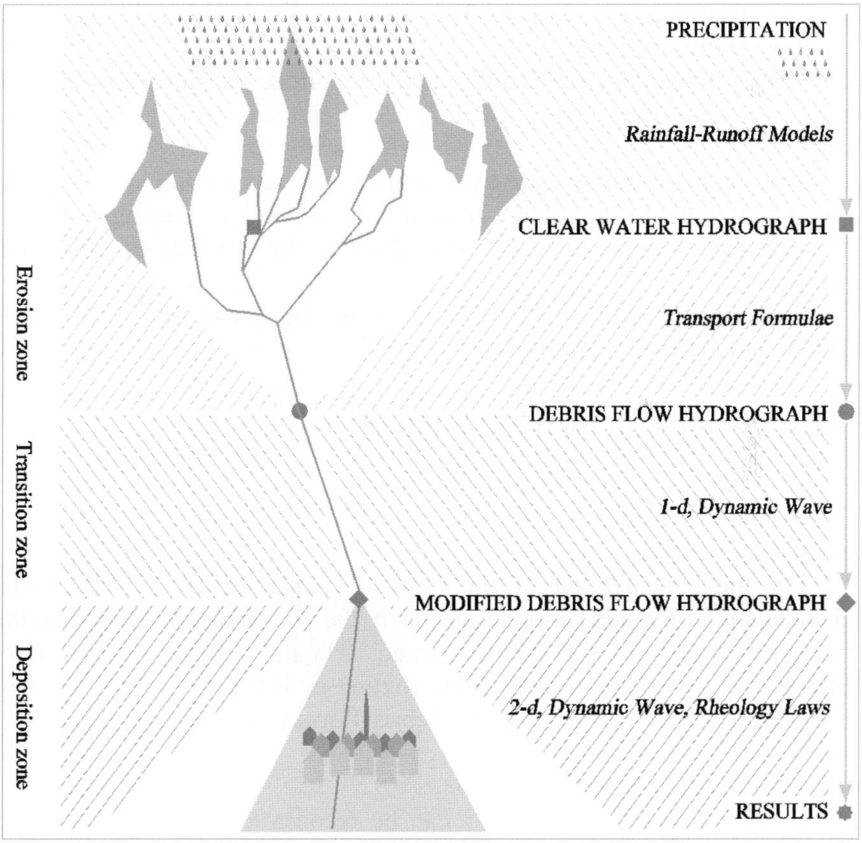

Figure 13: Flowchart of the Adopted Methodology That Takes into Account Debris Flow Initiation, Transport, and Deposition.

results. By means of extensive analysis, intensive field work and interviews with eyewitnesses, it has been possible to recognize essential characteristics of the Tschengls torrent, to assess its debris flow activity in a satisfactory detail and therefore to give recommendations for structural and nonstructural measures.

Figure 13 presents a flowchart of the adopted methodology. The modular and gradual structure of the physically based approach includes the formation, transportation and deposition of debris flows.

Analysis begins with a rainfall-runoff model. Precipitation data based on the local meteorological conditions are combined with the hydrological characteristics of the watershed to define clear water hydrographs at different points of interest.

Clear water hydrographs serve as input data for the next step of the analysis where debris flow hydrographs are developed, using different sediment transport formulae.

Debris flow hydrographs are routed through the transition zone using a 1D numerical model. Debris retention basins are considered in a simplified way.

The modified debris flow hydrographs represent the input for the 2D numerical model of the urbanized area. The simulation of different scenarios provides a series of results that allow to the proposal of mitigation measures.

A detailed description of the entire procedure can be found in Gostner (2002) and Gostner et al. (2003).

3.4.2. Situation in the Urbanized Area

In the numerical model (Figure 14), the depositions of a simulated debris flow event are shown. In the photo a portion of the deposits of the 1999 events can be seen.

For the investigation of the discharge capacity of the trapezoidal channel and of the deposition processes in the village and downstream, the numerical model FLO-2D (O'Brien et al. 1993 and O'Brien 2001) is used. FLO-2D is a two-dimensional flood routing model and allows the use of both clear water hydrographs and debris flows. The main results can be summarized as following:

1. The debris flows of 1999 can be reconstructed with the adopted methodology. Results of the modelling correspond well with reports

Figure 14: Numerical Model and Photo of the Deposition Zone with the Village of Tschengls (Upper Part) and the Alluvial Forest (Lower Part).

from eyewitnesses and field evidence. The first of the 1999 events corresponds to a debris flow with a high to medium probability of occurrence (return period between 30 and 100 years).

2. Maximum flow depths do not vary significantly and are enclosed in a narrow spectrum. However, the rheological behaviour of a debris flow is the determining factor for the maximum flow depth and not, as for clear water flow, the maximum discharge.
3. In general, the trapezoidal channel has a high discharge capacity due to the low roughness and the rather high slope of around 11 percent. The protection walls on both sides of the trapezoidal section contribute to the discharge capacity. Exceptions are two bridges that cross the torrent in the village. They restrict the discharge capacity of the torrent, and in several debris flow scenarios the maximum flow depth reaches the structure of the bridges.
4. The open area downstream of the village of Tschengls is sufficiently large for the deposition of debris flows, if the debris flow is routed across the village.

3.4.3. Proposal of Mitigation Measures

There are several measures that could mitigate the hydraulic risk of the Tschengls torrent; the most effective ones are briefly discussed in the following:

1. The side walls on both sides of the trapezoidal channel contribute significantly to the discharge capacity of the torrent. They confine the flow and allow it to pass through the village. Therefore it is necessary to examine their condition since they are around 130 years old and seem weakened in some places.
2. The study has shown that two bridges are the weak points in the village protection strategy since they restrict the channel capacity. The construction of mobile bridges would be an efficient measure. During a debris flow event they could be opened, enabling the debris flow to pass through. Mobile bridges exist in other countries and have already proven their effectiveness (Vischer and Bezzola 2000).
3. Emergency planning. Even if protection measures are well planned and executed, a residual risk remains. Thus it is of primary importance to plan for emergencies using temporary measures (for example, stop-logs at the openings of the side walls), warning devices and evacuation plans to prevent major damage in the case of catastrophic events.
4. Mapping of hazard zones should be done to prevent increased damage potential due to the urbanisation of the depositional area. This measure is very useful as there is still a large free surface available for the activity of the Tschengls torrent. In the village, the discharge section has to be maintained and the deposition zone should be utilized only extensively.

4. CONCLUSIONS

The case study of Tschengls presents an example of the way to deal with the natural danger of debris flows. The time series of debris flow events and the history of hazard management at the Tschengls torrent show the direct link between catastrophic events and subsequent protection works. Their realization was and is, of course, dependent on the availability of financial resources. In former times, protection measures were mainly based on empirical knowledge. The integrated analysis of the Tschengls torrent shows how modern research and better understanding of the physical processes involved in a debris flow event allow us to assess the debris flow activity of a torrent in satisfactory detail. Thus, it is possible to define more and more precise recommendations and to design structural and nonstructural measures. However, in order to provide reliable and useful results an integrated analysis based on scientific approaches must take advantage of all available sources including eyewitness accounts, written resources, field evidence, and event history.

FLOOD VOLUME ESTIMATION AND FLOOD MITIGATION: ADIGE RIVER BASIN

Salvatore Manfreda
Dipartimento di Ingegneria e Fisica dell'Ambiente
Università degli Studi della Basilicata
Potenza, Italia
manfreda@unibas.it

Mauro Fiorentino
Dipartimento di Ingegneria e Fisica dell'Ambiente
Università degli Studi della Basilicata
Potenza, Italia

Abstract: In the present work, we describe an extended flood risk analysis carried out in the Adige River basin in Italy. The methodologies adopted were used in a comparative approach that highlighted the limits and potentiality of some methods with respect to others. Principles presented may be considered of interest for general problems of flood risk management. The work carried out shows interesting results along with a broad number of specificities that may constitute a useful support for those who will apply hydrological analyses on large-size basins. The study basin covers a wide area of about 12,000 km^2. In such a case, a satisfactory analysis becomes complex because of the large number of phenomena involved in flood generation that need to be taken into account.

Keywords: flood risk, water management, flood volume estimation, Adige River

1. INTRODUCTION

Flood risk management is critical for territories that remain vulnerable despite the proliferation of advanced technologies. Population is exposed to higher risks as cities increase their boundaries, neglecting or sometime forgetting about natural river systems. Flood risk management represents a problem that is not easily addressed due to many political, social, and economic factors. In addition, prediction remains difficult because of the numerous mechanisms involved the generation of floods. Moreover, the use of relatively short gauging records, and the uncertainty in the flow rating curves, as well as the errors involved in the measurements of extreme events influence the reliability of flood prediction.

E. Wiegandt (ed.), Mountains: Sources of Water, Sources of Knowledge, 243–264.

Therefore, it is mandatory to base analyses on methodologies able to interpret hydrological dynamics, and that are sufficiently consolidated to be accepted and fully understood by politicians, who bear the responsebility for making decisions of high economic and social impact. Methodologies must be reliable and able to provide the most detailed information as is possible. Hydrological data should not be merely analyzed with statistical tools, but further investigation from annals, journals, and technical reports may provide a more comprehensive framework for hydrological studies.

The objective of the present work is to reduce the high level of uncertainty about the prediction of hydrological extremes by using multiple approaches to achieve a more reliable estimation of flood peaks and their corresponding flood volumes. The methodology uses both statistical models and rainfall-runoff simulations in order to quantify hydrological response. Elaborations based on systematic data are enriched by the use of historical research on the documented events.

Annual maxima series of rainfall and floods represent the most common database, containing information about the event time, which is useful to detect seasonal frequency of the events (see part 3). In some cases, local measurements are not sufficient to provide consistent frequency estimates. Regional models, based on the concept of spatial homogeneity of the populations of annual maxima, may improve statistical models' performance removing limitations due to short series, error in the measurements, etc. (see part 4).

Flood volume estimation is central to a quantitative study. In fact, this is the most important variable for the design of water works for flood mitigation. In the case of the Adige River, it was particularly complex to estimate because of the nonunique behavior of the basin. In this study we compared three different methodologies including a rainfall-runoff model with the specific task of estimating the flood volume (parts 5, 6, and 7). In the rainfall-runoff model, the hydrological processes are interpreted by using a conceptual model, which calculates the runoff and the subsurface runoff that contributes to the stream flow. The model, in combination with other methodologies, is applied to investigate the variation of the hydrograph at different return periods. The analyses allow a reliable prediction of flood risk in terms of flood peak and flood volume.

The comparison of different sub-catchments highlights substantial dissimilarity in their hydrological behavior. Such a condition becomes critical in the phase of defining possible solutions for flood mitigation. In this context, it is important to note that the main purpose of this research was to find ways to safeguard the city of Trento on the Adige River.

2. THE ADIGE BASIN AND THE VULNERABILITY OF THE CITY OF TRENTO

The Adige is one of the most important Italian rivers. Its basin has an area of 11,954 km^2 and the main river course is 409 km long. The river originates in the Province of Bolzano, crosses the Trentino Region, the city of Verona, and finally empties into the Adriatic Sea. The main tributaries stem from Alpine saddles and rims and are characterized by slight gradients. Secondary streams begin at higher altitudes and flow down into the recipient branches after a short stretch.

Because the focus of this chapter is to find ways to safeguard the city of Trento, characteristics of the whole river basin are relevant. At the station of Trento (in S. Lorenzo) the basin has an area of about 10,000 km^2 and, for the purpose of this study (see Figure 1), can be divided in 3 sub-catchments: Adige at "Bronzolo" (6,926 km^2), Noce (1,372 km^2) and Avisio at Lavis (934 km^2). The city of Trento is located, as shown in Figure 1, on the border of the main River and is close to the Avisio and Noce outlets. The city of Trento is thus in a highly vulnerable location given its downstream position along the main river.

Figure 1: Description of the Adige River Basin up to Trento and its Major Tributaries: Avisio, Noce, and Bronzolo's Watersheds. (Map Obtained Using a Digital Elevation Model of 240-M Resolution.)

2.1. A Brief Review of Past Extreme Events

Among the floods that occurred in the recent past, one major event (November 1966) deserves special attention as it affected vast areas and caused serious damage to the Provinces of Trento and Bolzano. The flood event of November 1966 was characterized by two antecedent phenomena: (1) the increase of temperature that caused snow melt above 2,500 m with the consequent increase of torrent level; (2) the contrast between two airflows, one warm coming from the south and one cold coming from the north, which joined to form a persistent cyclone zone in the north of the Alps, thereby provoking a severe storm over entire Northern Italy. Extreme floods were also recorded in the Arno River and the Brenta River, with the tide reaching its maximum levels in the lagoon of Venice.

Even if this event were considered the most intense flood event of the last two centuries for the Adige, its severity at Trento was moderated by two factors (see Table 1). The flood peak was reduced by the river bank breaches upstream from the city (according to Dorigo, 1967, breaches caused a reduction of the peak of about 144 m^3/s) and by the flood storage operated by the "S. Giustina" dam, which accumulated 12 mm^3 of water (Menna, in 1998, estimated that the reduction of the peak flow was of about 300 m^3/s).

Table 1: Hydrological Data Describing the Flood Event of November 1966.

November 1966	Basin area (km^2)	Peak discharge (m^3/s)	Virtual discharge considering river banks break and S. Giustina storage (m^3/s)	Antecedent rainfall during the previous 32 days (mm)	Total rainfall (mm)
Noce Rupe	1,372	575		146.3	168.8
Avisio at Lavis	934	1,048		166.0	184.4
Adige at Bronzolo	6,926	1,380			
Adige at Trento	9,763	2,321	2,321+ 444 ≅ 2,765		

The stream flow hydrograph may better describe the dynamic of the flood event from a hydrological perspective (see Figure 2). It is observed that the stream flow of the Noce at S. Giustina is the outflow of a dam, thereby minimizing the flood peak discharge. In fact, it increases slowly compared to other hydrographs and does not have the typical recession curve of natural systems. On the other hand, the hydrograph of the Avisio River highlights a short lag time and high peak flow. This condition is directly reflected by the hydrograph recorded at Trento. It may be considered the main factor responsible for the fast increase of the recorded flow at Trento. Comparing the peak discharges of each subbasin with their relative size, one may observe that the peak discharge of the Avisio was almost 50 percent of the peak recorded at Trento. This amount is surprisingly high if compared with the relative surface of the subbasin that is only 10 percent of the total basin (see Table 1).

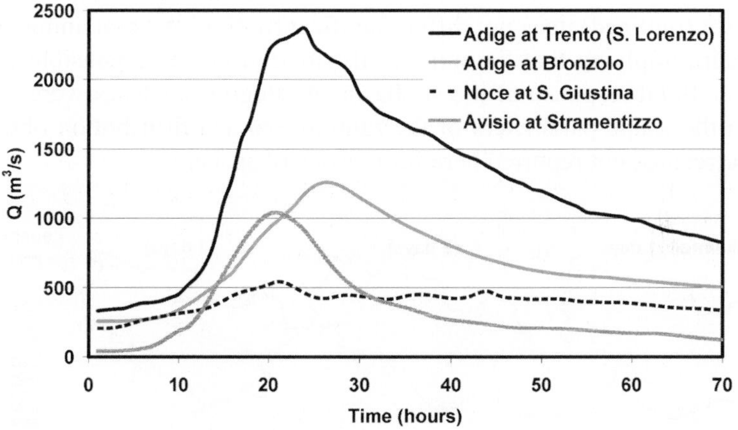

Figure 2: Historical Hydrograph Recorded, during the Flood Event of November 1966, at the Water Level Gauges of Trento, Bronzolo, S. Giustina, and Stramentizzo.

More recently, minor flood events occurred in the Adige River basin. These did not affect the city of Trento, but the risk was significantly high. At these occasions, large areas situated immediately upstream to Trento, were surrounded by water, producing an involuntary flood mitigation with respect to the major city. This behavior is recurrent and somehow affects the shape of the Cumulated Distribution Function (CDF) of floods in some stations of the Adige River upstream to Trento (see part 5).

3. RAINFALL DYNAMICS AND SEASONALITY EFFECTS

Rainfall is a fundamental starting point for flood studies. It has been analyzed in two ways: in terms of its frequency of extremes and in terms seasonality effects on precipitation. In particular, the spatial distribution was investigated with the aim of verifying if it displays any anomaly that may justify the high values of stream flow recorded on the Avisio sub-basin.

Rainfall spatial distributions may be extremely heterogeneous, constituting a critical point in the flood risk assessment. In this particular case, the mean rainfall maxima have been analyzed across the basin at durations of one, two, and three days (see Figure 3). The spatial distribution appears fairly homogeneous around values 40–80 mm for durations of one day. With increases in duration, differences becomes higher, showing that higher depth occur in the central part of the basin. Moreover, the Avisio subbasin (outlined) does not differ significantly from the remaining part of the basin, implying that the rainfall distribution is not responsible for the peculiar flood response of the subbasin. Analogous analyses were carried out on the shape parameters of the rainfall maxima distribution obtaining similar results, not reported here for reasons of space.

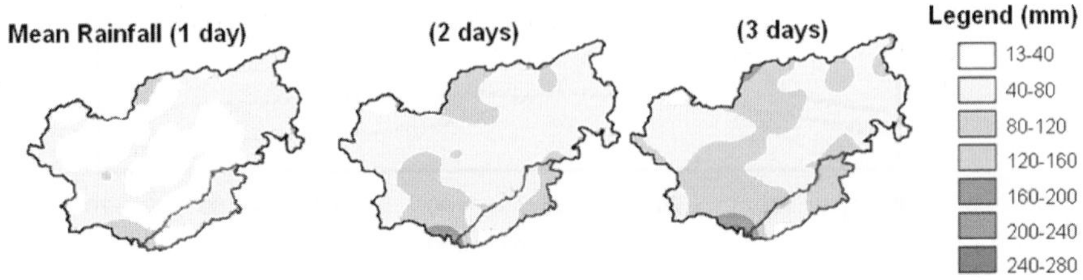

Figure 3: Maps of the Mean Values of the Rainfall Maxima Over Durations of 1, 2, and 3 Days. (Maps Obtained Using the Technique of Kriging with Exponential Semivariogram Based on Rain Gauges Data.)

Furthermore, we analyzed two different samples of rainfall: the first with ordinary and the second with extraordinary data. In Figure 4a and Figure 4b, we describe the frequencies of rainfall maxima at different durations ranked by season: autumn, winter, spring, and summer. The first graph considers a dataset with all the rainfall annual maxima, while the second contains only the annual maxima over a threshold equal to 1.5 times the mean of the ordinary component.

The analyses produced interesting results: (1) the annual maxima of short duration (1–3 h) occur essentially during summer periods (50–70 percent of the events), with both the complete record and the extraordinary maxima having the same behavior; (2) The annual maxima for greater duration (12–24 h) occur in 50 percent of the cases in the autumn, with peaks of 65 percent for the extraordinary component; (3) the duration of 6 h represents an intermediate duration between the two regimes. In general, it was found that short duration rainfall is more relevant during summer, while the longer durations are more relevant in the autumn. The rainfall annual maxima have a strong seasonal influence that becomes more and more significant with the increase of the event severity (threshold).

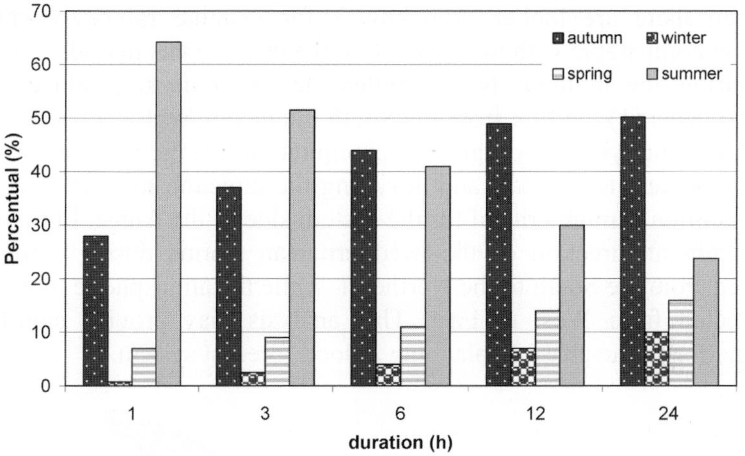

Figure 4a: Distribution of the Annual Maxima of Rainfall for the Duration from 1 to 24 H—Ordinary.

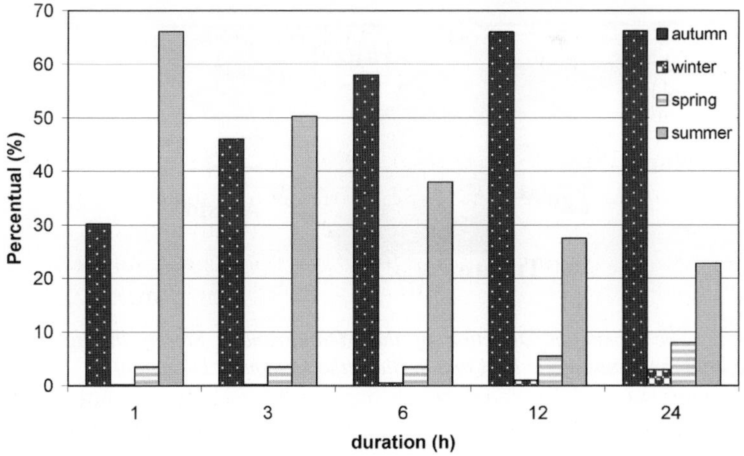

Figure 4b: Distribution of the Annual Maxima of Rainfall for the Duration from 1 to 24 h—Extraordinary Precipitation. (Obtained Introducing a Threshold on the Rainfall Data Equal to 1.5 Times the Mean Value of Rainfall Depth of the Ordinary Component.) Each Symbol Distinguishes the Percentage of Events Occurred in the Different Season for a Given Duration.

In principle, the catchments produce the maximum peak discharge under a constant rainfall intensity of durations equal or greater than the catchment lag time (τ). The parameter τ depends on many factors including the basin area. In the specific case of the Adige at Trento, τ assumes a value of about 20 h. In light of the basin characteristics and of the above results, one may observe that the most dangerous period is autumn, which is when there are higher probabilities for extreme rainfall events of duration comparable with the lag-time of the basin in this period.

During the autumn, two simultaneous movements produce precipitating cells: (1) the flux from the South interacting with the terrain relief and producing typically orographic precipitation; (2) the movement of the weather system to the East and deviating the direction to Northeast. This last condition is more critical for the eastern side of the Adige. In Figure 5, the dominant direction of the Mediterranean storms during autumn is oriented from the South to the Northeast, while the atmospheric movement is directed from West to East. This analysis may provide significant information in the phase of planning a flood forecast system.

Figure 5: The Dominant Direction of the Mediterranean Storms during Autumn (in Transparent the Atmospheric Flux and in Black the Movement Direction).

4. FLOOD FREQUENCY ANALYSIS

The primary objective of frequency analysis is to relate the magnitude of extreme events to their frequency of occurrence through the use of probability distributions (Chow et al. 1988). Data observed over an extended period of time in a river system is analyzed assuming that the flood peaks are independent and identically distributed. Furthermore, it is assumed that the floods have not been affected by natural or man-made changes in the hydrological regime in the system. This last assumption is not always realistic, especially for the most severe events that may produce flooding and that also reduce the peak flow downstream to the flooded area. For this reason we recommend a careful review of historical information regarding the recorded floods using contemporary scientific, academic, and engineering publications along with technical reports and any other available sources.

An efficient approach for the estimation of the flood peak and/or of the peak volumes associated with different probability levels or return periods derives from regional analysis. This approach reduces estimate uncertainties and overcomes the lack of hydrological data (Cunnane 1989). In Italy, such an approach is adopted by the VAPI procedure (Evaluation of Floods in Italy), which refers to a probabilistic model, known as two component extreme values or TCEV (Rossi et al. 1984). The expression of the probability distribution is the following:

$$F_X(x) = P[X \leq x] = exp(-\Lambda_1 \, exp(-x/\theta_1) - \Lambda_2 \, exp(-x/\theta_2)) \qquad (1)$$

where Λ_1, Λ_2, θ_1, and θ_2 are parameters with the same meaning of the Gumbel distribution parameters. The TCEV introduces the distinction between an ordinary component (1) and an extraordinary component (2).

Using the VAPI procedure, it was possible to characterize the flood probability distribution at each sub-catchment of the study area. For this, we used the records of about 13 stream gauges, thus obtaining probability distributions that fit the distribution of the recorded extreme floods well, especially in the case of small and midsized subbasins (two examples are given below). The entire basin area was considered homogeneous. The regionalization model was useful to predict the basin response at the higher return period and to interpret some incoherent aspects of the recorded data. In particular, the probability distributions obtained with the regionalization approach are able to interpret nonlinearities in flood peak distributions, including for cases where the data do not reveal such non linearities.

The case of the Avisio at Lavis is remarkable as it demonstrates the reliability of the regionalization model in hydrological analyses of extremes (Figure 6). In this case, the probability distribution of the recent records are less skewed with respect to the predicted CDF (continuous line) but, considering the recent data along with the historical data of the last two centuries, it is clear that the TCEV interprets the real basin behavior, even at the higher return period. This is even more remarkable if one realizes that the historical data were not used for the calibration of the model parameters. The analysis of flood distribution was based on systematic records obtained from the hydrological annals. Some additional information collected from old journals and other documents was also available about the most severe events that occurred in the last two centuries, between the years 1868–2000 (see triangle in Figure 6). Those are reported in the graph using the plotting position of Hazen (n−0.5)/N, where n is calculated considering these 5 events as the highest (with rank position ranging from 128 to 132) during the time period from 1868 to 2000 and N is the total number of events that is equal to the number of years (N=132).

On the other hand, the flood CDF refers to the stations upstream to Trento. Specifically Bronzolo and "Adige at Ponte Adige" show different behavior at the higher return periods. In those cases, the problem is not the limited extent of the records, but something else, as one can clearly see Figure 7. The probability distribution overestimates the peak flows for the higher return periods with respect to the plotting position of the recorded data. This discrepancy could be explained by looking at documentation on recorded flood events (see, e.g., Reichenbach et al. 1998; AVI project). In fact, as we expected, the marked peaks were associated with overspill and flooded area upstream to the stations due to the limited hydraulic capacity of the river cross section.

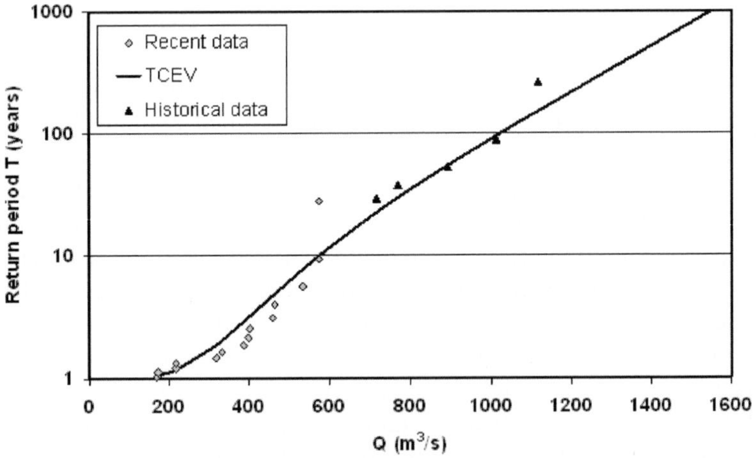

Figure 6: Probability Distribution of Floods for the Avisio at Lavis. The Continuous Line Represents the TCEV Distribution Estimated with Regionalization Model, Diamonds the Systematic Data, and the Triangles are the Historical Floods.

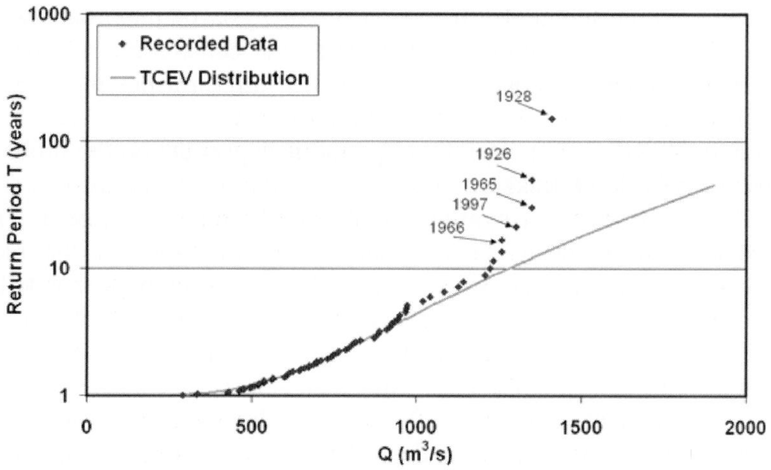

Figure 7: Comparison between the TCEV Distributions Estimated with Regionalization Model and Recorded Annual Maxima at Bronzolo.

Following the described approach, we estimated the flood peaks for all the small and midsized basins. Choosing a return period for the flood protection equal to 500 years, it followed that the peak flow of the Noce River is $Q_{T=500}$=600 m^3/s, while for the Avisio River $Q_{T=500}$=1,234 m^3/s. For the water level gauges of Bronzolo and Trento, we preferred the at-site model because it is able to account for the superimposition effects of different sub-catchment contributions (Adige at Bronzolo $Q_{T=500}$=2,000 m^3/s, and Adige at Trento $Q_{T=500}$=3,150 m^3/s).

5. THE HYDROGRAPHS OF THE ADIGE AT TRENTO

In flood management problems, the definition of hydrological risk includes several aspects. It refers not only to the peak discharges, but also to the flood volume and the shape of the flood hydrographs. The estimation of this latter variable is possible through the formulation of a synthetic hydrograph of a given return period obtained by statistically analyzing the flood volumes of recorded events. This is a frequently used technical practice.

If a long series of recorded hydrographs is available for a river section, the analysis of the flood volumes can be performed following different procedures. We decided to compare results of synthetic hydrographs, a hydrological model and statistical analyses to define the basin response at the higher return periods in terms of flood volumes.

The method adopted is based on the analysis of maximum average discharges of given duration and leads to the construction of the Flow Duration Frequency Reduction curve (FDF) that furnishes the maximum average discharge $Q_D(T)$ in each duration D for a given value of return period T (NERC, 1975). The return period is dependent on the probability distribution of flood peaks.

The construction of synthetic hydrographs for the Adige at Trento is carried out by using 11 recorded events. Analyses of the records (surprisingly) reveal two systematically different responses that fall into two distinct categories. The first has a mean increase in discharge of 60 $m^3/s/h$ (Figure 8a), and the second a faster increase of the discharge of about 240 $m^3/s/h$ (Figure 8b). This behavior is unexpected and may produce significant errors in the flood risk evaluation if not accounted for. We will address this problem in detail later, but now we proceed assuming both the hydrographs equally possible at any return period. As we will see later, this is a wrong assumption.

The synthetic hydrograph can be constructed, under the simplifying assumption of symmetry respect to the time of the peak, according to the expression proposed by Fiorentino (1985):

$$q(t) = Q_T e^{-\frac{2|t|}{k}} \quad \text{for } t \in [-\infty, +\infty] \tag{2}$$

where Q_T $[L^3/T]$ is the peak flow referred to the return period, t [T] is the time, k [T] is a shape parameter of the hydrograph. Assuming the hydrograph collapsed in the positive t axis, it follows that the previous formula can be rewritten for computational purposes as:

$$q(t) = Q_T e^{-\frac{t}{k}} \quad \text{for } t \in [0, +\infty] \tag{3}$$

Equation 3 is particularly useful to derive an analytical function of the flood volumes over a given threshold q_o (Fiorentino & Margiotta 1998):

$$V_{q_0} = Q_T \int_0^{t_0} e^{-\frac{t}{k}} dt - q_0 t_0 = kQ_T\left(1 - e^{-t_0/k}\right) - q_0 t_0 \tag{4}$$

where $t_0 = -k \ln(q_0 / Q_T)$ is the time duration in which the discharge $q(t)$ reaches the value q_o.

In the case study, the shape parameter k at Trento may assume two different values: the first refers to the slow events (k_{slow}=92 h) and the second to fast events (k_{fast}=41 h). Using equation 4, it is possible to evaluate flood volume corresponding to a given threshold and for a return period of 500 years. The threshold q_o represents the hydraulic capacity of the cross section of the river that, in our case, is equal to 2,200 m^3/s. The flood volume, obtained following this procedure, ranges from 53.1 to 23.7 mm^3 moving from a slow event to a fast one.

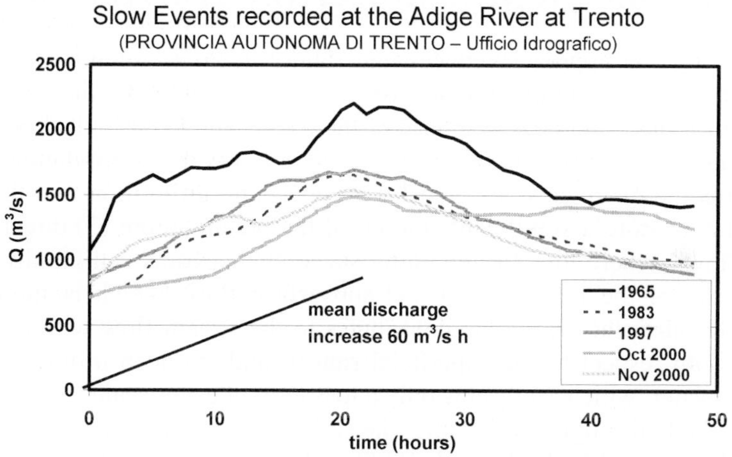

Figure 8a: Slow Events Recorded at the Water Level Gauge of Trento.

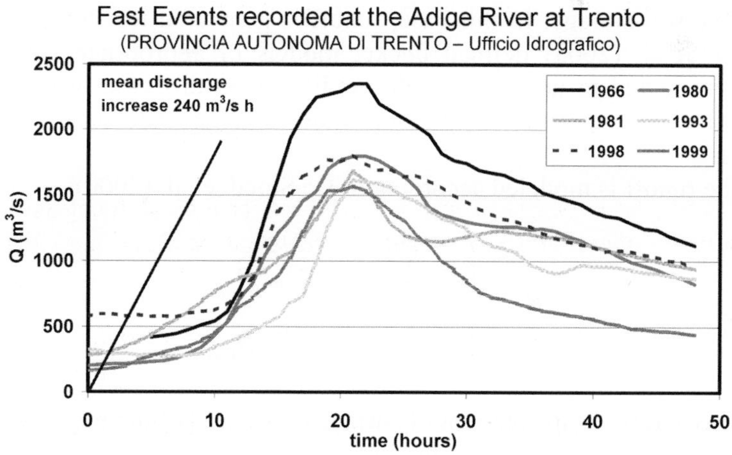

Figure 8b: Fast Events Recorded at the Water Level Gauge of Trento.

This method provides two significantly different results according to the two event dynamics. Neglecting the presence of two event typologies would lead to a value of the flood volume of about 35 mm^3. This value, as we will clarify later, still overestimates the flood volume of the basin at T=500 years. These uncertainties led to deeper analyses undertaken by using hydrological modeling.

6. HYDROLOGICAL SIMULATION AT EVENT SCALE

The simulation approach reduces uncertainty in flood volume estimation. To carry out our simulation, we use an event scale model that is able to reproduce the flood peak response of the basin. The hydrological elements relevant for a drainage basin, at this scale, are: (1) the precipitation input, which is the main cause of runoff; (2) superficial infiltration into the soil, which may store a significant amount of the precipitation; (3) direct overland flows, where discharging along successive streams can rapidly swell the flows of the main stream; (4) subsurface flow, which is runoff for shallow subsurface flow that contributes to the stream flow with a certain delay with respect to the superficial runoff; and (5) deep infiltration into groundwater, which represents only a loss term at event scale.

In the hydrological modeling, the soil state significantly influences the basin behavior (Manfreda et al. 2005) and it is therefore necessary to take into account its variability. To this end, Manabe (1969) suggested that the land surface water balance can be simulated by using a simple model of effective soil storage. In this case, runoff is generated for storage excess. Farmer et al. (2003) defined bucket models of appropriate complexity mainly oriented to ungauged basin prediction. In this case, we defined a bucket scheme to simulate the soil water storage state during extreme events.

The runoff is modeled according to De Smedt et al. (2000):

$$R_t = \begin{cases} C\dfrac{S_t}{S_{\max}}P_t & S_t < S_{\max} \\ P_t & S_t \geq S_{\max} \end{cases} \qquad (5)$$

where R_t [L] is the amount of surface runoff, P_t [L] the precipitation, S_t [L] the total water content of the bucket at time t, S_{max} [L] the maximum water storage capacity of the bucket, and C [–] the runoff coefficient. Equation 5 states that the runoff is proportional to the soil water content until the cell reaches the saturation state. After that point there is no more infiltration into the soil and all the precipitation becomes runoff.

The soil moisture storage is the quantity of water held, at any time, in the active soil layer. It varies in time depending on rainfall, interflow, and groundwater recharge, according to the following water balance equation:

$$S_{t+\Delta t} = S_t + I_t - R_{out,t} - L_t \tag{6}$$

where: $S_{t+\Delta t}$ [L] is the total water content of the bucket at time t+Δt, I_t ($I_t = P_t - R_t$) [L] the infiltration amount during the time-step Δt, $R_{out,t}$ [L] the subsurface outflow in Δt, and L_t [L] the leakage in Δt.

The model accounts for the subsurface production assuming that the subsurface flow constitutes a fraction of the water exceeding a given threshold. The subsurface outflow is evaluated by the following equation:

$$R_{out,\ t} = max\{0, c(S_t - S_c)\} \tag{7}$$

where S_c [L] is the threshold water content for subsurface flow production, and c [1/T] is the subsurface coefficient. The groundwater recharge is evaluated according to (Eagleson 1978):

$$L_t = k_s \left(\frac{S_t}{S_{max}} \right)^{\beta} \Delta t \tag{8}$$

where: L_t [L] is the groundwater recharge in Δt, k_s is a parameter that interprets the permeability at saturation [L/T], β is a dimensionless exponent. Hydrological losses such as vegetation interception and evapotranspiration are neglected, because they are less relevant at the event scale.

The discharge is computed by using a linear relationship to the total generated runoff, and by considering a constant delay time to reach the basin outlet (Figure 9). Therefore, the discharge at the outlet is evaluated as the sum of the following components:

$$Q(t) = Q_s + Q_{sub} + Q_b = \alpha_s W_s + \alpha_{sub} W_{sub} + Q_b \tag{9}$$

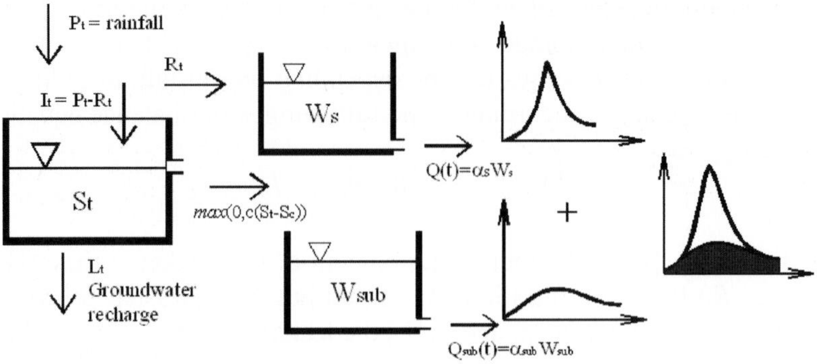

Figure 9: Lumped Model Scheme Used to Interpret the Catchment Response.

where: Q_s [L^3/T] is the discharge due to the superficial runoff; Q_{sub} [L^3/T] is the subsuperficial; Q_b [L^3/T] is base flow contribution, which is assumed constant; W_s [L^3] is control volume of the generated runoff; W_{sub} [L^3] is control volume of the generated subsurface runoff; α_s [T] is runoff recession constant; and α_{sub} [T] is the subsurface runoff recession constant.

Model capabilities are evaluated on the base efficiency and error functions such as: Efficiency (*EFF*), Absolute Average Error (*AAE*), and Root Mean Square Error (*RMSE*). For brevity we report only the expression of the efficiency:

$$EFF = \frac{\sum_{i=1}^{n}\left(Q_{oi} - \overline{Q_o}\right)^2 - \sum_{i=1}^{n}\left(Q_{oi} - Q_{ci}\right)^2}{\sum_{i=1}^{n}\left(Q_{oi} - \overline{Q_o}\right)^2} \tag{10}$$

where: Q_{oi} is the *i*th ordinate of the observed discharge, Q_{ci} is the *i*th ordinate of the simulated discharge, $\overline{Q_o}$ is the mean value of the observed discharge, *n* is the number of registrations.

The stream flow was interpreted as the sum of contributions coming from two sub-catchments: the Avisio and the remaining part of the Adige. Each of them was interpreted using the described bucket-scheme. In this way, we could also define the contribution of the Avisio sub-catchment to the peak flow of the Adige at Trento.

The model presented here has a simple structure, but at the same time it provides good results as confirmed by the event simulation reported in Figure 10. In this graph, we plot the recorded and simulated hydrographs of the Adige at Trento and Avisio at Lavis. The simulation provides efficiency higher than 90 percent. The model has been tested on four other rainfall-runoff events, and in all the cases the simulations provided are satisfactory. The model parameters are given in Table 2.

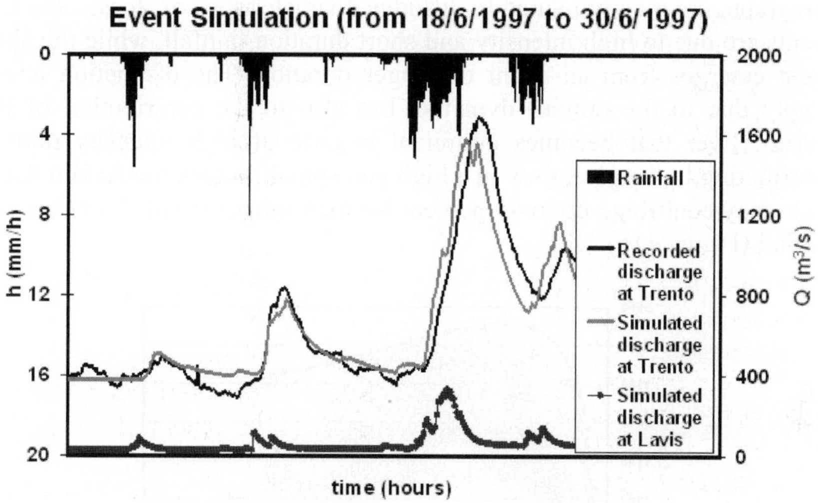

Figure 10: Model Simulation of a Recent Event that Occurred in June 1997 (EFF = 92.4%, RMSE = 109.9 m^3/s).

Table 2: Parameters of the Bucket Model.

	Adige	Avisio
C (–)	0.45	1.00
S_c (mm)	85	23
S_{max} (mm)	101	100
c (h^{-1})	0.05	0.011
Ks (mm/h)	22	48
β (–)	5.36	4.84
α_s (h^{-1})	0.041	0.16
α_{sub} (h^{-1})	0.025	0.025
Q_b (m^3/s)	420	0

After a complete analysis of calibration and validation, the bucket model was used to estimate the contribution of the Avisio sub-catchment to the peak flow at Trento and to verify the form variation of the hydrograph for a given return period. In particular, using rectangular rainfall pulses as input of the hydrological model, we could evaluate the basin and subbasin responses caused by rainfall of different durations. Rainfall depth of specified return period is evaluated by using the intensity duration–frequency relationship (IDF), estimated at basin scale.

In Figure 11a, we report the shape variation of the hydrographs at Trento, obtained using rectangular pulses with different durations and for a return period T=500 years. In light of the obtained results we could deduce that the two different hydrograph shapes, described in the previous paragraph, are mainly related to different characteristics of storms: the fast events are due to high intensity and short duration rainfall, while the slow event emerges from an event of longer duration. The distinction is not simply due to the rainfall dynamic, but also to the contribution of the Avisio River that becomes dominant in case of high intensity rainfall contributing to the peak flow in a high perceptual. In fact, the Avisio River basin may contribute up to 45 percent for high intensity and short duration rainfall (Figure 11b).

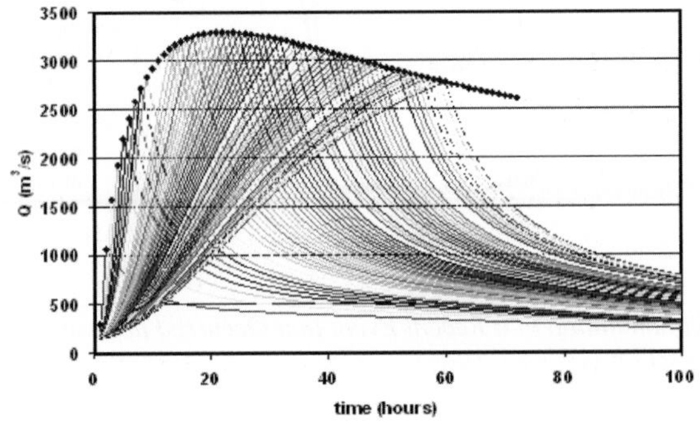

Figure 11a: Variation of the Hydrographs and Peak Flow Obtained by the Bucket Model Using Rectangular Pulses of Precipitation of Different Duration for a Specified Return Period (T=500).

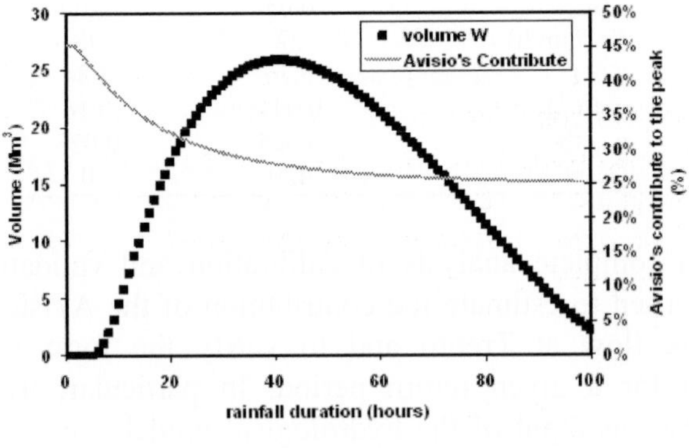

Figure 11b: Variation of the Volume Over the Threshold 2,200 m^3/s for a Precipitation of a Specified Return Period at Different Duration and Avisio's Contribution to the Adige's Peak at Trento.

The increase of rainfall duration causes an increase of the peak flow. After a certain value of rainfall duration (D = 20h) the peak starts to decrease with respect to its maximum value of 3,150 m^3/s. The maximum volume over the threshold of 2,200 m^3/s for each hydrograph is about 26 mm^3 (see Figure 11b). This value does not agree with either of the previous results, implying that the more dangerous event is something in between the fast and the slow event. At the same time, the results clearly indicate that only fast events may reach the peak flow of 3,150 m^3/s while slow events with the same probability have a lower value of the peak flow. Such a condition evidences that the estimation of the flood volume with synthetic hydrograph corresponding to a slow event is incorrect and represents a gross overestimation.

7. A STATISTICAL APPROACH FOR FLOOD VOLUMES ESTIMATION

The presence of different estimates of the flood volume may be incorporated by introducing a third model in order to facilitate a comparative analysis with different procedures. In this view, we pursued a direct estimate of the flood volume over a given threshold. This method is the only one, which provides a direct estimation of the flood volume.

Defining the stochastic variable W:

$$W = \int_0^t \left(q(t) - q_0\right) dt \tag{11}$$

where W [L^3] is the flood volume over the threshold q_0 [L^3/T], $q(t)$ [L^3/T] is the discharge during the time t. The variable can be estimated from the recorded hydrographs of the flood events. For every threshold we define flood volumes series of data from recorded events. For this purpose, it is necessary to consider a threshold sufficiently high to take into account only the runoff production but at the same time not a very high threshold, which could make the series too short. In our case, we fix such a threshold limit equal to 1,500 m^3/s, obtaining the series reported in Figure 12. The cumulative distribution function (CDF) used is the Power Extreme Value (e.g., Villani 1993):

$$F(X) = exp\left(-\Lambda\, exp\left(-\frac{X^{\gamma}}{\beta}\right)\right) \tag{12}$$

where γ is the shape parameter, $\beta = E[X^{\gamma}]$ is the position parameter, and Λ = expected number of flood events exceeding the threshold q_0/year.

Parameters γ and β are estimated through maximum likelihood fitting and are assumed independent from the threshold value. The parameter β was also verified for other threshold values higher than 1,500 m^3/s. The estimated values of the parameter at the station of Trento were $\gamma = 0.66$ e $\beta = 3.59$ $Mmc^{0.66}$.

Using these hypotheses it is possible to extrapolate the probability distribution of the flood volumes over different thresholds. The threshold has a physical meaning—it represents the limited hydraulic capacity of the river cross sections—and the volume over the threshold represents the over spill of water.

The mean number of events is the number of times in which the threshold is exceeded, which means:

$$\Lambda = 1/T(q_0) \tag{13}$$

where $T(q_0)$ is the return period of the discharge q_0 deduced by the CDF of the maximum annual peak flow of the considered station (Trento).

This method allowed the estimation of flood volumes using a pure statistical approach based only on the recorded data, using both the peak flow series and the recorded hydrographs. In Figure 12, the probability distribution of the flood volumes for different thresholds is drawn. As is

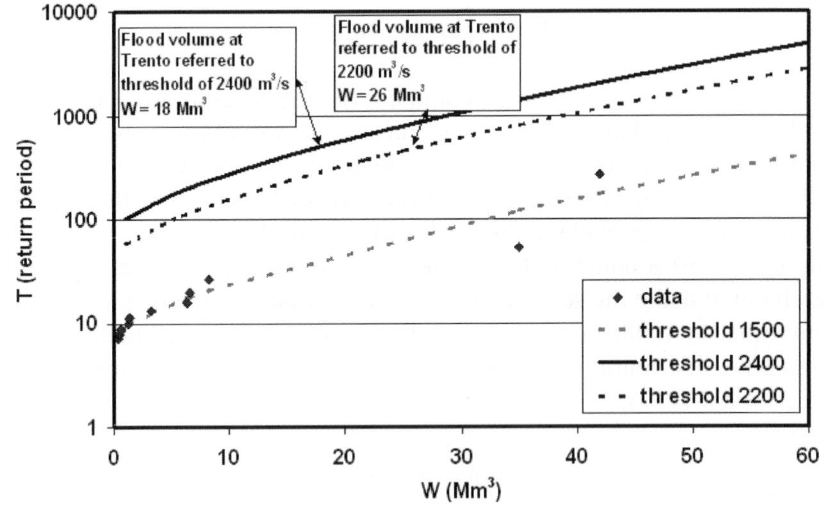

Figure 12: Probability Distribution of Flood Volumes over Different Thresholds for the Adige at Trento.

clear from the graph the estimated volume (with T = 500 years) over the threshold 2,200 m^3/s is 26 mm^3. This result is coherent with the one obtained from the previous analyses and corroborate each other.

8. FLOOD MANAGEMENT STRATEGIES: GUIDELINES

Flood risk mitigation is based on real time actions such as flood warning, flood forecasting, flood reservoir management, emergency planning, and long term actions such as land planning and zoning, structural protection measures, and property insurance. Risk management may be subdivided into three levels of actions: operations, planning, and design. The last two may be considered dynamic processes that account for variations in "sensitivity" toward risk through time (Plate 2002). Risk education is also a valuable way to evaluate differences between expert knowledge and people's behavior (e.g., during crises and when assessing the needs for land planning). The overall objective of flood mitigation management is to integrate all the methodologies in order to reduce potential losses.

The flood risk, over the study area, can be managed by using an integrated system, which consists of a combination of a forecasting system and structures for flood mitigation. The protection actions can be further subdivided into three: (1) Setup of a flood forecasting system that permits the prediction of flood risk and the evaluation of the progress of floods, thereby enabling the responsible authorities and involved populations to take personal, material, and organizational decisions to reduce the detrimental consequences of the imminent flood. These decisions may range from routine responses (e.g., change of dam operation instructions) to preventive instructions to emergency measures (e.g., announcing a generalized alert). (2) Nonstructural action that may reduce the risk of ordinary events by increasing, for instance, the actual capacity of the river system. (3) Structural actions that can be necessary during extraordinary events to control flood peaks by storing floodwater. The flood control works should be arranged, when possible, considering water storage capacity of existing structures or realizing new structures if necessary.

9. CONCLUSIONS AND FINAL REMARKS

The Adige represents a stimulating case study that raises interesting questions and highlights numerous particular cases. In our opinion, the description of the outcomes of this work may be extremely useful for hydrologists facing similar problems. For the sake of brevity, we presented only the most significant results, which does not reduce the relevance of the present work.

The study introduces innovative strategies for flood volume estimates and underlines possible strategies for flood management. The methodologies presented here can be usefully applied to general problems of flood risk mitigation, in which the flood volume estimation is crucial. The work allows the reduction of uncertainties by using multiple analyses that highlight the limitations of procedures based on the use of synthetic hydrographs. The hydrological simulation and direct statistical approach provide a more accurate estimate of the basin response at the higher return periods. Particularly interesting is the coherence of the results obtained following the last two procedures described below.

In the first procedure, flood peaks data were studied using a regionalization model along with the TCEV model. The regionalization approach was particularly successful in small and medium size subbasins. Furthermore, the use of historical documents was found to be extremely useful in the analysis of model results and in the understanding of real behavior of the basin. Rainfall maxima were found to be fairly homogeneously distributed over the basin despite the remarkably different subbasins hydrological responses. This implied that the main differences in the peak flow distribution were related to the presence of less permeable soils in the Avisio basin rather than differences in the rainfall. Furthermore, ranking rainfall maxima according to seasons allowed the detection of the autumn season being the period with higher probability of events which are potentially dangerous for the area.

In the second procedure, the hydrological response of the basin was subdivided into two classes of events according to rainfall characteristics. Temporal dynamics of rainfall dictated the distinction between fast and slow events; but also the contribution of the Avisio subbasin was detected as responsible for the production of the so-called fast events.

While studying hydrologic risk, one may face a broad number of uncertainty factors. Rigorous analyses, if not physically based, may lead to unreliable results. For this reason, hydrologists must approach the problem with a critical sense in order to obviate rough errors without which overestimation or, even worse, underestimations of the flood events is possible.

HYDROLOGICAL ASSESSMENT FOR SELECTED KARSTIC SPRINGS IN THE MOUNTAIN REGIONS OF BULGARIA

Tatiana Orehova
Department of Hydrogeology
Geological Institute at Bulgarian Academy of Sciences
Sofia, Bulgaria
orehova@geology.bas.bg

Elena Kirilova Bojilova
Department of Hydrology
National Institute of Meteorology and Hydrology
Sofia, Bulgaria

Abstract: Karstic water is an important source of water in the rural areas of Bulgaria. In this study, we estimate the impact of climate variability on the regime of karstic springs of two mountainous regions of the country. Since 1981 Bulgaria has experienced a continuous decrease in rainfall combined with an increase in air temperature. As a result, ground water levels and spring discharge have decreased. Data from three karstic springs were used. The springs refer to karstified Proterozoic marbles. Their watersheds are situated in the Pirin and Rhodopes mountains located in the southwestern part of Bulgaria. The infiltrated snowmelt water is the main source of spring recharge. The springs are included in the National Hydrogeological Network. Time series of spring discharge were studied, with a special focus on the drought period during 1982–1994, which was compared to the 1960–2001 observation period. The 1982–1994 drought period in Bulgaria also considerably influenced the evaluated springs. The strongest reduction in spring discharges was registered during the period 1985–1994. After 1996, the yearly average discharges have tended to reach their multi-annual average values. However, reduced values of spring discharges were observed again in 2000 and 2001. The quantification of the effect of a documented long drought period is of great significance for the prediction of the effects of future climatic change on groundwater resources.

Keywords: karstic springs, spring discharge, hydrological assessment, drought, multi-annual variations

E. Wiegandt (ed.), Mountains: Sources of Water, Sources of Knowledge, 265–280.

1. INTRODUCTION

Since 1981 in Bulgaria, there has been a continuous decrease in rainfall combined with an increase in air temperature, resulting in reduced river flow. The ensuing drought has elicited great interest because of its relation to global climate change, which is expected to threaten water resources (World Data Center, Trends' 93 1994; Arnell 1999).

The drought period may be considered as a model for future global changes. A recent study of water resources during the drought period (Gerassimov et al. 2001) describes the general characteristics for this period for the whole of Bulgaria. The study concerns the three main hydrological zones in Bulgaria: (1) the zone with direct discharge to the Danube River; (2) the zone with direct discharge to the Aegean Sea; and (3) the zone with influence of the Black Sea.

In addition to the study of precipitation and river flow, some analysis of groundwater variation has also been undertaken (Gerassimov et al. 2001; Orehova and Bojilova 2001). In the present study, more explicit information is given for the Aegean Basin, namely for the watersheds that are situated in the Pirin and Rhodopes mountains located in the southwestern part of Bulgaria. This basin has its source in the mountains, often referred to as "water towers." With regard to water resources, the importance of the mountainous regions is primarily based on enhanced precipitation due to the orographic effect. Colder temperatures at higher altitudes result in lower evapotranspiration rates, with mountainous areas providing the greater part of the fresh water flowing downstream. However, these water towers are subject to droughts, thus producing unfavorable impacts on water resources.

The aim of the present study is to characterize the general behavior of a few selected springs. In this regard, the influence of the drought during 1982–1994 on the regime of selected karstic springs was estimated. For this reason the studies of variations in the spring discharge were made for the longer 1960–2001 period.

2. GENERAL CHARACTERISTICS OF THE DROUGHT PERIOD

The general characteristics of the drought period are presented in the monograph *Drought in Bulgaria: A Contemporary Analogue for Climate Changes* (Gerassimov et al. 2004).

2.1. Available Data

Eight representative hydrological stations for the Aegean Basin were used (Gerassimov et al. 2001 and 2004). Observation data on water levels for rivers were available since 1909, and regular water discharge measurements started in 1936.

Data series for precipitation and air temperature were obtained at the base of all rain gauge and meteorological stations in the territory of Bulgaria. They were extrapolated using correlation and regression analysis, taking into consideration the altitude of stations. The number of stations for precipitation and air temperature is given in Table 1.

Table 1: Used Data Series with Annual Values for Precipitation and Air Temperature (after M. Genev et al. 1998).

Area	Precipitation stations (P), number	Temperature stations (T), number
Aegean Basin	114	69
Bulgaria total	300	169

2.2. Analysis of Multi-Annual Variations of Air Temperature, Precipitation, and River Flow

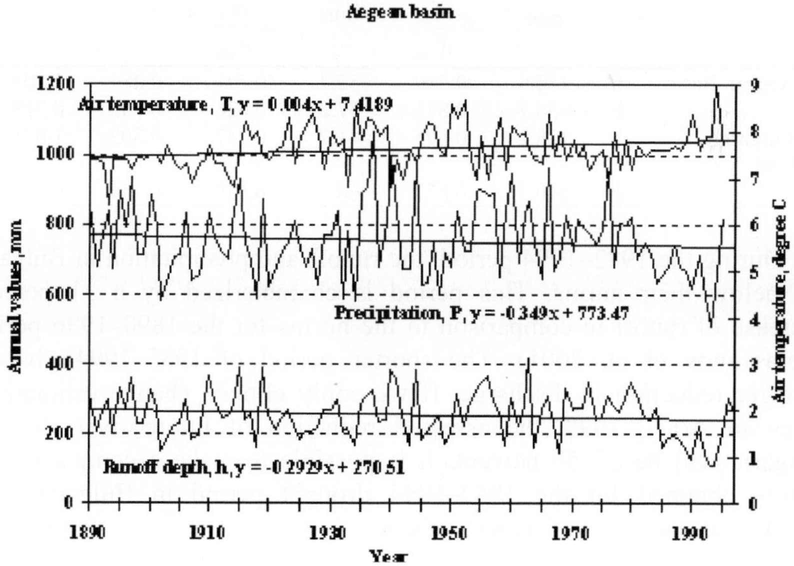

Figure 1: Mean Annual Values of Air Temperature T, Precipitation P, and River Flow h (as water layer) for the Aegean Basin (after Gerassimov et al. 2001).

Data extrapolation and weighting made it possible to calculate the mean annual temperature, precipitation, and river discharge for the 1890 to 1995 period for each of the three Bulgarian hydrological basins.

A strong decrease in precipitation and runoff was registered in the Aegean hydrological zone (see Figure 1).

2.3. Definition of the Drought Period for Precipitation and River Flow

The drought period 1982–1994 for discharges and precipitation is defined and analyzed by Gerassimov et al. 2001. The statistical structure of the period is represented by two basic parameters—mean value $\overline{X}$ and standard deviation σ_X—and is given in Table 2. The statistical parameters are compared with values from the longest period of data between 1890 and 1995. Table 2 shows that the chosen estimators ($\overline{X}$ and σ_X) are considerably lower from their values over the 106-year period.

Table 2: Statistical Structure of the Drought Period 1982–1994: $\overline{X}$, σ_X, Cv and Their Deviation in Relation to the Period 1890–1995 (Kx, KT, Kcv) (after Gerassimov et al. 2001).

Area	h, P	$\overline{X}_{13}$ mm	$K_X = \frac{\overline{X}_{13}}{\overline{X}_{106}}$	σ_{13} mm	$K_\sigma = \frac{\sigma_{13}}{\sigma_{106}}$	Cv_{13}	K_{Cv}
Aegean Basin	h	182.6	0.719	52.7	0.791	0.289	1.103
	P	658.7	0.873	67.2	0.626	0.102	0.718
Bulgaria total	h	138.2	0.695	41.7	0.747	0.302	1.075
	P	640.2	0.877	66.5	0.639	0.104	0.732

During the 1982–1994 period, the runoff and precipitation in Bulgaria are below their norms. This period is characterized by a 31 percent decrease of runoff in comparison to the norms for the 1890–1996 period (Gerassimov et al. 2001). The shorter period of 1985–1994 gives a stronger reduction in discharge. If we apply climate change estimations given in Arnell 1999, the assessed reduction of future river flow in Bulgaria will be 25–50 percent. It is possible to make a comparison of results obtained for the 1982–1994 drought period in Bulgaria with Arnell's scenario as well as with others.

3. DESCRIPTION OF THE STUDY AREAS

For the purpose of this study, two study regions were selected (Figure 2). They are situated in the Pirin and Rhodopes mountains located in the southwestern part of Bulgaria. Both study areas belong to the Rila-Rhodope Geomorphological Region. The climate of the northwestern part of Bulgaria is temperate with Mediterranean influence (Koleva and Peneva 1990).

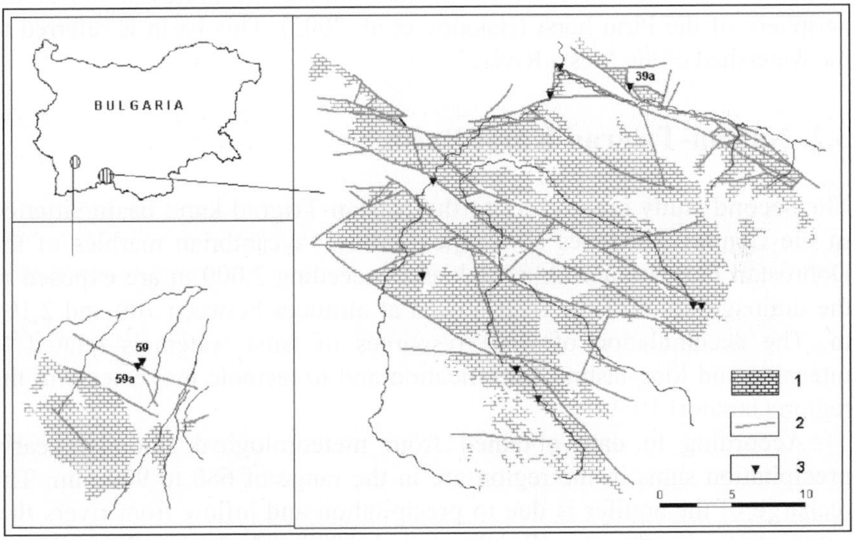

Figure 2: Map of the Razlog and Nastan-Trigrad Karst Basins (prepared by A. Benderev): (1) Marbles; (2) Faults; (3) Main Springs.

3.1. Razlog Karstic Basin

The first study area refers to the Pirin mountain, which in its modern boundaries corresponds to the Pirin neotectonic horst (Zagorchev 1995). In northern Pirin, Precambrian marbles of the Dobrostan Marble Formation (Rhodopian Supergroup) are developed, with a thickness of more than 1,000 m (Zagorchev 1995, 2001). Marbles construct the highest part of the mountain from 950–1,000 to 2,914 m above sea level (a.s.l.). Mostly vertical caves are developed there as a result of karstification of marbles (Bakalov et al. 2002).

The Razlog karstic basin (Boyadjiev 1964; Antonov and Danchev 1980) is drained by several springs, of which the Jazo and Kjoshka are the most important. These are located in the preserved area of the Pirin

Mountain National Park and are included on the list of the UN World Natural Heritage sites.

Due to high karstification of marbles, there is no surface runoff in the region. Scant vegetation within the alpine belt and low temperatures impose low evapotranspiration (<30 percent). Large values of yearly precipitation (800–1,000 mm in high parts of the mountain), provide considerable recharge to the ground water. Besides this, karst waters are fed by inflow from rivers. The karst water flows to the north and northeast towards Razlog kettle. The spring issues are manifested among the proluvial fan at the periphery of the Pirin horst (Bakalov et al. 2002). This basin is referred to the watershed of the Mesta River.

3.2. Nastan-Trigrad Karst Basin

The second study area refers to the Nastan-Trigrad karst basin situated in the Central Rhodopes (Boyadjiev 1964). Precambrian marbles of the Dobrostan Formation with a thickness exceeding 2,000 m are exposed in the drainage basin of the river Vucha at altitudes between 700 and 2,190 m. The accumulation of large resources of karst waters is related to intensive and long-lasting karstification and to tectonic movements in the region (Jaranoff 1959).

According to data obtained from meteorological stations, yearly precipitation sums in the region are in the range of 680 to 900 mm. The recharge of the aquifer is due to precipitation and inflow from rivers that enter into the karst terrain (Benderev et al. 1997). The Nastan-Trigrad karst basin is drained by many springs. For the purpose of this study, the Beden spring (N 39a) was selected. The recharge of karstic water is mainly due to snowmelt. The general characteristic of the three karstic springs from both study areas (Razlog and Nastan-Trigrad karst basins) is presented in Table 3.

Table 3: General Characteristics of the Karstic Basins. (for the climatic period 1961 – 1990).

Parameter	Unit	Jazo N 59	Kjoshka N 59a	Beden N 39a
Spring altitude	m a.s.l.	913	941	786
Mean discharge*	m^3/s	1.142	0.417	0.819
Location		Pirin	Pirin	Rhodopes
Village		Razlog	Razlog	Beden
Geological age		Pt	Pt	Pt
Lithological composition		marbles	marbles	marbles

4. SPRING VARIABILITY ANALYSIS

4.1. Information Database

The selected springs (see Table 3) are included in the National Hydrogeological Network of the National Institute of Meteorology and Hydrology, Sofia. The period beginning with the first observation (between 1959 and 1964) for the respective station was investigated in this study. The frequency of measurements was noted 12 times for each year using a current meter. Water stage and water temperature at the mouth of the spring were measured mainly by observers; recorders were operating only at limited number of stations. Using a rating curve, the daily data of water stage were transformed into spring discharge. The groundwater quality was determined four times per year for selected springs. Basic components like nitrates were defined.

Whereas in the springs of Kjoshka and Beden no direct anthropogeneous impact was observed, the water of the Jazo spring was used for domestic, industrial, and irrigation purposes. For the latter, the gauging station is located after water has been supplied to the three sectors. The total amount of water used before the gauging station is estimated to be 21 percent of the total spring discharge. For this study, water usage has been assumed to be constant through the years. However, the domestic and industrial uptake of water started prior to the first observation period while the water uptake for irrigation purposes started after the gauging station was installed.

4.2. Regime of Spring Flow

The regime of the selected springs is described by time series data obtained from the National Hydrogeological Network. The period under study begins from the first observation recorded for the respective station.

As a rule, maximal discharges occur in spring or summer due to snowmelt. The delay in the extremes for springs 59 and 59a is due to the location of their watersheds in the high mountain Pirin.

Beden spring has maximal discharge in April–May and minimal discharge in September–October. Kjoshka spring has a well-defined seasonal cycle with maximum discharge in June and minimum in February–March. At the Jazo spring the maximum and minimum occur one month later and its seasonal cycle is smoother. It shows weak seasonal variation over the year.

Seasonal variability of the spring discharge for the selected springs in relative units Q/Qy (where Q is monthly discharge, Qy—mean yearly value) is presented in Figure 3. Average values of discharges for the 1961–1990 period were used.

Results obtained by Bojilova (1994 and 2001) for the analysis and generation of monthly and seasonal discharges for the studied karstic springs in Bulgaria (Jazo, Kjoshka, Beden) show applicability of stochastic models to reconstitute interannual distribution of the karstic flow. E. Bojilova (2004) applied the method of composition to the Beden and Jazo springs to extend the probability curve of the empirical distribution of both springs. The method makes possible the estimation of quantiles in the range of very low probability of occurrence.

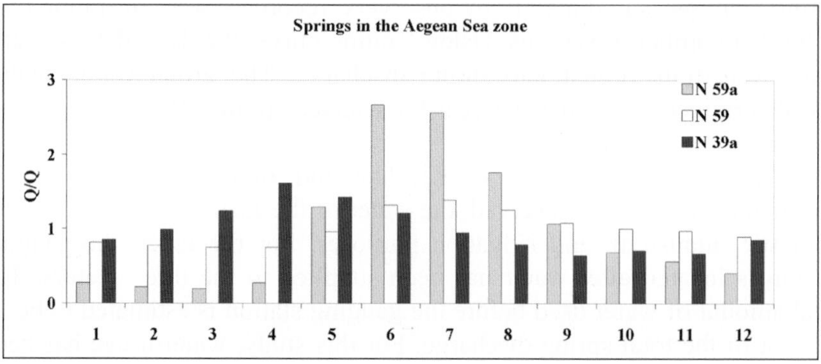

Figure 3: Monthly Regime for Selected Karstic Springs in the Aegean Sea Zones (Orehova, ICHE 2002).

4.3. Regime of Groundwater Temperature

Orehova (2001) made a comparative analysis of variations in water temperature for the chosen springs. Time series of water temperature and spring discharge for the period 1993–1997 were analyzed, showing that each spring has its own temperature behavior.

The frequency distribution of the water temperature is presented in Figure 4. The distribution obtained is bimodal for all springs. Comparative temporal variations for different springs are given in Figure 5. The Jazo and Kjoshka springs are located in the same region but their temperature difference is about two degrees. The altitude of the outflow for Kjoshka spring is higher that for Jazo spring (see Table 3), therefore it is supposed that Kjoshka drains the upper part of marbles, and consequently has a lower temperature and a well-defined seasonal cycle.

The temporal variation of water temperature (minimal monthly temperature) and spring discharge (average month value) is presented in Figures 6–8. The time series are presented in deviations:

$$\psi = \frac{X - \overline{X}}{\sigma_X}, \tag{1}$$

where $\overline{X}$, σ_X are average values and standard deviations respectively.

There is a clear correlation between water temperature and discharge (for some cases the best correlation is obtained for the logarithm of the discharge). The lowest temperatures are observed when spring discharge is at a maximum (see Figures 6 to 8). This corresponds to snow melt and to a massive input of water at low temperature into the aquifers. The recharge events always lead to a decrease in spring water temperature in relation to background temperature.

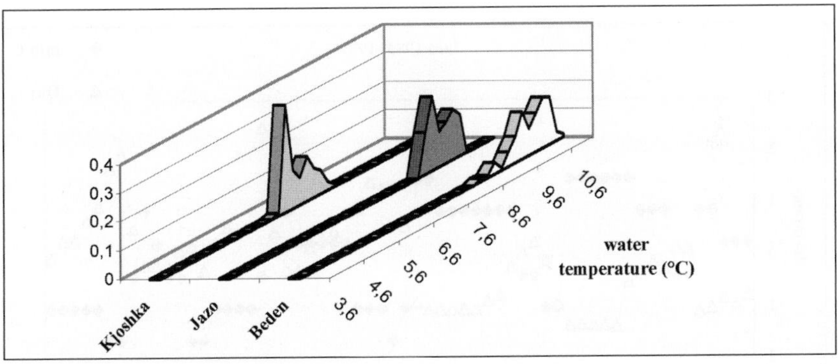

Figure 4: Frequency Distribution of Temperature for the Chosen Karstic Springs.

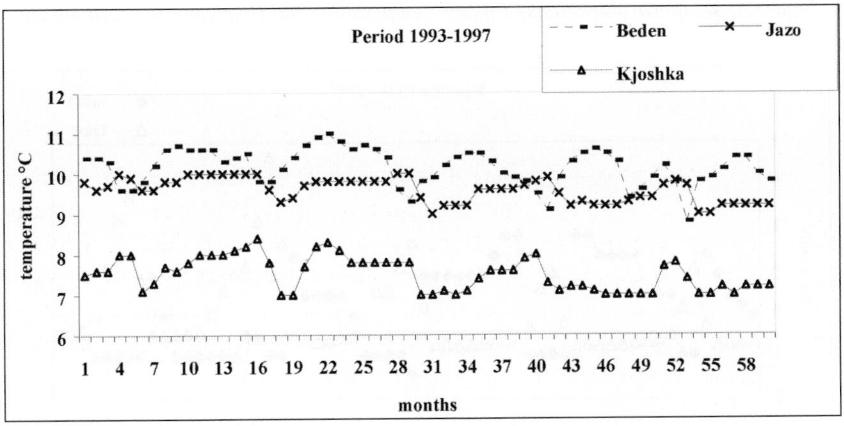

Figure 5: Comparative Temporal Variations of Water Temperature for All Springs.

As discharge falls, the water temperature rises. All studied springs reach maximum temperatures during baseflow period.

The following conclusions can be made (according to Orehova 2001):

- The joint analysis of spring discharge and water temperature confirms some known characteristics of karst springs: the lower temperatures occur during the periods with high discharges.
- All studied springs reach maximum temperatures during baseflow period.
- The time series of the two springs situated in Pirin mountain are asynchronous to springs from the other regions. This is due to the fact that their watersheds are located at much higher altitudes and therefore the snowmelt occurs some months later.

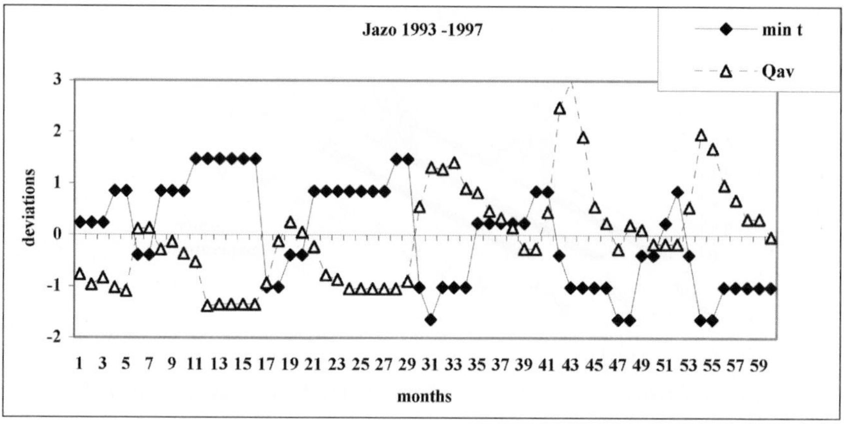

Figure 6: Temporal Variations of Water Temperature and Discharge for Jazo Spring.

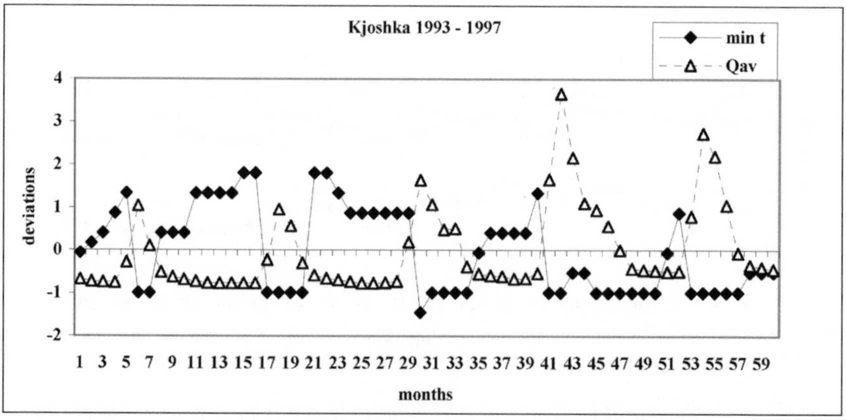

Figure 7: Temporal Variations of Water Temperature and Discharge for Kjoshka Spring.

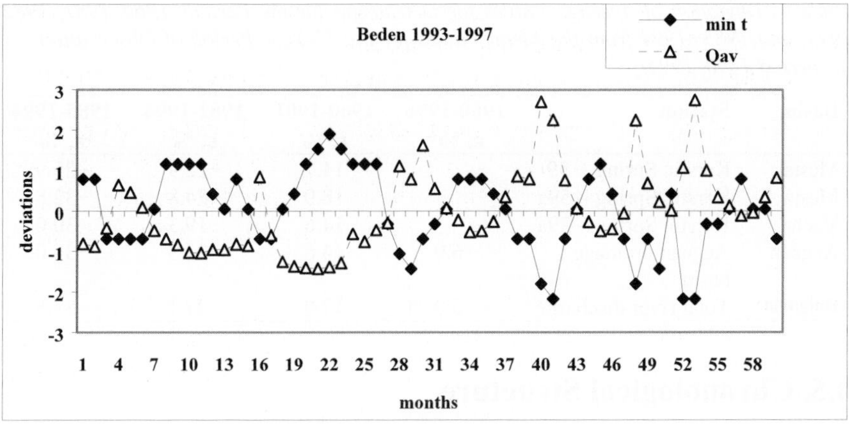

Figure 8: Temporal Variations of Water Temperature and Discharge for Beden Spring.

Water temperature gives valuable information on karst systems. The two springs from Razlog karst basin appear near one another but at different altitudes (see Table 3). Thus their comparative analyses of the discharge and water temperature are useful. Orehova (2001) showed that significant discharge variation and lower temperature give evidence that the Kjoshka spring drains the upper part of marbles, and Jazo their deeper part. Bakalov et al. (2002) also drew the same inference, subjoining lower total mineralization of the upper spring (235 mg/l for Kjoshka and 253 mg/l for Jazo on average for the period 1980–1991).

4.4. Quantitative Assessments of the Effect of the Drought Period to Groundwater

For a quantitative assessment of the 1982 to 1994 drought period the mean values for the periods 1960–1996, 1960–1981, 1982–1994 and 1985–1994 have been calculated. The percent deviation is obtained by the following equation:

$$\varepsilon = \left(\frac{\overline{X}_n}{\overline{X}_N} - 1 \right) 100\% \qquad (2)$$

where n refers to the short period and N for the whole period.

Deviations for shorter periods in comparison to the longest one are presented in Table 4, as well as data for the Aegean Basin and total river discharge for Bulgaria (Orehova and Bojilova 2001a).

Table 4: Deviation of Average Values for Discharges for the Periods 1960–1981, 1982–1994, and 1985−1994 from the Mean Values for the 37-Year Period of Observation. (for the period 1890–1995)

Basin	Station	1960-1996 ε, %*	1960-1981 ε, %	1982-1994 ε, %	1985-1994 ε, %
Mesta	Karstic Spring – 59		14.7	−22.9	−25.9
Mesta	Karstic Spring – 59a		18.0	−24.8	−32.8
Vacha	Karstic Spring – 39a		14.6	−19.3	−30.0
Aegean	Aegean drainage basin	−6.0	14.6	−25.3	−34.0
Bulgaria	Total river discharge	−3.9	17.0	−27.7	−35.8

4.5. Chronological Structure

The time series of spring discharge are analyzed using the method of double-mass curves. Figure 9 is an example of an application of this method.

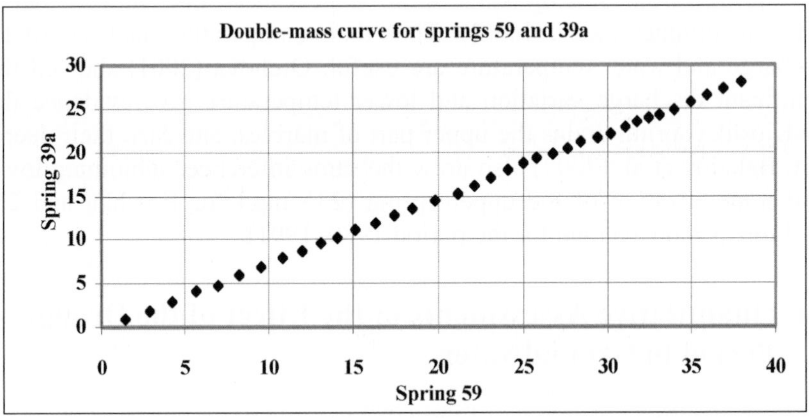

Figure 9: Double-Mass Curve for Springs 59 and 39a (m^3/s).

The double-mass curve is smooth and without significant change in the direction. From this analysis we can infer that there is no visible anthropogenic impact on the regime of karstic springs.

The chronological structure of the investigated periods is presented in Figures 10 to 12. The graphs represent the annual discharge for springs and Aegean drainage basin in deviations using equation (1). For the purpose of this study, the existing data for Aegean Sea basin for the period 1960–1997 were used.

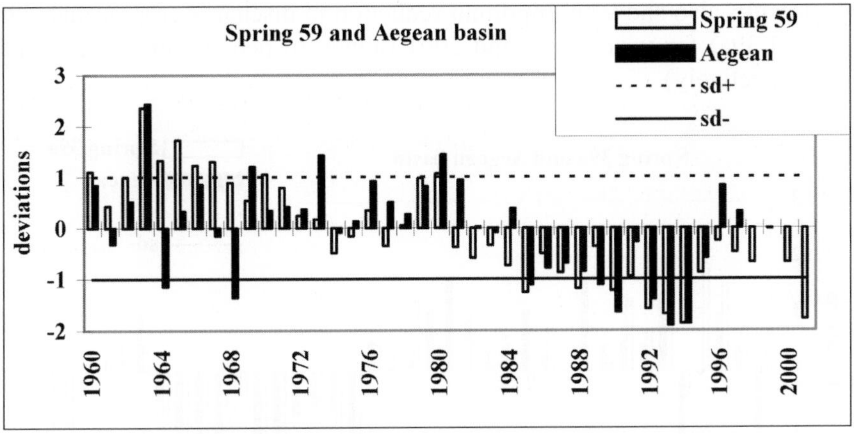

Figure 10: Discharge for Spring 59 and Aegean Basin in Deviations ψ.

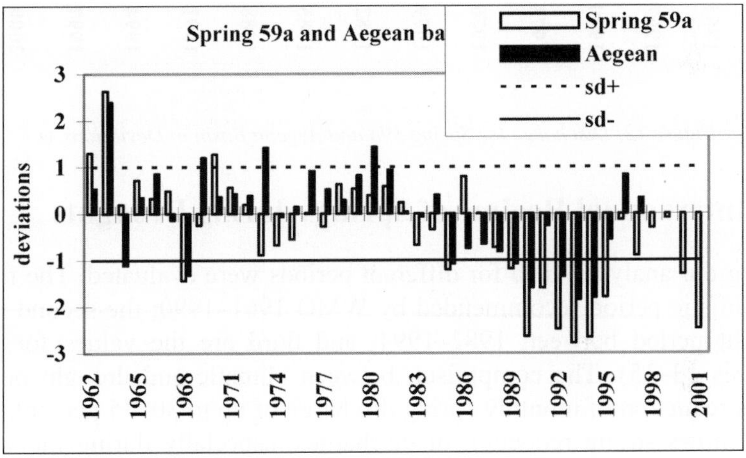

Figure 11: Discharge for Spring 59a and Aegean Basin in Deviations ψ.

The information presented in Table 4 and Figures 10 to 12 provides the following results:

- The drought period 1982–1994 and especially the short one between 1985–1994 are characterized by considerable reduction in karstic spring discharges (up to 25–30 percent);
- The chronological structure of the karstic spring discharge is similar to that of the river runoff in the Aegean zone;
- The minimal values of spring discharges are registered in 1993–1994. Since 1995–1996 the yearly average discharges tend to reach their multi-annual average values.

- Results also show an important reduction in discharge (for springs 59a and 39a) for years 2000 and 2001 (about 30 percent and 40 percent, respectively).

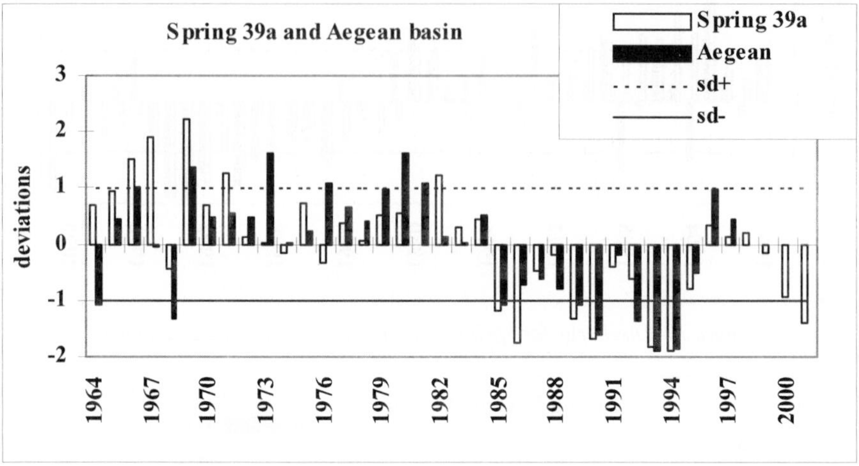

Figure 12: Discharge for Spring 39a and Aegean Basin in Deviations ψ.

4.6. Interannual Regime of Springs during Droughts

During our analyses, data for different periods were evaluated. The first is the climatic period recommended by WMO 1961–1990; the second is the drought period between 1982–1994; and third are the values for 2001 (Figures 13–15). The comparison between climatic and drought periods shows reduction of monthly spring discharge of up to 30–35 percent. Year 2001 shows strong reduction of discharges, especially during the winter months (Orehova 2002a). According to Andreeva and Orehova (2004), 2001 had a mild winter. Besides higher temperatures, mild winters are characterized by lower precipitation, and thus lead to reduced recharge to the aquifers.

For surface runoff, M. Genev assessed the probability structure of drought in Bulgaria, evaluating the extended period 1982–2001. He established that during the period under investigation a very low probability of occurrence was observed. According to his study, it is not possible to clearly say where exactly is the end of the drought period observed on the Balkan Peninsula (Genev 2004).

Our recent research confirms previously reached conclusions that the reduction of precipitation during mild winters has a strong negative influence on groundwater recharge (Andreeva and Orehova 2004). The obtained results can be used for the estimation of long-term spring variability.

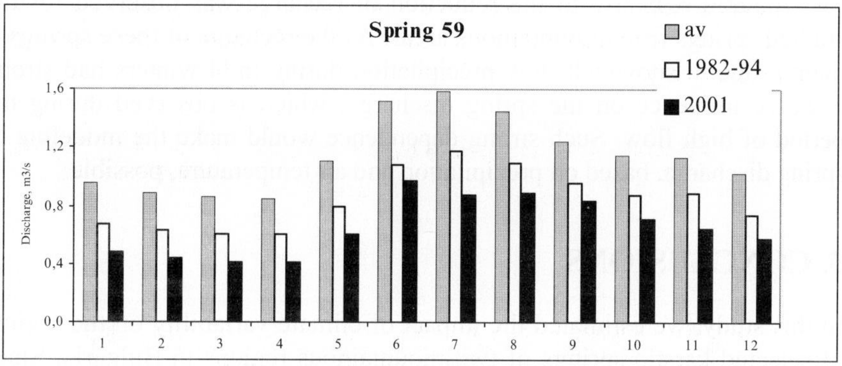

Figure 13: Monthly Values for Jazo Spring.

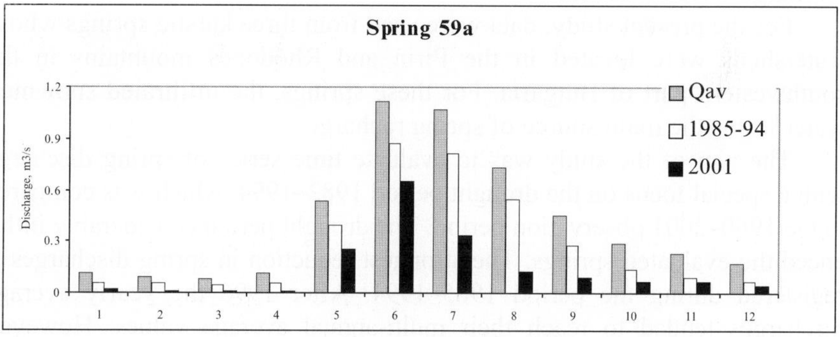

Figure 14: Monthly Values Kjoshka Spring.

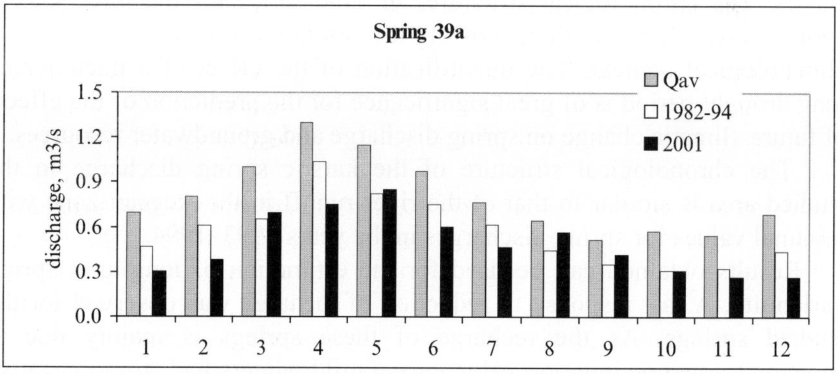

Figure 15: Monthly Values for Beden Spring.

A rapid response to the reduction of recharge was observed for the studied springs from mountainous areas. As the recharge of these springs is mainly due to snowmelt, low precipitation during mild winters had strong negative influence on the spring discharge, which is observed during the period of high flow. Such strong dependence would make the modeling of spring discharge, based on precipitation and air temperature, possible.

5. CONCLUSIONS

In this study, we estimated the impact of climate variability on the regime of selected karstic springs of two mountainous regions in Bulgaria. Since 1981 a continuous decrease in rainfall combined with an increase in air temperature was registered. As a result reduced river runoff and spring discharge were observed.

For the present study, data were used from three karstic springs whose watersheds were located in the Pirin and Rhodopes mountains in the southwestern part of Bulgaria. For these springs, the infiltrated snowmelt water forms the main source of spring recharge.

The aim of the study was to evaluate time series of spring discharge with a special focus on the drought period 1982–1994, which was compared to the 1960–2001 observation period. The drought period considerably influenced the evaluated springs. The strongest reduction in spring discharges is registered during the period 1985–1994. After 1996 the yearly average discharges tended to reach their multi-annual average values. However, reduced values of spring discharges were observed again during 2001.

Spring discharges were analyzed using the method of double-mass curves. The chronological structures of time series for the three karstic springs resemble each other, showing the similarity in their geological and climatological context. The quantification of the effect of a documented long drought period is of great significance for the prediction of the effects of future climatic change on spring discharge and groundwater resources.

The chronological structure of the karstic spring discharge in the studied area is similar to that of the river runoff in the Aegean zone with minimal values for spring discharges in the years 1993–1994.

Results obtained can be used for the estimation of long-term spring variability. A fast response to reduction of recharge was observed for the studied springs. As the recharge of these springs is mainly due to snowmelt, low precipitation value during mild winters had strong negative influence on the spring discharge. The strong dependence would give rise to the possibility of modeling spring discharge based on precipitation and air temperature.

POLICY IMPLICATIONS FOR EFFICIENT AND EQUITABLE WATER USE

WATER AND MOUNTAINS, UPSTREAM AND DOWNSTREAM: ANALYZING UNEQUAL RELATIONS

Urs Luterbacher
Graduate Institute of International Studies
Geneva, Switzerland
luterbac@hei.unige.ch

Duishen Mamatkanov
Institute of Water Problems and Hydropower
Kyrgyz National Academy of Sciences
Bishkek, Kyrgyzstan

Abstract: This chapter analyzes unequal access to water with a special focus on the Central Asian situation. It emphasizes the crucial issue of property rights and examines what happens when these are distributed in ways that lead to major inefficiencies and conflict. The chapter first presents a game theoretical investigation of this type of conflict situation and then ways in which the conflict might be solved if parties continue having relative risk aversion. In the case of Central Asia this type of solution would lead to mutually beneficial outcomes if credibility problems were lifted by using international institutions to guarantee the observance of contracts that contain prescriptions to share benefits associated with a change in the property rights structure.

Keywords: water, unequal access, property rights, Central Asia, game theory risk preferences

1. INTRODUCTION

Access to water at both the domestic and international levels is often characterized by inequality. Sometimes inequalities stem from different capabilities in terms of wealth and technology. Frequently, however, they are combined with geographical characteristics that give countries and regions differential access to water sources. This is particularly the case with upland-lowland situations, which are often combined with upstream–downstream relationships. The question raised by unequal situations is—as is often the case with natural resources—one of property rights. The issue

E. Wiegandt (ed.), Mountains: Sources of Water, Sources of Knowledge, 283–303.

is ultimately whether "ownership" can be clearly specified or whether users must share the water.

At the domestic level, the answer to such a question is relatively straightforward because in most cases some sort of property rights (or absence thereof) is specified. The problem is more complicated at the international level because there is no unambiguous jurisdiction with power of enforcement that can constrain the behavior of states. Thus power relations and power bargaining will often determine the outcome over a disputed access to natural resources, including water. Nevertheless, even at the international level, states often have established either contractual or customary arrangements over time. New problems in state relations often come about because such relations are questioned under the pressure of new economic or technological developments.

2. THE CENTRAL ASIAN SITUATION

Central Asia provides an emblematic context for addressing these questions because of both the level of tension water rights issues have generated and their underlying causal factors. What characterizes the Central Asian case is the evolution of political structures in which internal relations in the region, historically managed by a strong central authority, have become—since the demise of the Soviet Union—international relations between upstream and downstream countries where other asymmetries also play an important role. Reports from Central Asia regularly alert the international community to worsening ecological conditions, the dire social and economic status of its population, and the ensuing potential for serious civil and interstate conflict. The situation is particularly complex and delicate because familiar problems of over-extensive irrigation agriculture and population increase have become mixed with interstate politics as a result of the collapse of the Soviet Union. As a consequence, "a very complex water management problem became a very complex *transboundary* water management problem" (Veiga da Cunha 1994, p. 6).

River water resources, especially from the Amu Darya and Syr Darya Rivers, play essential roles in the economies and societies of the Central Asian states of Kyrgyzstan, Tajikistan, Turkmenistan, Uzbekistan, and Kazakhstan, some of which have become dependent in varying degrees on irrigated crops for survival. For example, cotton—the most important irrigation crop—is the major source of income and employment in Turkmenistan and Uzbekistan. Its production was encouraged under the

Soviet system as a source of hard currency. Overuse of water resources for irrigation is responsible for the drying of the Aral Sea, whose surface and volume have declined by 35 percent and 58 percent respectively since the mid-1980s. Water is wasted for cotton production in areas otherwise not suited for this culture in Uzbekistan and Kazakhstan. It is provided for free most often so there is no incentive to preserve it. The entire irrigation system relies on 32,000 km of canals, which are poorly maintained and full of leaks—including, for instance, the Karakoum canal, which constitutes a 1,340 kilometer open-air water way in the Turkmenistan Desert. Needless to say, losses to evaporation under such conditions are tremendous. The water that is used in agriculture can not be exploited by the upstream countries of Kyrgyzstan and Tajikistan, who rely on hydropower for 50 percent of their electricity production.

These competing water uses are aggravated by demographic pressures. Population grew by 140 percent between 1959 and 1989 (Horsman 2001, p. 71) and is projected to increase between 35 and 50 percent in most of the states between now and 2050 (Population Reference Bureau 2002).

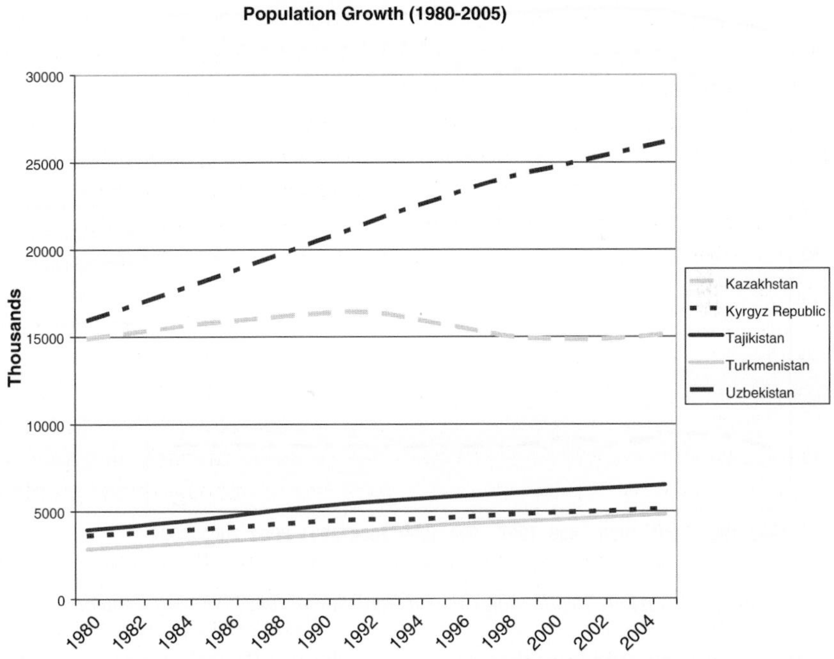

Graph 1: Population Growth (World Bank Social and Economic Indicators).

Although tensions over water allocation are not new, they have taken on new significance since the collapse of the Soviet Union. Managed until 1992 from Moscow by a centralized administration, water systems have suddenly come under the control of separate sovereign states that have no history of agreements or coordination structures. This poses important allocation problems because of the nature of water resources and the weakness of new state institutions in the Central Asian republics.

Because upstream states Kyrgyzstan and Tajikistan need water for hydroelectric production as well as irrigation—Kazakhstan and Uzbekistan use water mostly for irrigation—they have held up release of water or threatened to charge for delivery downstream in order to pressure downstream users to compensate for energy production lost when water is released for downstream irrigation. It is important to note that despite their control over the source of water, upstream states are implicated in allocation schemes that oblige them to provide water downstream.

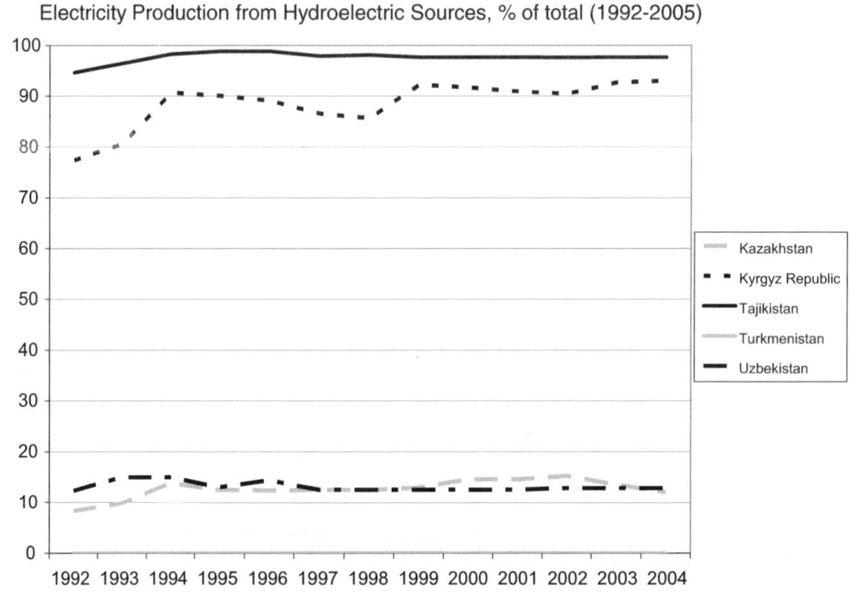

Graph 2: Electricity Production (World Bank Social and Economic Indicators).

It is common wisdom that resources whose allocation proves problematic because of the difficulty of assigning clear, unambiguous property rights are generally managed through common or centralized property institutions. This was the case under Soviet rule when the whole region was controlled from Moscow and agriculture, industrial, and energy production were part of national policy. The central state was able to enforce exchanges of water and energy between upstream and downstream users.

At present, the previous patterns are maintained, after being reaffirmed in the Almaty Agreement of 1992, but they are not perceived to be equitable. The arrangements reflect the favored status that downstream countries had achieved during the Soviet era. Kyrgyzstan and Tajikistan would like to expand irrigation agriculture and, especially, electricity production. However, even their dominant upstream position does not permit them to achieve their goals because of their political weakness in front of the downstream users' control over coal and gas and the energy produced by these fuels (Horsman 2001, pp. 74–75). Indeed, after independence, Uzbekistan and Kazakhstan introduced market prices for gas and coal. Kyrgyzstan could not pay these higher prices and their response was to increase electricity production in order to augment revenues. This meant that the amount of water available for downstream irrigation in Uzbekistan and Kazakhstan was reduced. As a consequence, agreements were not respected. In breaching their commitments, upstream states thus became vulnerable to reprisals from downstream states that have refused to provide energy in the form of natural gas and coal in return for water.

The case of Central Asia illustrates two important aspects of water resource allocation. The first is the nature of the asymmetries associated with its allocation. Upstream republics can control quantities of water sent downstream, but they are subject to reprisals because they do not control other critical resources like gas and coal. The second aspect is the shift from the status of regions in a single state to that of independent republics, which highlights the weakness of international regimes to regulate transboundary waters. No authority is empowered to impose a distribution system.

However inequitable an upstream–downstream relation may be, it will be stable unless the downstream user has other resources or power with which to pressure those upstream that control access to water. The asymmetric distribution of resources in the Central Asian republics is an example of an unstable relationship between users of a same natural resource with unclear property rights and crosscutting powers or access. In this case, there is no obvious solution that is both equitable and efficient. Management schemes must therefore be negotiated.

In Central Asia the asymmetry in access to waters could be compensated by industrial and agricultural developments that could, in the end, benefit all countries concerned. There might also be favorable spillover effects, where two countries or regions could share in the advantages created by the development of water resources in one region by specializing in complementary activities. This might consist of developing industry in the area less suitable for agricultural development and taking advantage of the cheaper electric power made available by dam construction in an upstream country. This dam development need not be limited to river run off low pressure constructions. High altitude countries can develop the kind of high pressure dams common in Alpine areas that are much less harmful for the downstream regions. These also have the advantage of providing large amounts of electric power under peak load conditions. They can then serve the industrial needs of firms located relatively far away from their particular location and thus be useful beyond the boundaries of a given country. Central Asia could be an example of such cooperation in power generation. The high altitude but relatively poor countries of Kyrgyzstan and Tajikistan could benefit from such schemes. In these cases, existing international organizations such as the World Bank would be appropriate institutions to devise policies that favor such positive spillovers and thus help to resolve otherwise intractable asymmetric water conflicts.

However, such a cooperative outcome is far from obvious. It requires a proper strategic analysis of asymmetric situations such as the ones we have been evoking above. We have to find out what the dilemmas are and which decisions and attitudes can get the parties out of the current dispute.

3. THEORETICAL CONSIDERATIONS

Access rights to water, even when inequalities are present, often take the form of shared property arrangements. It used to be difficult, at least in the past, to stop river flows so that waters from a basin were shared between different owners or different states. There are clear advantages to common property, notably risk sharing. The example of pools of water under properties defined at the surface is relevant. For each individual owner of the surface properties, digging a well might not be worth it because of the risks associated with the prospect of not finding any water under a particular property. Yet, as shown by Dasgupta and Heal (1979, p. 383), risk sharing in a common property arrangement tremendously increases the possibility of deriving benefits from digging wells in a coordinated fashion. In fact, the greater the number of participants in the risk sharing operation, the lower the costs associated with the enterprise and thus the higher the benefits for each individual owner. Similar reasoning can be made with respect to irrigation: it represents a kind of insurance scheme for agricultural producers who can then make use of it in case of drought. Thus, even risk-averse individual owners have an incentive to enter such an insurance scheme, which renders the costs of risk bearing negative (Dasgupta and Heal 1979, p. 386).

Within common property structures for water, the combination of the resource's inherent characteristics, technological features, and the institutional configurations related to its management underlie arrangements to prevent some groups or individuals from over-using water resources at the expense of others. Frequently the solution is to have a collective or central management structure. This form of management has its own special problems, the major one being the familiar "free rider" problem. Collective users have incentives to maximize the benefits of the common resource without paying their fair share of the costs. However, there are significant differences between water availability in the form of an underground or surface pool—like a lake—and in the form of a river basin with sequential access by different riparian users. Pool type access can be illustrated by the graph below, which shows that without socially imposed limits, the end result is overexploitation and dissipation of the resource.

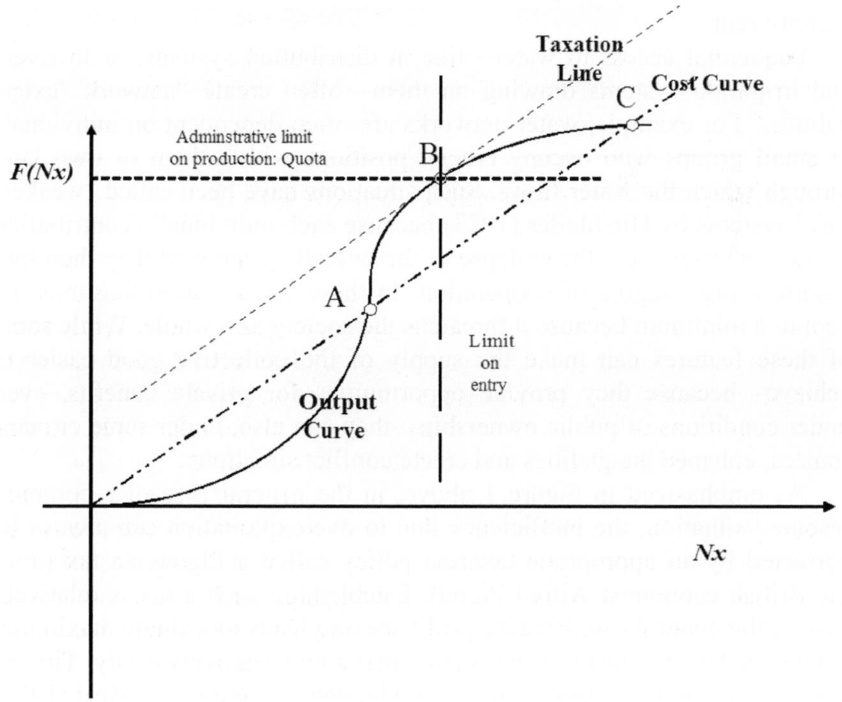

Figure 1: Optimal and Unsustainable Resource Use.

As this graph illustrates, production of a given natural resource type good like fresh water (the y axis) depends on the number of producers willing to share in its supply (the x axis). In order to ensure benefits in production, there must exist a relatively large number of producers, leading initially to increasing returns to scale. As more and more producers join, however, diminishing returns set in. This is illustrated by the S shaped aspect of the production curve. If each producer incurs a unit fixed cost, then the total cost of production can be represented by a (here the mixed dashed and dotted) straight line. This straight line crosses the production curve at two points: at A, where the number of producers is sufficient to initiate profitability; and at C, where too many producers have completely dissipated profits or surpluses. This also corresponds to the dissipation of the scarcity rent associated with the natural resource. Quite clearly the optimum lies at B, which maximizes surplus production over costs. This optimal situation can be maintained by limiting the number of producers entering the process, either by exclusion or by taxes, which increases the fixed costs of each entrant. A solution to the problem involving taxes is presented by the other (dashed) straight line, running parallel to the cost line and tangent to the production curve in B. A correct taxation of each entrant limits their number and assures maintenance of the scarcity rent.

Sequential access to water—like in distribution systems, or in rivers and irrigation systems drawing on them—often create "network "externalities." For example, water networks are often dependent on individuals or small groups who occupy crucial positions within them or own land through which the water flows. Such situations have been called "weakest link" systems by Hirshleifer (1983) because each individual's contribution is essential to prevent the collapse of the overall system and they therefore require a high degree of cooperation. In these cases, free-riding must be kept to a minimum because it threatens the society as a whole. While some of these features can make the supply of the collective good easier to achieve—because they provide opportunities for private benefits, even under conditions of public ownership—they can also, under some circumstances, enhance inequalities and create conflict situations.

As emphasized in Figure 1 above, in the general case of a common resource situation, the inefficiency due to overexploitation can always be corrected by an appropriate taxation policy called a Pigouvian tax (after the British economist Alfred Pigou). Establishing such a tax is relatively easy in the general case because profit seeking leads to a single maximum. A tax can then be used to reach such a maximum relatively easily. The tax keeps too many firms from entering production. Dasgupta and Heal (1979) show that such an outcome does not obtain when asymmetries, such as upstream–downstream relations with differentially defined property rights, are present. Imagine for instance the following: A downstream firm has to base its production on water that is also used by an upstream firm. Two parameters are important here. On the one hand, the degree of pollution generated by the first firm may significantly cut the possible profits available to the second firm; on the other hand, each firm has some kind of property rights as well as legal rights and obligations with respect to each other in terms of clean water. If the first firm has an unlimited right to pollute and the second firm no rights to clean water, the second firm will eventually be driven out of business in the absence of some Coasian (Coase 1960) type negotiated arrangement between the two. Conversely, if the first firm is constrained to produce without any pollution, it might have to cease its activities. However, things change when some property rights are assigned to one side or the other. If a maximum pollution level ê is allowed, the production possibilities of either firm are no longer convex and neither taxation nor buying or selling of rights will bring uniquely defined equilibrium solutions (Figure 2).

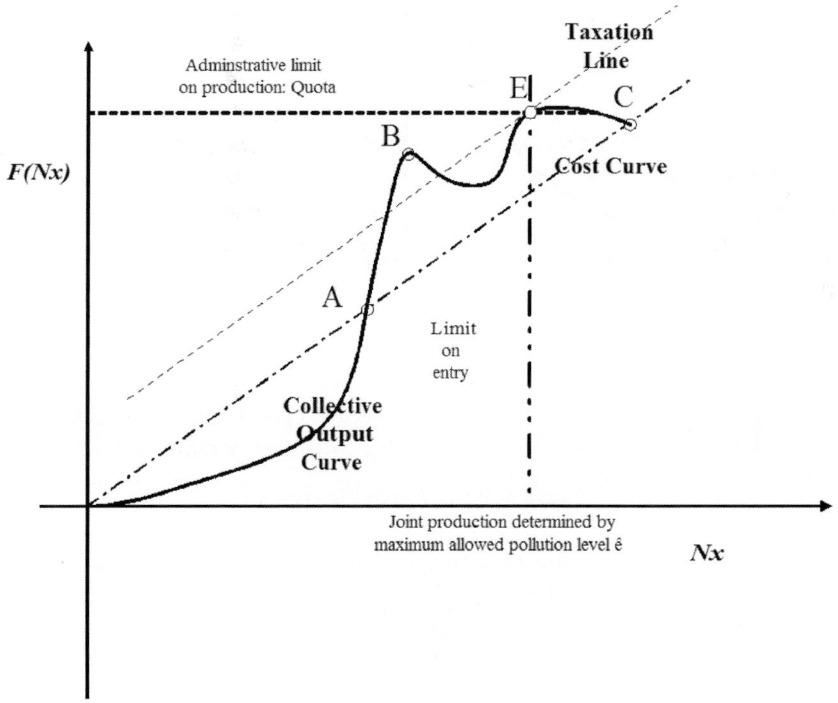

Figure 2: Water Use Disequilibria.

In Figure 2, point B would represent a better level of profit maximization than point E. But joint production will be stuck at the relatively suboptimal level E because property rights are determined in such a way that one firm uses the resource up to its maximal allowed level of depletion and thus keeps the other one from using the same resource more efficiently by restricting usage more for the first firm. Conversely, it is possible to imagine a situation where established property rights severely limit the access to resources of the firm that is the most efficient, thus reducing overall efficiency of resource use. This can be illustrated by the following graph (Figure 3):

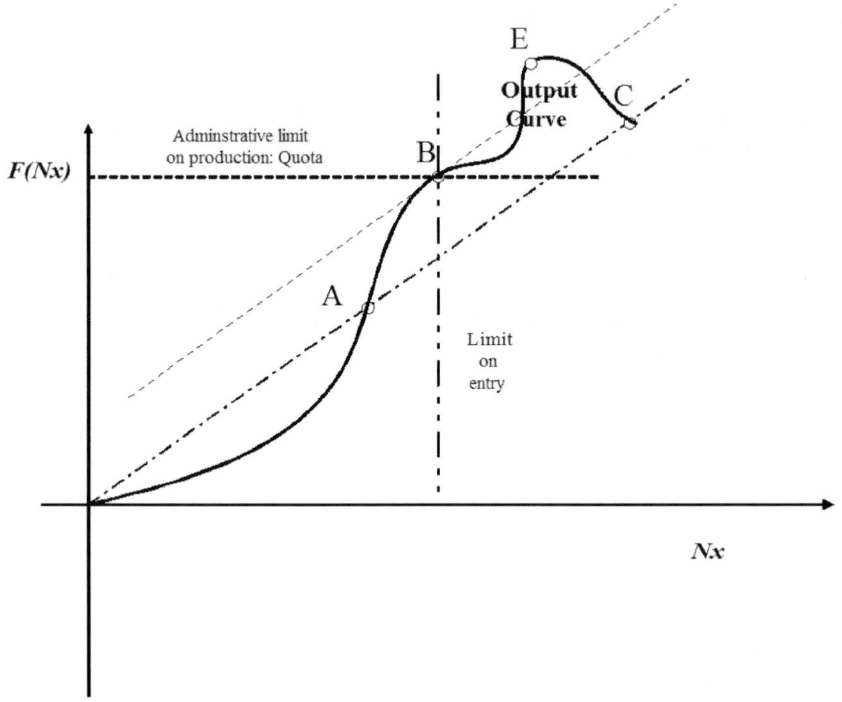

Figure 3: Water Use Inefficiencies.

One could ask why the two firms could not come to a Coasian type agreement where one of them could relinquish some of its property rights in exchange for a share in the higher profits generated by the one with the best productivity. The problem with such an arrangement is often one of credibility. Indeed, once property rights have been relinquished, or claims abandoned, there is no guarantee that the firm benefiting from this operation will then share the benefits if some superior authority does not insure the execution of the contract. This credibility issue is particularly acute if countries are involved instead of firms. Generally, in the international system no supreme authority exists to enforce contracts except in cases where states recognize an overarching authority such as an international court. This process can be described with the help of a sequential decision-making process with imperfect information between two decision makers, like the one presented in Figure 4. Here, two decision-makers, 1 and 2, decide sequentially about strategies to follow. Quite clearly, if information is complete and mutual responses fully anticipated on the basis of end payoffs, the equilibrium outcomes are generated by the sequences R L' or L R'(bold lines in Figure 4). However, only the sequence R L' is what is called subgame perfect, i.e., it can only be undermined by an incredible threat strategy, here a possible commitment by decision-maker 2 to play R' no matter what 1 does. This attitude is, however, incredible because once 1 has played R, 2 can only respond by playing L' to maximize his end gains (1 instead of 0). It is worth noticing however that the sequence L L' might bring in the end the best results for both decision-makers. Indeed, if somehow 1 and 2 can agree to share later on in the payoff of respectively 5 and 1, they could achieve a much better final outcome for both of them (assuming that 1 is able to redistribute from his own share of 5 a value to 2 that is higher than the 2 she would get if she reaches her best possible outcome). How can such a result obtain in a rational way, i.e., as deriving from the interactions between self interested

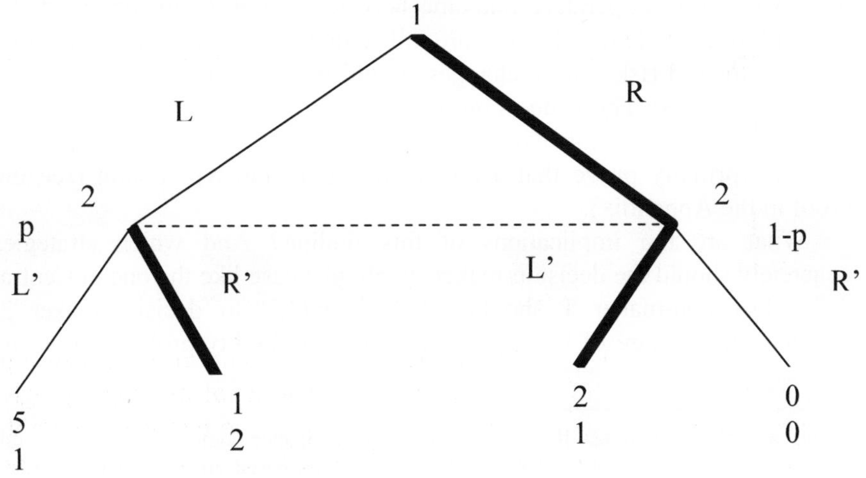

Figure 4: Game Tree with Imperfect Information.

parties? As we will try to show, the sequential decision-making process changes completely when the perspective of imperfect information is introduced. Assume that 2 does not know what 1 has chosen and attributes an equal probability of occurrence to the choice of both L and R by 1, i.e., $p=1-p$, which is consistent with a risk dominance analysis (Harsanyi and Selten 1988). Then 2 should determine her decision-making preferences on the basis of expected utility analysis.

However, here this decision-making procedure does not lead to any particular course of action since both L' and R' have the same expected utility (namely 1) for decision-maker 2 and therefore no simple risk dominance can be established. Even though both L' and R' are equivalent, it can be argued that 2 will act differently, depending on her attitude toward risk. If 2 is risk averse she will always chose L' because she will get 1 for sure; if she is risk preferring she will chose R', which gives her a chance of a higher reward. Decision-maker 1 will therefore get 5 if he anticipates that decision-maker 2 is risk averse; 1 if she is risk preferring.[1] If we assume that attitudes toward risk may change depending on the circumstances, the strategic problem for decision-maker 1 is to make sure that decision-maker 2 is risk averse. One is facing some kind of dilemma here because if 1 pushes to hard to extract concessions, 2 may feel threatened and become risk preferring. Decision-maker 1 has to try to anticipate 2's risk attitude and this may imply promising concessions rather than insisting on concessions by the other side. Once 2 has settled on a risk averse attitude, she will cooperate and make concessions of her own and a cooperative outcome between 1 and 2 will obtain if the assumption is made that 1 and 2 are utility maximizers and have changing attitudes toward risk. These changes should be occurring as a function of gains and losses: These decision-makers are risk averse with respect to gains and risk preferring with respect to losses. If all these are satisfied, one can formally prove that a cooperative outcome will obtain (see the proof in the Appendix).

What are the implications of this finding? And which strategies concretely should the decision-makers apply in a case like the one in Central Asia? Decision-maker 1 should offer a contract to decision-maker 2, guaranteeing a share of his gains to her. The credibility problem could be

---

[1] This analysis only holds if one uses the risk dominance logic that implies a full compatibility with backward induction. Other approaches based on the notion of perfect Bayesian equilibrium (for an illustration, see Gibbons 1992, pp. 175–183) would insist that for decision-maker 2, p can only be 1 because it corresponds to an equilibrium belief. The equilibrium is the same if we assume that 2 is risk preferring; but this is not necessarily a correct assumption.

lifted, on the one hand, if decision-maker 1 does not act too aggressively, thus making 2 risk preferring and, on the other, if the contract is guaranteed by a third party like an international institution. In our case, as already mentioned, this would enable upstream republics, in particular Kyrgyzstan, to develop their hydroelectric potential with a promise to share the benefits with the downstream republics. Calculations presented elsewhere (Luterbacher, Kuzmichenok, Shalpykova, and Wiegandt, forthcoming) show that the hydroelectric potential of the region is largely superior to the current consumption of the countries of the whole region. Sharing profits from hydroelectricity would thus be a win-win situation, especially if one includes in addition the environmental benefits from such an operation.

4. CONCLUSIONS

We have tried to show in this chapter that asymmetric water conflicts do not have to escalate to open hostilities and that often they can be resolved through the use of special cooperative strategies at the state level. A game theoretical approach can be used to show that such strategies can be implemented in a way that makes sense analytically and that leads to a win-win situation: namely, the development of a hydroelectric potential whose benefits would be shared among all Central Asian states and whose redistribution scheme would be guaranteed by an international institution such as the World Bank. The remaining problem is that such win-win outcomes, despite their attractiveness, are not always achieved because they get opposed by powerful domestic interests. The political analysis of international trade has shown that commercial openness will be countered domestically by groups who are bound to loose from it. These are often powerful enough, despite being a minority, to derail any successful interstate negotiation. It is quite clear that agricultural interests in the downstream republics might incur losses from a hydroelectric buildup in upstream countries because they might have to face higher irrigation costs. There are however two ways in which such tendencies can be countered. First, the same international guarantors can put pressure on countries to resist such moves and threaten to cut international credit if these win-win projects are not implemented. Second, domestic interests that export industrial and fossil fuel could be mobilized to support new projects. In both cases, international institutions can play a crucial role in eliminating these sources of international conflict. Such tasks should however be in close conformity to their ultimate objective: to help organize more harmonious relations among states.

APPENDIX: UTILITY FUNCTION

Are parties to a bargaining process risk averse or risk preferring? This is basically an empirical question. However, a theoretical model that is constructed in order to be applied to concrete situations has to be able to account for both attitudes. So far analysts have dismissed the appearance of risk preference attitudes as relatively rare, even though it provides the one who displays it with increased bargaining power. Risk preference attitudes by themselves are probably infrequent. They may however appear quite often within "mixed" attitude representations. Experimental psychologists and even observers of animal behavior have noticed that risk preference often appears after risk aversion when a decision-maker is faced with the prospect of losses (Stephens 1990). Risk aversion and risk preferring behavior are regularly seen together, and various attempts have been made to explain their joint appearance. The principal analyses of hybrid risk attitudes are Battalio et al. (1990), Battalio et al. (1985), Camerer (1989), Fishburn and Kochenberger (1979), Friedman and Savage (1948),[2] and especially Kahneman and Tversky (1979) and Tversky and Kahneman (1992). In particular, Fishburn and Kochenberger (1979) show that the majority of individuals have an everywhere increasing utility function $U(x)$, where x is a measure of gains and losses in terms of the valued item alluded to above. *Most individuals are thus risk averse over gains and risk preferring over losses.* This notion can serve as a theoretical justification for the contention elaborated by Hirshleifer (1991)—that the poor have a comparative advantage in appropriation, which is obviously a more risky way to acquire wealth than capital accumulation through savings. Further, like Kahneman and Tversky (1979, 1992), Fishburn and Kochenberger show that the utility of no change is 0 (i.e., $U(0) = 0$), and that U is more steeply sloped over losses than over gains (i.e., $U'(-x) > U'(x)$ *for all* $x > 0$). A systematic discussion of these findings is given in Neilson (1993).

A natural extension of these considerations is to represent an average decision-maker's utility function by an everywhere increasing S curve in x, which adequately expresses the mix of risk aversion under gains and risk preference over losses.[3] Without loss of generality, we can then present the following risk averse/risk preferring (S shaped) utility curve (Graph 3):

---

[2] Friedman and Savage use a perspective on the utility function in their article that differs markedly from ours.

[3] The S curve analysis and its application to conflict has been initiated by Dacey (1996a, 1996b, 1996c, 1998) and Dacey and Gallant (1997). The formulation used here below for the critical risk ratio is based on losses, whereas the formulation used in Dacey is based on gains. These formulations are logically equivalent. The formulation employed here is the one used in Harsanyi (1977).

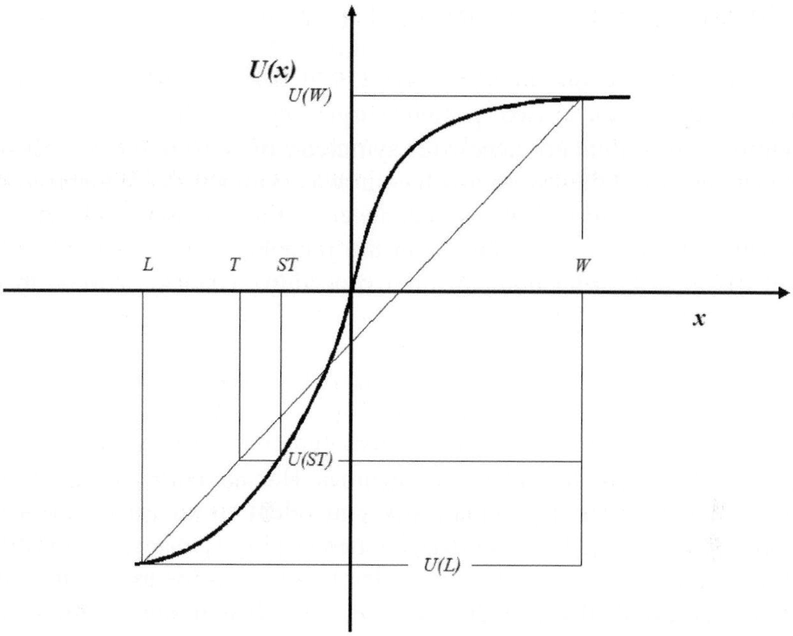

Graph 3: Risk Averse/Risk Preferring Utility Function.

We present graphically here a risk averse/risk preferring curve that spans an interval on the x axis from W (winning) to L (losing). A sure thing value ST of x is also presented on this axis. One should note that this sure thing value is susceptible to change as a result of bargaining with another agent. In other words, ST may represent an "offer" by the other side. These values of x are projected via the S curve onto the y axis where they give respectively $U(W)$, $U(ST)$, and $U(L)$. T is the projection of $U(ST)$ via the cord $U(L)$—$U(W)$ onto the x axis. T defines the interval $ST-T$ or $T-ST$—namely, the gain an individual seeks by taking risks or is willing to forgo by not taking them. We clearly have under risk aversion $T>ST$, and under risk preference $T<ST$. The switch from risk preference to risk aversion can then be described in terms of these inequalities.

What are the advantages of this model? It can give straightforward answers as we will show, as of when conflict initiation is preferred over staying at the bargaining table, when an agreement will be struck and why sometimes bargainers might get stuck in conflict.

Traditional bargaining theory[4] has been presented within two apparently different, but ultimately common frameworks. The first and older conception is due to John Nash (1950, 1953). Nash showed that a unique solution to a two person bargaining problem obtains under conditions of (1) joint efficiency; (2) symmetry of gains to the two actors if the game situation they were involved in was symmetric; (3) linear invariance of the solution; and (4) independence of the solution from irrelevant alternatives.[5] The unique solution to the joint bargaining problem is the result of the following maximization procedure: Choose actions (in our case) for *i* and *j* so as to:

$$Max\ [U_i - U_i(L_i)][U_j - U_j(L_i)]$$

where U_i and U_j are the respective utility functions of players *i* and *j* and $U_i(L_i)$ and $U_j(L_j)$ are disagreement outcomes. Harsanyi (1977) then showed that Nash's theory is mathematically equivalent to an earlier theory of bargaining developed by Frederick Zeuthen (1930). Harsanyi demonstrates that the Zeuthen theory expresses the bargaining process as a sequence of moves that eventually converge to the Nash bargaining solution. This demonstration is based on the notion of a critical risk ratio. As noted above, the critical risk ratio measures the probability of defecting or choosing the conflict outcome. It is:

$$r_{ij} = \frac{U_i(x_{ij}) - U_i(x_{ji})}{U_i(x_{ij}) - U_i(c)}$$

Here x_{ij} represents what agent *i* expects from agent *j* in the bargaining process, whereas x_{ji} is what he gets as an offer from *j* and *c* means the value of no agreement or conflict between the two agents. Obviously from the above fraction, r_{ij} is *0* if the offer from the other agent corresponds exactly to what he wants. On the other hand, r_{ij} is *1* if the offer from the other side does not differ from the value of the no agreement or conflict situation. Thus r_{ij} varies between *0* and *1*. A symmetric consideration holds for agent *j*. Harsanyi further postulates that, within a bargaining process,

---

[4] Because the Nash solution sometimes involves the use of cooperative strategies, it is often considered to be only a part of cooperative game theory. However, the Nash theory is *not* confined to cooperative games. Harsanyi, for instance, has a lengthy discussion about its pertinence for noncooperative games (Harsanyi 1977, pp. 273–290). This point is also emphasized by Hargreaves Heap and Varoufakis (1995, p. 113).

[5] A thorough presentation of the Nash postulates is presented in Harsanyi (1977, pp. 144–146) and also in Binmore (1998, pp. 94–98).

the player with the lower critical risk ratio concedes to the player with the higher critical risk ratio. When the critical risk ratios of both players are equal, both make a concession. We will call these the Harsanyi-Zeuthen rules. Harsanyi shows that if both agents behave in this way the bargaining process will inevitably converge to the Nash solution. More recently, the bargaining approach advanced by Rubinstein (1982, 1985) has been considered more convincing than the Harsanyi-Zeuthen-Nash approach.

On the surface, these two approaches look very different. While the Harsanyi-Zeuthen-Nash theory can be interpreted as a sequence of bargaining moves, the particular sequential nature of the bargaining process is not taken into account. The Rubinstein conception is explicitly built on a process of alternating offers and counter-offers at different moments in time, according to the following script: Agent C makes an offer at time 1 to agent R for a division of a certain good. The amount of the good is assumed to be fixed so that if the offer made by C is x, then $(1-x)$ would be left to R. The bargaining process is characterized by time discounting: as time goes on, the value of the good shrinks at a different rate for each agent. This discounting and the sequential nature of the bargaining process favors the agent who makes the first offer because rejecting an offer is costly for the other agent. Rubinstein (1982) shows that if the first agent anticipates in his first offer the discounting of the other agent with respect to successive offers and counter-offers, then his initial offer will be accepted. In fact, the Rubinstein conception can be reduced to a special case of the Harsanyi-Zeuthen bargaining process with the introduction of discount rates. What the Rubinstein approach[6] shows with respect to bargaining is that if agents have different discount rates, they also have different attitudes toward risk and the curvature of their respective utility functions will be affected. Thus the Rubinstein conception pleads even more in favor of analyzing various attitudes toward risk in bargaining.

How does our conception, based upon the S shaped utility curve, fit with the perspective of the bargaining literature? We can observe the following, which holds independently of the geometry of the S curve: If a conflict is considered a gamble and this gamble is chosen over a sure thing or a (narrowly defined) status quo situation we have:

$EU(gamble) = p\,U(L) + (1-p)U(W) > U(ST) = EU(sure\text{-}thing\ act)$, which is equivalent to:

[6] The Rubinstein perspective is well described in Osborne and Rubinstein (1990).

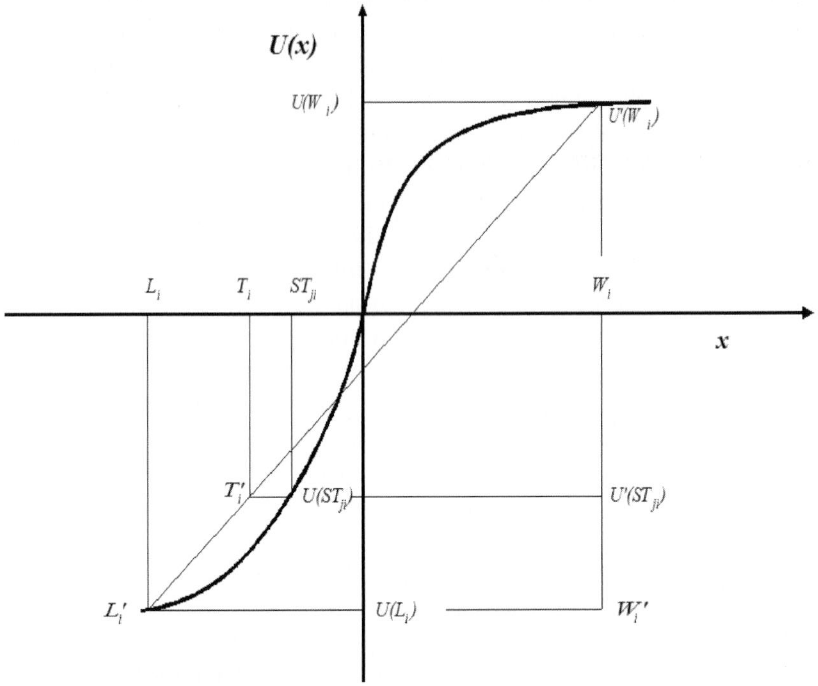

Graph 4: The S Shaped Utility Curve and Bargaining.

$$\frac{U(W)-U(ST)}{U(W)-U(L)} > p$$

This, of course, is the Harsanyi critical risk ratio. We can now give this ratio an interpretation in terms of the geometry of the S curve. If the S curve is defined in terms of a bargaining situation between *i* and *j*, we get: We can now establish the following:

Basic Lemmas:

Lemma 1: The equality $\frac{U(W_i)-U(ST_{ji})}{U(W_i)-U(L_i)} = \frac{W_i - T_i}{W_i - L_i}$ holds everywhere in the domain spanned by the S shaped utility curve and the cord $U(W_i)$—$U(L_i)$. Proof: Consider the geometry of the S shaped utility curve and the interval spanned by the cord $[U(W_i), W_i]$—$[U(L_i), L_i]$, as for instance in Graph 4. Within the triangle $U'(W_i)$—W'_i—L'_i, there is another similar triangle $U'(W_i)$—$U'(ST_{ji})$—T'_i. Obviously, the interval $U'(W_i) - W'_i$ is equal to the interval $U(W_i) - U(L_i)$, the interval $U_i'(W_i) - (U'(ST_{ji})$ is equal to the interval, the interval $U(W_i) - U(ST_{ji})$, $U'(ST_{ji}) - T'_i$ is equal to the interval $W_i - T_i$, and the interval $W_i' - L_i'$ is equal to $W_i - L_i$. Thus we can establish that by similar triangles $\frac{U(W_i)-U(ST_{ji})}{W_i - T_i} = \frac{U(W_i)-U(L_i)}{W_i - L_i}$ then can be rearranged as $\frac{U(W_i)-U(ST_{ji})}{U(W_i)-U(L_i)} = \frac{W_i - T_i}{W_i - L_i}$ which completes the proof.

Lemma 2: The observable critical risk ratio $\frac{W_i - ST_{ji}}{W_i - L_i}$ is always higher than the "subjective" critical risk ratio $\frac{U(W_i)-U(ST_{ji})}{U(W_i)-U(L_i)}$ under risk aversion and always lower than the subjective critical risk ratio under risk preference. Proof: Given the equality $\frac{U(W_i)-U(ST_{ji})}{U(W_i)-U(L_i)} = \frac{W_i - T_i}{W_i - L_i}$ established by Lemma 1 and that under risk aversion $T_i > ST_{ji}$, the difference $W_i - ST_{ji} > W_i - T_i$ means that $\frac{W_i - ST_{ji}}{W_i - L_i} > \frac{U(W_i)-U(ST_{ji})}{U(W_i)-U(L_i)}$.
Similarly, under risk preference we have $ST_{ji} > T$ and thus $W_i - ST_{ji} > W_i - T_i$ and then $\frac{W_i - ST_{ji}}{W_i - L_i} < \frac{U(W_i)-U(ST_{ji})}{U(W_i)-U(L_i)}$.

Existence of cooperative equilibria in the form of Nash bargaining solutions and of conflict equilibria:

Let us assume with Harsanyi the following: (1) Two bargainers will follow the Harsanyi Zeuthen principles. (2) They will not make concessions beyond the Nash bargaining solution. (3) No concession will be smaller than a minimum size $\varepsilon > 0$. We will add our own (reasonable) assumption that in a bargaining game, two bargainers estimate each other's subjective critical risk ratio (or probability) $\frac{U(W_i)-U(ST_{ji})}{U(W_i)-U(L_i)} = r_{ij}$ through the more objective estimator $\frac{W_i - ST_{ji}}{W_i - L_i} = \hat{r}_{i,j}$

Then:

Theorem 1: If both bargainers are risk averse, they will through mutual concessions always reach the Nash bargaining solution. Proof: All we have to show is that the estimation of the subjective risk ratio through the objective risk ratio is consistent with the Harsanyi Zeuthen principles. Then we can apply the Theorem demonstrated by Harsanyi (Harsanyi 1977:152–153) saying that: Under assumptions 1, 2, and 3 above, two bargainers will eventually reach the Nash bargaining solution. We proceed as following: The bargainers will estimate their mutual r_{ij}'s through $\hat{r}_{ij}$. Under risk aversion we always have $\hat{r}_{ij} > r_{ij}$ and also, of course, $\hat{r}_{ji} > r_{ji}$. Thus, whenever, $r_{ij} > r_{ji}$, $\hat{r}_{ij} > \hat{r}_{ji}$. This means that bargainer *j* will then make a concession to bargainer *i*. If these inequalities are reversed *i* will make a concession to *j*. Both of these conclusions are thus consistent with the Harsanyi Zeuthen rules. The case $r_{ij} = r_{ji}$ remains to be examined: Under risk aversion, $\hat{r}_{ji} > r_{ji}$, thus we can conclude $\hat{r}_{ji} > r_{ij}$, and hence *i* will make a concession to *j*. However, for the same reason $\hat{r}_{ij} > r_{ji}$ and hence *j* will make a concession to *i*. This is consistent with the Harsanyi Zeuthen rule that whenever $r_{ij} = r_{ji}$, both bargainers make concessions. Thus our construction is completely consistent with the Harsanyi Zeuthen rules and we can apply the Harsanyi theorem.

Theorem 2: If two bargainers are risk preferring, they will end up in an equilibrium in conflict. Proof: We need to show that the Harsanyi Zeuthen rules do not lead to any concessions under risk preference for the two bargainers. The critical risk ratio $\frac{U(W_i)-U(ST_{ji})}{U(W_i)-U(L_i)} = r_{ij}$ is again estimated via $\frac{W_i - ST_{ji}}{W_i - L_i} = \hat{r}_{ij}$ with the difference that $\hat{r}_{ij} < r_{ij}$. Thus $r_{ij} > r_{ji}$ does not mean that *j* will make a concession to *i* since $\hat{r}_{ij}$ eventually $\leq r_{ji}$. We can establish initially that whenever $r_{ij} = r_{ji}$ since $\hat{r}_{ij} < r_{ij} = r_{ji}$ no bargainer will offer a concession. This will also occur whenever $\hat{r}_{ij} < r_{ji} < r_{ij}$. We thus have to deal only with the case $r_{ij} > \hat{r}_{ij} > r_{ji}$. Suppose this condition holds. Then, we would have under risk preference, a space where j would offer ST_{ji} to i because this offer would appear superior to the conflict outcome. For such a space or interval to exist we would have always an interval $[a,b]$ such that for all $x \in [a,b]$, $u_A(x) \geq W_A - L_A$ and $u_B(x) \geq W_B - L_B$. However, under one-sided risk preference, $u_A(x') \leq x'$ and thus eventually $u_A(x') \leq W_A - L_A$, which is a contradiction and so the interval does not exist for A and if one reasons similarly neither does it for B. So no such space exists ever. A similar reasoning can be made for i. Hence the estimator $\hat{r}_{ij}$ is $\leq r_{ji}$ and thus no concessions are made on either side, which lead to bigger demands by the two bargainers and thus to a convergence $ST_{ji} \rightarrow L_i$ and $S_{ij} \rightarrow L_j$.

Theorem 3 (Asymmetry): If one bargainer is risk averse while the other one is risk preferring, two equilibria can occur: (1) The risk preferring bargainer presents a take it or leave it request to the risk averse bargainer, who accepts it (equilibrium in surrender). Proof: As already established above, the risk preferring bargainer never makes a concession. Therefore, only one offer is made. The risk averse bargainer overestimates the critical risk ratio of the risk preferring bargainer and is prepared to make the concession provided that his being worse of through it does not push him below the chord in terms of the geometry of the S curve, i.e., if he is sufficiently risk averse. (2) The risk preferring bargainer presents a take it or leave it request to the risk-averse bargainer. The other bargainer refuses it and fights. Proof: This offer puts the risk-averse bargainer in a position where he becomes risk preferring. No bargainer makes any concession as established above and either noncooperation prevails or fighting starts.

CREATING A POLICY ENVIRONMENT FOR SUSTAINABLE WATER USE

Karina Schoengold
School of Natural Resources and Department of Agricultural Economics
University of Nebraska
kschoengold2@unl.edu

David Zilberman
Department of Agricultural and Resource Economics
University of California, Berkeley

Abstract: This chapter discusses the historical context of water use, rights, and development, much of which has contributed to the current water shortage in many arid regions of the world. In addition, it attributes many perceived water crises to poor water management instead of an insufficient water supply. Poor water management is discussed in several contexts, including the ecological implications of past water development, social and public health ramifications, and a general underestimation of all costs associated with water development. Recommendations are provided to improve the management of water resources in the future. These recommendations include recognizing that uncertainty about costs and benefits of water management choices must be taken into account by decision-makers, and that incorporating community management of water resources can lead to improved water management. Other recommendations include improving water pricing and allocation systems, such as through a switch from queuing systems to tradable permits; as well as using water pricing mechanisms that incorporate the negative externalities of water quality deterioration. Finally, the chapter provides description of how to improve the cost-effectiveness of policies to improve water quality and quantity through better targeting.

Keywords: water shortages, water management, water pricing, cost-effective water policies

1. INTRODUCTION

Water resources have been the source of environmental services and economic benefit for millennia. Water provides multiple benefits, in both consumptive and nonconsumptive uses. Examples of consumptive uses include drinking water and agricultural production; nonconsumptive uses

E. Wiegandt (ed.), Mountains: Sources of Water, Sources of Knowledge, 305–326.

include water recreation and habitat provision for aquatic species. Historically, an adequate supply of water has not been a constraint in many locations, and institutions and policies have been established to encourage utilization of water. Governments who were rich in resources, but poor in capital, established systems that were liberal in dispensation of water rights for beneficial use. Under this type of system, water is not used where it provides the greatest benefit. Increasing population growth, growing concern for environmental quality, and financial constraints are leading to the realization that water use has to be curtailed and water has to be managed in a sustainable manner for the long run.

As the word "sustainable" can be ambiguous, its meaning in this context should be explained. It requires that water use and resource development take into account the limitations of natural systems, and that the existing stock of the resource not be depleted. Surface water sources are replenished regularly, but the amount of water provided is limited and subject to variability. Groundwater aquifers provide another source of fresh water. While some groundwater aquifers are replenished, others are nonrenewable and their use should be carefully monitored. These supply constraints need to be recognized in any sustainable water management plan. A key element in sustainable water use is the development of new technologies to use water more efficiently and to reuse residual water more effectively.

Many regions have a perception of water crisis because existing water resources are not sufficient to meet growing needs. In most cases, the real problem is a *water management crisis*. Incentives for efficient and socially responsible management of water are lacking. Water projects that cannot be justified economically, and are damaging environmentally, are being built. Users are paying well below the value of the water they use, and are encouraged to consume water. Polluters of water bodies are often not penalized. In looking at the sustainability of water resources, it is not enough to only require a sufficient quantity of water, it is also important that supply be of sufficient quality.

In this chapter, we discuss past trends and their implications for water use. We also discuss necessary components of any water policy that considers future needs. To achieve sustainable water use, water policies and institutions must be reformed. Concerns about adequate water supply can be addressed through better management of existing sources. In this spirit, we will present incentives and policies to improve water project design, water conveyance and pricing, micro level choices, and water quality.

2. PAST TRENDS

2.1. Population Trends, Food Security, and Water

A growing world population is putting increasing pressure on freshwater supplies. Global population has increased from 2.5 billion in 1950 to 6 billion in 2000. As much as half of the world's population lacks adequate water for basic sanitation and hygiene (Sullivan 2002). By 2050, the world population could be 10 billion or higher. In addition to population pressures, there has also been an increase in water use per capita. The combination of these forces has led to worries about the adequacy of water supplies in the future.

We find both good and bad news in addressing this problem. First, the good news—in many ways, irrigation has allowed the world to overcome the potential food supply problems associated with population growth. Worldwide, irrigated land has increased from 50 mha (million hectares) in 1900 to 267 mha today (Gleick 2000). While 17 percent of cropland is irrigated, it produces 40 percent of the food supply, and the value of output per acre is six times the value of output on rain-fed cropland (Dregne and Chou 1992). In addition, there is some evidence that the high productivity of irrigated agriculture has slowed the rate of deforestation. Agriculture is one of the primary reasons for deforestation in developing countries, and increased yields (a change at the intensive margin) have decreased the need to expand the total land under cultivation (a change at the extensive margin). This is particularly important in mountainous regions. The steep slope of this land not only makes it difficult for food production, it also increases its value in conservation. Protection of forests in mountainous regions protects the soil, and decreases erosion and runoff into water bodies.

Now some bad news—there are limits to increasing irrigation water sources. Worldwide water consumption in 2000 is 4–5 times that of 1950 levels. Most of the obvious sources of water have been developed, and many that remain are marginal at best. The costs of developing water for irrigation are increasing and are not uniform across regions. Postel (1999) reviews the result of a World Bank study that shows the cost of irrigation has increased substantially since the 1970s. The study of more than 190 Bank-funded projects found that irrigation development now average \$480,000/km^2. This cost varies by location—the capital cost for new irrigation capacity in China is \$150,000/km^2, while the capital costs in Africa are \$1,000,000–2,000,000/km^2. Mexico's irrigated area has actually declined since 1985 due to lack of capital.

There is also an increased understanding of the importance of freshwater for environmental services, such as ecosystem health, as well as the environmental costs of water projects, such as habitat destruction. Soil salinity is a problem on irrigated arid lands, both reducing productivity and forcing land out of production. In many places with insufficient surface water supplies, groundwater is used as a substitute. While the availability of groundwater has benefited the global food supply, its use as an input has progressed in an unsustainable manner. As much as 8 percent of food crops grown on farms use groundwater faster than the rate at which aquifers are replenished. In 1973, only 3 percent of India's groundwater tables were below 10 m. By 1994, this figure was 46 percent (Postel 1999). Using irrigation in a more efficient manner will be necessary to protect water sources while still meeting goals of food security.

2.2. Social Concerns and Water Development

Social concerns associated with water development include waterborne diseases and displacement of native populations. There have been a number of large dams whose construction contributed to local public health problems, including increased incidences of diseases such as malaria, diarrhea, cholera, typhoid, and onchocerciasis (river blindness). However, there is evidence that many of these cases have been the result of poor planning, and not a necessary effect of dam construction. Often, increased vector breeding occurs in fields and not in the dams and canals (von Braun 1997). Incorporating public health concerns into the planning of a new water project can reduce the impact of the project.

Another negative result of water development has been the displacement of native populations. Between 1950 and 1999, 40–80 million people were displaced as a result of water development. In addition to their physical displacement, it has also often resulted in forced lifestyle changes. Compensation for these forced resettlements has been minimal. Resettlement plans regularly fail to take into account the loss of a viable livelihood in addition to the loss of physical land, often leaving resettled populations worse off than before dam construction. For example, one study found that 72 percent of the 32,000 people displaced by the Kedung Ombo Dam in Indonesia were worse off after resettlement (World Commission on Dams (WCD) 2000). The construction of the Liu-Yan-Ba Dam on the Yellow River in China forced the resettlement of 40,000 people from fertile valleys to unproductive wind-blown highlands. This has led to extreme poverty for many of the resettled people (ibid).

3. WATER PROJECTS

The last 50 years have witnessed the construction of thousands of water projects worldwide. Many of these projects have been built in marginal areas, without consideration of the true cost of construction. It is important that water projects rely on social benefit-cost analysis. In many cases, the estimated costs of construction have been too low. A recent study of 81 large dams by the World Commission on Dams found that the average cost overrun was 56 percent. These costs include only capital costs, and if environmental degradation was also included, the "true" costs would be much greater. In addition, ex-ante predictions of the benefits of water projects have often been overly optimistic. This combination of factors has resulted in observations that the internal rate of return to most water projects is well below the expected rate of return, although most of the return rates are still positive (World Commission on Dams 2000). Only projects with a positive net present value, when all costs and benefits are included, should be considered for construction.

In the past, capital subsidies and low estimates of environmental values have led to oversized projects. There has often been an attempt to include environmental benefits of water development (such as recreation in a reservoir) in a cost-benefit analysis. However, many environmental costs such as habitat destruction or extinction of species were considered immeasurable, and were ignored. Methods of accurately measuring these costs are being developed and improved, and they should be used whenever possible. In evaluating different projects, issues of uncertainty, optimal project design, and future costs must be considered.

3.1. The Importance of Uncertainty

One important issue in water development is that of uncertainty, particularly when making irreversible decisions. Water development requires choices about the location and size of a water project. These choices are made under uncertainty regarding future technology, population size, and environmental preferences. For example, the construction of a large dam will permanently alter the surrounding ecosystem. Traditional cost-benefit analysis looks at the net present value of a project to determine whether or not the project should be built. This type of analysis ignores a third possibility—the option of waiting. If the value people place on the benefits of this ecosystem is uncertain, then waiting to build the project can allow further information to be learned about these benefits. If the benefits of water development are uncertain, the uncertainty can be decreased as further knowledge becomes available.

Arrow and Fisher (1974) and more recently Dixit and Pindyck (1994) develop models that suggest that in these cases the decision-maker may consider delaying the decision about optimal project design to collect more information about the costs and benefits of project construction. They not only look at the question "to build or not to build"; they also consider the importance of when to build. Delaying building a project by one or two periods may lead to the loss of benefits in these periods but will lead to a future gain as more information is taken into account. They show that if the gains from acquiring new information are greater than the foregone benefits of current construction, it is better to delay construction of a new project. The gain from not making an immediate decision is referred to as "option value." In particular, in cases when there is uncertainty about productivity of water as a result of availability of a new technology or as a result of uncertainty about environmental impacts of water diversion activities, the "option value" of waiting may be quite high and there may be significant gain from delay. Because of this, a positive net present value of a benefit-cost analysis is a necessary, but not a sufficient, condition for construction.

Zhao and Zilberman (1999) extend this analysis to consider projects where restoration is costly but feasible. This is more realistic for water development. Dams are being removed from many sites worldwide, and natural habitats are being restored. They find that in some cases, it might be better to construct a new project even if there is a chance it will lead to costly restoration in the future. This could happen if the expected benefits of a project are larger than the expected future restoration costs.

For policymakers, it is necessary to consider not only the expected benefits and costs of developing a water project, but also the potential to learn more information when those benefits and costs are uncertain. If the gains from waiting and learning are high, it might be best to delay the decision until better information is known.

3.2. Project Design

In the design of water projects, it is not enough to only consider the physical construction of the project; it is also important for engineers, economists, and biologists to work together in the project design process. Project design should include both physical aspects and managerial specifications. At a system-wide level, management and storage capacity can be considered substitutes. Designing a management system in conjunction with physical infrastructure may reduce the size of the project and its environmental side effects.

Traditionally water systems have been managed by large bureaucratic agencies, leading to much inefficiency. As discussed by Easter (1986), there has been a shift in recent years from the development of new water projects to better management of existing projects. This has led to an increased reliance on water user associations (WUAs). A WUA is usually comprised of landowners in a small geographical region who are charged with the distribution of water and the collection of costs of provision. Because they manage the conveyance and costs themselves, they have incentives to find ways to do so efficiently. One element that is essential for the success of a WUA is well-defined property rights to water. In cases where rights are ambiguous, monitoring and charging water users an appropriate cost becomes very difficult. The successful use of WUAs for water distribution and cost recovery is used in parts of India, Mexico, and most of Madagascar (among other countries). The growing use of WUAs is a positive step toward improved management, but there is still room for much improvement.

Another aspect of project management is developing uses for recycled/reclaimed water. Often water is not fully consumed in its first use, and can be reused. For example, certain crops are tolerant of water with high salinity levels, and runoff water from other less tolerant crops can be used for some irrigation needs. Treated wastewater from sewage plants can be reused for many industrial water needs. As more technologies are developed, further possibilities will be found for reusing and recycling water in an effort to make the limited supply last.

Water system management should also include sufficient upkeep in maintenance. The lack of maintenance is seen in both municipal water and irrigation infrastructure. Poor management of irrigation systems leads to conveyance losses of up to 50 percent (Repetto 1986). When users of conveyance structures invest in maintenance, they get some benefit because of improved water supply. However, water users downstream also incur a positive externality, which is often not taken into account in maintenance decisions. Economic theory shows that private users under-invest in canal maintenance. This has a few implications for water users. First, canals will be shorter than optimal if based on private interests. Second, there will be overapplication of water upstream, and under-availability of water downstream. Third, improving conveyance will improve the well-being of downstream farmers' more than upstream farmers. There is a need for governments to take charge in developing optimal canal maintenance and pricing. Water prices should not be uniform; they should take into account a user's location on a canal

(Chakravorty et al. 1995). These charges need to take into account transportation costs and inefficiencies in conveyance—the cost of water downstream should be greater than the cost of water upstream.

In the development of new water projects, policymakers must consider all elements of the project, including capital, management, and conveyance. These elements need to be planned jointly, as failure to do so could lead to a higher than necessary reliance on one part of the system.

3.3. Dynamic Costs of Water Development

Many of the costs of developing a water project are not immediate. There are environmental problems that have occurred over time as the amount of land being irrigated has expanded. These costs include increased salinity levels in fresh water sources, and waterlogging and salinization of soil. They also might include future construction of drainage systems to keep land usable for agriculture or decreased productivity in traditional fisheries.

For example, the Aral Sea used to be a thriving site for the fishing industry, employing 60,000 individuals. Between 1962 and 1994, diversions to provide water to grow cotton have reduced the volume of water in the sea by 75 percent. The fishery industry has been entirely wiped out, with many fish species disappearing (Calder and Lee 1995). These costs are more likely to arise in arid locations that are susceptible to water logging and salinization of the soils. In arid regions, there is little rainfall to dissolve the salts in the soil. When too much water is applied without proper drainage, the evaporation in arid climates can quickly lead to high levels of salt in the soil, reducing the yield potential of the land. Estimates are that 20 percent of the irrigated land worldwide is affected by soil salinity, and that 1.5 mha are taken out of production each year as a result of high salinity levels.

In evaluating potential water projects and water project design, it is important to consider future costs, where these costs are broadly defined to include future environmental damage as well as social costs.

4. WATER ALLOCATION AND PRICING

Water allocation systems are a primary source of inefficiency in water use. Existing water rights systems are mostly designed as a "queuing system," where an order of seniority is established among water users. Most of these systems also have banned the trade of water between users, or the sale of individuals' water rights. A prior appropriation rights system gives a permanent water right to the first person to divert water from a source, while a riparian rights system gives water rights to landowners whose land is adjacent to a water source. It has been shown that a transition to trading, where water is priced according to its opportunity cost, will increase social benefits. The original development of this argument can be attributed to Coase's seminal work entitled *The Problem of Social Cost*. Coase (1960) argues that if transaction costs are zero and property rights are well-defined, then allowing trade of those rights leads to a first-best (most efficient) outcome. These gains from trade increase with water scarcity. However, several of the required assumptions for this argument do not hold with the water industry. For water the assumption of zero transaction costs is incorrect. Water is a difficult commodity to move, and trading is only beneficial when the gains from trade exceed the transaction costs. In addition, this outcome assumes perfect enforcement, which is generally not the case with water use and application.

4.1. Gains from Water Trading

In many places, trade in water is not necessary in normal years. However, during times of drought, the benefits of water trading increase. For example, from 1987 to 1991 California experienced one of the worst droughts in recent history. By 1991, the California Department of Water Resources established the California Water Bank, which bought water from users for $125 per acre-foot and sold them at $175 per acre-foot (prices are at the Delta). During this year, the water bank purchased 825,000 acre-feet of water, with 166,000 acres of agricultural land fallowed to provide a portion of this water. The water bank continued during the dry years of 1992 and 1994, although with lower prices due to the lower severity of the drought. Studies have shown that the California Water Bank provided a significant economic welfare gain to the state (Zilberman 1997).

Another example that shows the potential gains from the introduction of water trading is Australia. Like in Chile, Australia has moved to a water-trading regime, and has decoupled ownership of land from the right to use water. The shift from traditional water rights stemmed from a growing realization that greater flexibility was needed in water rights that water resources are necessary in the natural habitat. A 1994 bill separated water rights from land ownership, and established a water allocation for environmental services and the development of water markets. The results of the change in Australia have been positive, and estimates are that the annual gains from the shift to tradable water rights are $12 million in Victoria, and $60–100 million in New South Wales (ACIL 2003). Despite these gains, there are still some barriers that have been identified as an impediment to the highest possible returns to tradable water rights. One of these impediments is a limitation on the lease of water-use rights. Water rights can be permanently sold in all states of the country, but some states still have a restriction on short-term (i.e., one year) leases of those rights. Another aspect that has been identified as a limitation on the benefits of trading is the lack of an options market in water resources. The elimination of these barriers to a fully functioning water market will only increase the benefits already realized in Australia.

In many places, for trading to become a viable option requires an expansion of existing canal systems. When infrastructure required for the transportation of water does not exist, it makes trades between users close to impossible. Even if the use of trading is small relative to total water use, it could still lead to a large gain in total welfare, especially in times of shortage.

Trading in water, particularly in permanent water rights, can induce long-term investment and development, and reduce uncertainty about future water supply. Trading can also allow governments or conservation groups to purchase water rights from individual users for environmental benefits and conservation.

4.2. Design of Water Trading Programs

In designing a water trading regime, several issues need to be addressed. One of these is the ownership of water rights—the distribution of water rights might be based on a number of criteria, including current use, need, or willingness to pay. The modalities for sale of water rights also need to be clearly specified—whether these rights can be sold on a permanent basis, or only on an annual basis. In 1981, Chile had a major reform in their water law, decoupling water rights from ownership of the land. While there have been many trades since the reform, there have been far more transactions to define ownership of water rights. At the time of the reform, there was a lot of uncertainty about the ownership of much of the water used. Much of the energy since the reform has gone into defining water rights, and some areas have seen 10 times as many water rights approvals as water sales (Bauer 1998). Clearly, well-defined water rights are a necessary condition for welfare-improving water sales.

As mentioned before, third-party effects need to be considered. When using water for irrigation, not all of the applied water is actually used by the plant. Some of the water is runoff, and some is stored in the ground below. Using drip irrigation or better management can increase the water-use efficiency of applied water, but there will always be some loss. Third parties might actually benefit from the availability of the water runoff, and will be affected if a farmer sells their entire water quota. To address third-party effects, individuals should only be allowed to sell the effective water they use, not the total applied water. Several choices need to be made when introducing a water trading program.

One decision is the size of the area to use for a trading region. Limiting trading to within a water basin will decrease third-party effects and lower transaction costs. However, it will also limit the number of potential trading partners of an individual. Another question is if individuals should be allowed to sell their water rights permanently, or to only lease them on a year by year basis. Uncertainty about the future, and particularly the future value of those rights, will lead some to prefer a short-term lease. However, this leads to higher uncertainty for the purchaser, who will be less likely to invest in efficient technology if they are unsure about future water supply. The use of permanent water sales are more appropriate in places with a chronic water shortage, while yearly sales are preferred to deal with drought situations in locations which generally have sufficient water.

Another decision is the type of trading system. Brill et al. (1997) discuss the difference between active and passive water trading schemes. Active trading occurs when a water district assigns water allocations to farmers based on some benchmark, and charges farmers the average price for their allocation. Farmers are then allowed to trade between themselves, so that those with a higher marginal value of water can buy a portion of the allocation of those with a low marginal value of water. Passive trading occurs when a water agency sets both an initial allocation of water per farmer and a price, with the price chosen for the water market to clear. Allocations are set so that the sum of individual allocations equals the total demand, with individual farmers allowed to use either more or less than their allocation, and either pay an additional fee or receive a rebate accordingly.

The details of a chosen trading system are influenced by various considerations. Certain choices are better for existing rights holders, while others will be better for governments or water agencies. Political economy considerations and the feasibility of the system must also be a factor in the choice.

4.3. Water Pricing Systems

There is a problem with many current water pricing systems. They are often aimed at cost recovery, and not at promoting efficient water use. Despite this goal, many water systems do not come close to recovering their costs. Recovery of operation and maintenance costs range from a low of 20–30 percent in India and Pakistan to a high of close to 75 percent in Madagascar (Dinar and Subramanian 1997). The most common means of pricing are per-hectare fees. This leads to inefficient water use, since the marginal cost of applied water to users is zero. Using volumetric pricing with the water priced at the marginal cost of delivery leads to efficient water use because the marginal cost reflects the opportunity cost of water. The marginal value of water in one use should equal the marginal value in another. Concerns about equity can be addressed through tiered pricing, while still retaining volumetric pricing and some level of efficiency. Often, there is no effort to recover fixed costs of water development, particularly in places where capital costs have been financed by international agencies. Attempts to recover fixed costs is usually done by applying a "hook-up" fee, where an individual has to pay a set fee for access to water from a source.

As mentioned earlier, efficiency can be gained with volumetric pricing of water. For efficiency, it is not sufficient to have an invariant marginal price for water. The value of water changes both by season and location. Ideally, we would like to impose a water pricing structure with variation in prices by both of these variables. While this sounds complicated, it is used in many places. Prices can change by time, location, or crop. These varying prices should reflect differences in the costs of supply, specifically conveyance costs and environmental side effects. In many places, volumetric pricing is not feasible due to the high costs of monitoring. It is possible to mimic volumetric pricing by imposing per-hectare fees that vary by the season/crop/irrigation choice.

Regardless of whether volumetric or per-hectare fees are used, it is crucial that these prices reflect the environmental side effects of water use. The use of greener/cleaner application technologies (such as drip instead of gravity irrigation) should be rewarded.

4.4. Groundwater Management

Groundwater suffers from an open-access problem, as an aquifer is accessible to anyone who builds a well. Each individual user affects the quantity of water available to others, but has no incentive to take that into consideration in their pumping decisions.

Groundwater resources are being exploited and being used at unsustainable rates in many areas of the world. For example, India increased pumping of groundwater by 300 percent from 1951–1986. This increase has continued in recent years. In using groundwater, farmers should pay both the direct pumping costs of obtaining that water, and a user fee to reflect future scarcity. Another issue is the subsidized costs of pumping. Many places have very low costs of electricity (the main cost of pumping groundwater). If the cost of electricity is subsidized, the perceived cost of pumping to farmers is below the actual cost. As with surface water, tiered pricing can be used to address equity issues between individuals. One concern is that a higher price of surface water will lead to increased dependence on groundwater resources, as water users substitute groundwater for surface water.

Because of the externality imposed on other water users, the elimination of electricity subsidies still leads to a sub-optimal groundwater price. The theory of exhaustible resources dictates that the price of groundwater should equal the sum of the cost of extraction and the user cost, with the user cost equal to the opportunity cost (Hotelling 1931; Devarajan and Fisher 1981). The user cost measures the loss of future benefits because of depletion and the increase in future pumping costs associated with depleted stock. A first-best solution would be to impose a tax equal to the user cost on every acre foot of groundwater extracted (Shah, Zilberman, and Chakravorty 1993; Howe 2002). However, the monitoring and enforcement of a tax such as this would be impossible given the costs and currently available technology. As discussed in Shah, Zilberman, and Chakravorty, a second-best solution would be to base the tax on the irrigation technology and crop choice.

As mentioned above, it is difficult to accurately monitor groundwater use. Despite this, it is necessary to have some monitoring of groundwater sources. It is possible to either do this directly, with test wells at various locations, or indirectly, by estimating water use based on land allocation choices. The fact that these are substitutes for each other requires a joint management plan of surface and groundwater sources.

4.5. Conjunctive Use of Water Sources

In managing water resources, there are several important differences between surface water and groundwater. Surface water supplies are replenished every year, but they are subject to large variation from year to year. Some groundwater sources can be recharged, while others should be considered a nonrenewable resource. For example, parts of the Ogallala aquifer in the Western United States has not had any recharge in over 1,000 years, and use of existing water is depleting the available reserves. For water management, conjunctive use of surface and groundwater can limit the uncertainty of surface water availability and the overdraft of groundwater. Conjunctive use of the two sources can decrease the risk associated with a stochastic surface water supply. In some cases, the separate management of groundwater and surface water resources has led agricultural producers to substitute groundwater for surface water after laws are passed to protect the environment through the reduced use of surface water. Therefore, the conservation goals of such policies are not achieved, as users substitute one unsustainable pattern of water use for another.

Arvin Edison Water and Storage District (AEWSD), located in California's Central Valley, provides a model of beneficial conjunctive use. AEWSD utilizes underground water banking in their water management plan. In years when they receive large quantities of water, they store some of it underground, providing a net gain in stored water. During dry years, when the water supply is insufficient to meet demand, they can pump this stored water for the growers in the district. Tsur (1997) estimates the value of this supply stabilization by the district to be $488,523 per year, a value equal to 47 percent of the total value of groundwater.

These outcomes show that there is a need to regulate surface water and groundwater jointly, as they are substitutes for each other in many places. Failure to do so can lead to inefficiency in water use.

5. WATER QUALITY AND ENVIRONMENTAL MANAGEMENT

For water systems to be sustainable it is not enough to have a large quantity of water available, it also must be of adequate quality. It should be noted that the definition of "adequate" quality varies by end use. The quality requirements of drinking water are much greater than that of water use in many industrial or agricultural applications.

5.1. Externalities

In economics, an externality is an unintended benefit or a cost imposed on a third-party. For example, producers get some economic benefit from the right to pollute water bodies. However, other users of the water (both for consumptive and nonconsumptive uses) are harmed, suffering a negative externality because of that pollution. If there is no regulation, polluters will ignore these externalities in their decision of how much to pollute. There are several ways that pollution can be decreased to a level that considers these externality costs.

Some of the most often discussed policy interventions are the use of taxes, subsidies, and quotas. Taxes impose a fee per unit of water use or pollution, equal to the damage incurred by third parties. Firms or agricultural producers will then take this cost into consideration when they decide how much water to use in their production decisions. Another possibility is to give firms a payment (a subsidy) for each unit of water they do not use. In this type of payment, the choice of a baseline level is difficult, but crucial for successful implementation. Using historical levels of water use punishes firms that were relatively efficient in their water use before the regulation. A similar solution is to subsidize the purchase of water-saving technologies. While these two policies can lead to the same outcome in a static framework, the long-term effects of a subsidy or a tax are different. A subsidy increases firms' profits, giving others an incentive to enter the market. A tax decreases total profits, and can lead to exit from the market.

Another possibility is the use of quotas, where a maximum level of pollution or water use is allowed, permits are distributed to potential users, and then those users are allowed to trade for permits between themselves. Examples of this include the 1990 Clean Air Act Amendments in the United States, which developed a tradable permit program for sulfur dioxide emissions of power plants. The choice of an appropriate policy depends on a number of factors, including implementation and monitoring cost, transaction costs, and political will.

Water quality management should be based on polluter-pay principles that control disposal of contaminants to bodies of water. When pollution can be linked to a polluter, and its marginal damage is known, a pollution tax equal to the marginal damage will result in an optimal outcome. However, it is often difficult to monitor individual pollution emissions, and to calculate the marginal damage of that pollution. Incentives may be used to encourage adoption of monitoring technologies and pollution abatement technologies to control runoffs and erosion. A major environmental problem is that of uncontrolled water movement, and the resulting soil erosion and flooding that occurs as a result. Better management of forest resources to limit soil erosion is a necessary part of water quality management.

In any policy designed to limit pollution, it is important to minimize the use of direct control. By allowing flexibility among users, efficiency and compliance are improved. The nature of a pollutant affects the choice of the regulatory mechanism. There are two distinctions that must be made with pollution. The first is the difference between point and nonpoint source pollution, while the second distinction is that of uniformity or nonuniformity in damages.

5.2. Point and Nonpoint Source Pollution

In discussing water pollution, it is necessary to distinguish between point source and nonpoint source pollution. The difference is mostly one of technology: point source pollution can be traced to the polluter using currently available, cost-effective technology, while nonpoint source pollution cannot be traced to a single polluter, but only measured in aggregate. To control pollution, a first-best policy is to tax pollution directly. As point source pollution is directly observable, it can be controlled with a single tax per unit of emissions. This requires monitoring each firm. While this can be costly, several solutions are available to help with the cost.

One type of program can include self-monitoring, where firms are responsible for reporting their emission levels to a regulatory agency, and are subject to a fine if they report inaccurately. One example of a program that uses self-reporting is the Environmental Protection Agency's (EPA) Permit Compliance System. This program enforces standards for water pollutants. Shimshack and Ward (2005) show that there is little evidence of strategic behavior by individual firms, and that self-reporting is usually accurate. Another possibility is for the government to take a role in subsidizing monitoring costs. This decreases the cost to individual firms and regulatory agencies.

While it is ideal to monitor and tax pollution directly, with nonpoint source pollution this may be impossible. Shortle and Horan (2001) describe some of the difficulties and developments that have been made regarding the control of nonpoint source pollution. For example, if pesticides are used by farmers and enter the water system, it might be impossible to accurately measure the quantity of runoff associated with each farmer. One method that can be used is a tax of an associated input. Using the example above, while it is difficult to accurately track pollution from pesticide use, placing a tax on pesticides when they are sold is relatively easy to enforce, and will lead to the desired outcome of less pesticide runoff. If different technologies are available to an industry, and some are more polluting than others, taxing either the use or purchase of the "dirtier" technology is another option. Another possibility is to subsidize certain behavior, such as the use of integrated pest management (IPM) or low-tillage crop production. Similarly the protection of forest resources, particularly in mountain regions, reduces soil erosion and increases water quality. Subsidizing those who protect these resources can reduce water pollution. In cases where these programs are not feasible, one option is to impose a lump-sum tax on any operating business. Despite the lower cost of implementation, this option is not a good choice when there is a lot of heterogeneity among firms. If firms are fairly homogeneous, however, it can work well.

In determining the appropriate policy to reduce water pollution, decision-makers must make a distinction between point and nonpoint source pollution, and develop policies that are appropriate to the kind of pollution. This could mean either taxing a pollutant directly, or taxing another input into the pollution generating process.

5.3. Uniform and NonUniform Damages

An additional distinction must be made between pollution damages that are uniform and those which are not. If damages are uniform, the total level of emissions is what matters in determining appropriate regulatory policies. If damages are nonuniform, the policies of pollution control should vary spatially. An example of a pollution problem with uniform damages is greenhouse gas emissions, where an individual firm's emissions are not as important as the combined total emissions from all firms in determining damages. Generally, damages from water pollution vary spatially. For example, a polluting factory that is upriver from farmers and urban communities affects all downstream users who require a clean water source. Because of this, it is not optimal to charge a constant tax on pollutant emissions for all users—those who are in locations where the pollution causes greater damage should pay a higher tax on their emissions.

When damages are uniform, a tradable permit program can be an effective way of limiting total pollution. When damages are not uniform, trading can only be allowed in a small region, or should be otherwise restricted.

5.4. Environmental Services

A public good has some element of non-rivalry and non-excludability in use. This means that one individual can enjoy the benefits of the good without reducing the benefits to another, and that it is difficult to restrict people from these benefits. Environmental services are benefits that accrue from either the preservation or improvement of natural resources. For example, the preservation of a forest sustains biodiversity, acts as a carbon sink, and reduces soil erosion. While environmental services tend to be underprovided by the private sector, they are a growing portion of environmental policy interventions worldwide, with several examples of successful programs in existence.

An example of a program used for the purchase of environmental services is the Conservation Reserve Program (CRP) in the United States. This program pays farmers to set aside crop land to be preserved as natural habitat. Other examples of this type of program are funds to buy wetlands in California or to protect the Pyramid Lake in Nevada. The U.S. Fish and Wildlife Service used a purchasing program of land to protect Pyramid Lake, where the purpose was to acquire the water rights necessary to protect the lake itself. In 1997, Costa Rica established a system of payments for forest environmental services, with hydrological services such as the provision of water for human consumption, irrigation, and energy production explicitly recognized as one of those services. The goal for this program is to eventually be self-sufficient with its funding. This has not been achieved yet, with hydroelectric power plants being the primary beneficiary group to contribute money to the program (Pagiola 2002). Another example of the successful use of an environmental services purchasing plan is seen in Sukhomajri, India (Kerr 2002). In this program, authorities built a system of irrigation water pools to benefit a local village. In exchange for protecting a fragile region prone to erosion, which resulted in the siltation of a recreational water body; the villagers received ownership of the irrigation water provided by the project development. The development of a water market between villagers allowed both landowners and landless peasants to share equally in the benefits of the project, providing an excellent example of how these programs can benefit everyone involved.

As discussed by Viviroli and Weingartner in this volume, mountain regions provide more than their share of freshwater resources. They also find lower variation in these water supplies than those originating in lowlands. Since mountain regions are often the source of water for downstream areas, keeping water resources clean provides a benefit to downstream users. In designing programs to purchase environmental services, the choice of which areas to target is important. As shown in Babcock et al. (1997), this is not always the least expensive land or the most ecologically diverse land. These two indicators are often not independent of each other—it could be the case that the most ecologically diverse land also has the richest soil, and therefore is the most lucrative for agricultural production. Their analysis shows that it is important to try to target spending to get the highest value of environmental benefits per dollar spent. This can be seen in Figure 1.

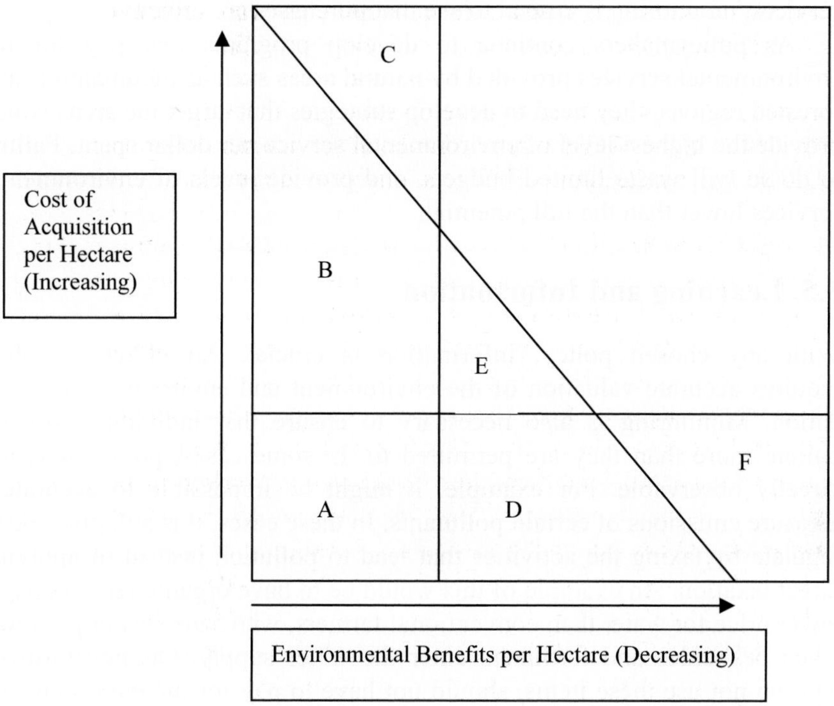

Figure 1: Distribution of Environmental Benefits and Cost of Acquisition per Hectare (Babcock et al. 1997).

In the past, many programs have either tried to focus on acquiring the largest possible area of land, or on saving only the most ecologically important. In Figure 1, the first goal would target regions A, D, and F; while the latter goal would target regions A, B, and C. Babcock et al. argue that with limited funding a program should target regions A, B, E, and D.

Mountain regions provide many environmental services. For example, protecting forests in mountainous regions decreases soil erosion and runoff. Combined with the fact that mountainous regions are often marginal for other activities (agriculture or urban development), mountain regions are a logical target for purchase through conservation programs, and should be targeted in a program where the goal is an improvement in water quality. For any type of program that purchases environmental services, monitoring is crucial to see that purchases are effective.

As policymakers continue to develop programs that pay for the environmental services provided by natural areas such as mountainous and forested regions, they need to develop strategies that target the areas which provide the highest level of environmental service per dollar spent. Failure to do so will waste limited budgets, and provide levels of environmental services lower than the full potential.

5.5. Learning and Information

With any chosen policy, information is crucial. An effective policy requires accurate valuation of the environment and environmental degradation. Monitoring is also necessary to ensure that individuals do not pollute more than they are permitted to. In some cases, pollution is not directly observable. For example, it might be impossible to accurately measure emissions of certain pollutants. In these cases, it is still possible to regulate by taxing the activities that lead to pollution instead of applying direct taxation. An example of this would be to have organic farmers pay a lower price for water than conventional farmers, who cause water pollution when pesticides and fertilizers enter the water supply. Organic farmers, who do not use these items, should not have to pay for the environmental costs of their use.

Unlike the case of perfect information, the seminal work of Weitzman (1974) showed that when environmental benefits and costs are uncertain, the choice between taxes and tradable permits leads to different results. Much has been written since this paper was published, but a major theme is that uncertainty matters in policy development. A program of adaptive management, one that allows policies to change as more is learned, is necessary. Designing effective water policies requires constant learning and updating of knowledge of natural phenomena and human behavior. Development of new monitoring technologies can help to track water pollutants and assign appropriate penalties/rewards. The use of satellite imagery can aid in the spatial analysis of different policies. An example of an agency that currently uses this technology is the Argentinean government. They recently started a program that uses satellite imagery to observe the crops and acreage that farmers have in production, and they are using this information to collect appropriate taxes (Smith 2003). Similar programs could be used to monitor individuals providing environmental services, or to assign appropriate taxes for pollution.

One crucial component of continued learning is policymaker education and interdisciplinary dialogue. In designing policies, we need to recognize the limitations of our knowledge.

6. CONCLUSIONS

In this chapter we have argued that humanity can do much more with the water we have already appropriated from nature. Our challenge is to introduce water policies and management strategies that will lead to sustainability of water resources. We argue that some current practices are sustainable while others are not, and that water resources may be depleted without intervention. The deterioration of groundwater aquifers throughout the world is a repeated example where subsidizing the price of water in the present may lead to resource depletion in the long run. The key elements of reform are the introduction of appropriate pricing and design of effective institutions and incentives that will promote conservation and responsible use of water, and reduce water contamination and pollution.

The policymaking process consists of many separated acts of decisions, and achieving sustainability will require reducing the likelihood of unsound policy choices. Therefore, water development plans need to be scrutinized by a social cost-benefit analysis. This analysis will consider the public and private, as well as the market and nonmarket, costs and benefits of proposed policies. Improved incentives can be used not only to address issues of a sufficient quantity of water resources, but also the quality of the water supply. Taxing polluters for the social cost of their activities will decrease pollution and ensure that pollution only occurs where its benefits outweigh the social costs. Public programs need to be developed to pay for conservation, environmental services, and other public goods. For these programs to work, the development of new technology for monitoring and enforcement are essential and need to be well utilized.

While we have a growing knowledge about policy design, we are also aware that implementation of reform is challenging intellectually and politically. Good science and economics are necessary but not sufficient for achieving water reform which is first and foremost a political act. A successful reform requires political will, leadership, and timing. Reform may be embodied in policy packages that are not ideal but represent major improvements. Reform has to be designed so that possible losers will not master the power to block it. So a reform may include compensations and resource transfers. The design of the reform may need to take into account not only the power distribution between various consumers and consumer groups, but also between government agencies, to overcome barriers by potential losers of reform. History has shown that frequently it is necessary for a crisis situation to occur (such as a severe drought) to develop the political and economic conditions that will lead to changes of policies. It is important that policymakers be well informed so that the opportunity afforded by a crisis can be used to make the best reform decisions.

The implementation of sustainable policies is still hampered by a lack of sufficient information. There is a need to know the parameters of both damage and benefit functions that show the relationship between people's behavior, policy choices, and the natural environment. There is a need to move to more interdisciplinary research efforts that incorporate biological, economic, and engineering knowledge to develop integrated decision-making frameworks resulting in ecologically sound and economically efficient solutions. Improved water policies have a limited capacity to lead to sustainable outcomes, as other policies are also of paramount importance. Water resource use depends on other factors and social trends. Many processes and factors generate pressures on water resources and threaten sustainable use. These include population growth, subsidization of agriculture and other industries, neglect of the risk posed by climate change, insufficient control of urban sprawl, and mining of the ocean or natural resources. Achieving sustainable water policies must thus be part of an across-the-board effort to attain resource sustainability.

BIBLIOGRAPHY

ACIL Tasman

2003 Water Trading in Australia Current and Prospective Products: Current Trends and Prospective Instruments to Improve Water Markets. Prepared for the Water Reform Working Group.

Adams, William M.

1993 Indigenous Use of Wetlands and Sustainable Development in West Africa. The Geographical Journal 159,2:209–218.

Aemmer, Fritz

1958 Determination of the Capacity of Hydro-electric Stations when Planning their Inclusion with Large Thermal Stations in the Operation of Interconnected Systems. Presented at UNIPEDE World Conference, Paper III, 3, Lausanne.

Afework Hailu

1998 An Overview of Wetland Use in Illubabor Zone, South-West Ethiopia. The University of Huddersfield, Huddersfield.

Afework Hailu, Alan Dixon, and Adrian P. Wood

2000 Nature, Extent and Trends in Wetland Drainage and Use in Illubabor Zone, South-West Ethiopia. Report for Objective 1: The University of Huddersfield, Huddersfield.

Allain, Sophie

2001 Une procédure de gestion intégrée et concertée de l'eau comme celle des SAGE est-elle un outil de développement territorial? Montagnes Méditerranéennes 14:25–29.

Allan, J. Tony

1996 Water, Peace and the Middle East. London: Tauris.

Altinbilek, Dogan

2004 Development and Management of the Euphrates-Tigris Basin. International Journal of Water Resources Development 20,1: 15–33.

Ames, Alcides
1988 Glacier Inventory of Peru. Huaraz: CONCYTEC.

1998 A Documentation of Glacier Tongue Variations and Lake Development in the Cordillera Blanca, Peru. Zeitschrift für Gletscherkunde und Glazialgeologie 34,1:1–36.

Ames Marquez, Alcides and Bernard Francou
1995 Cordillera Blanca glaciares en la historia. Bulletin de L'Institut Français d'Études Andines 24,1:37–64.

Amundsen Eirik S., Andersen Christian, Sannarnes Jan Gaute
1992 Rent Taxes on Norwegian Hydropower Generation. The Energy Journal 13,1:97–116.

Amundsen Eirik S., Tjøtta Sigve
1993 Hydroelectric Rent and Precipitation Variability: The Case of Norway. Energy Economics April:81–91.

Andreeva, Teodosia and Tatiana Orehova
2001 Climate Variability and its Influence on Groundwater in Central Bulgaria during the Last Decades. Comptes rendus de l'Académie bulgare des sciences 54,11:39–44.

2004 Influence of Mild Winters on Groundwater in Bulgaria. BALWOIS – Conference on Water Observation and Information System for Decision Support, 25–29 May 2004, Ohrid, FY Republic of Macedonia.

Anonymous
1941a El InformeTécnico: 8 millones de metros cúbicos de agua, incrementados por toneladas de piedra y agua se precipitaron desde las alturas sobre Huaraz. El Comercio 5,16 Diciembre.

1941b Huaraz sufrio grandes daños por el desborde del río Quilcay. El Comercio,14 Diciembre.

1945a En torno al vaciado de las lagunas de la Cordillera Blanca. El Departamento.

1945b Recuperación de sus tierras por los propietarios de la antigua Avenida Raymondi. El Departamento 3,21 Junio.

1945c Una Catástrofe mas en Ancash. El Pueblo (Seminario Informativo Independiente) 2,21 Enero.

1951 Ansiosa Espectativa Urbanistica. El Departamento 2,23 Agosto.

1954 La catástrofe de Huaraz. La Hora Año 1, 53,13, Diciembre, 2.

1955 Amenazas Hidrológicas. La Hora Año II, No. 212,4 Julio, 2.

1956a Asociación de Propietarios de la Zona Aluviónica. El Departamento 2,12 Abril.

1956b Obras públicas que requiere Huaraz. El Departamento 3,2 Agosto.

1956c Urbanisación de la zona del aluvión. El Departamento 2,7 Enero.

1959a Desagüe total de laguna Tullparaju para salvar Huaraz de aluviones. El Comercio 1–2,12 Diciembre.

1959b Estado de obras de defensa causa zozobra en Huaraz. El Comercio 1–14,12 Diciembre.

1970 Piden que el Gobierno profundice estudio sobre reubicación de la ciudad de Yungay. El Comercio 10,15 Noviembre.

Antonietti, Thomas
1993 L'esthétique du tourisme. Manifestations de l'industrie des loisirs à Crans-Montana et à Zermatt. In Antonietti Thomas et Marie-Claude Morand (dir.). Mutations touristiques contemporaines. Valais 1950–1990 3:63–90.

Antonov, H. and D. Danchev
1980 Groundwater in Bulgaria. Sofia: Technika. 360p. (in Bulgarian).

Antúnez de Mayolo, Santiago
1941 La caida de agua del Cañon del Pato. (Extracto del la Revista de la Escuela Nacional de Artes y Oficios de Lima, No. 32, Oct. 15 de 1941), Lima, Doc #. Ancash 00038, Instituto Riva Aguero. Vol. No. 32. Lima: Instituto Riva Aguero.

Arnauld, Antoine and Pierre Nicole
1962 La Logique, ou l'art de Penser, Commonly called The Port Royal Logic; available in English as The Art of Thinking, 1964,

Indianapolis: Bobbs-Merrill, Library of Liberal Arts, and as Logic or the Art of Thinking, translated and edited by Jill Vance Buroker, 1996, Cambridge: Cambridge University Press.

Arnell, Nigel, Chunzhen Liu, Rosa Compagnucci, Luis Veiga da Cunha, Keisuke Hanaki, Carol Howe, Gabriel Mailu, Igor Shiklomanov, and Eugene Stakhiv
2001 Hydrology and Water Resources. In Climate Change 2001: Impacts, Adaptation, and Vulnerability. McCarthy, James, O. Canziani, N. Leary, et al., eds. Cambridge: Cambridge University Press. pp. 191–234.

Arnell, Nigel W.
1999 The Effect of Climate Change on Hydrological Regimes in Europe: A Continental Perspective. Global Environmental Change 9:5–23.

Arora, Vivek K. and George J. Boer
2001 Effects of Simulated Climate Change on the Hydrology of Major River Basins. Journal of Geophysical Research – Atmospheres 106, D4:3335–3348.

Arrow, Kenneth J. and Anthony C. Fisher
1974 Environmental Preservation, Uncertainty, and Irreversibility. Quarterly Journal of Economics 88:312–319.

Association of Power Exchanges (APEX), ed.
2001 Creating Powerful Markets. Noordwijk, The Netherlands.

Ayllón Lozano, José
1955 En torno del control de las lagunas. La Hora Año 1, No. 118,4 Marzo, 3.

Babcock, Bruce, P.G. Lakshminarayan, Junjie Wu, and David Zilberman
1997 Targeting Tools for the Purchase of Environmental Amenities. Land Economics 73,3:325–329.

Bağış, Ali Ihsan
1997 Turkey’s Hydropolitics of the Euphrates-Tigris Basin. Water Resources Development 13:567–581.

Bakalov, Pavel, Stefan Shanov, Iren Ilieva, and Aleksey Benderev
2002 Physic-Geographical Characteristics for Creation of Caves in the Region of Pirin Mountain. In Pirin – Caves and Precipices, Speleo Expedition (in Bulgarian). Sofia: Iskar Press. pp. 11–28.

Bandyopadhyay, Jayendra, John C. Rodda, Rick Kattelmann, Zbigniew Kundzewicz, and Dieter Kraemer
1997 Highland Waters – A Resource of Global Significance. In Mountains of the World: A Global Priority. Messerli, Bruno and Jack D. Ives, eds. New York and London: Parthenon. pp. 131–155.

Barbier, Edward B.
1993 Sustainable Use of Wetlands – Valuing Tropical Wetland Benefits: Economic Methodologies and Applications. The Geographical Journal 159,1:22–32.

Barker, Mary L.
1980 National Parks, Conservation, and Agrarian Reform in Peru. Geographical Review 70,1:1–18.

Bartle, Alison
2002 Hydropower Potential and Development Activities. Energy Policy 30:1231–1239.

Bartle, Jim
1985 Parque Nacional Huascarán, Ancash, Perú. Lima: Nuevas Imágenes S.A.

Battalio, Robert C., John H. Kagel, and Don N. MacDonald
1985 Animals' Choices over Uncertain Outcomes: Some Initial Experimental Results. American Economic Review 75:597–613.

Battalio, Robert C., John H. Kagel, and Komain Jiranyakul
1990 Testing between Alternative Models of Choice under Uncertainty: Some Initial Results. Journal of Risk and Uncertainty 3:25–50.

Bauer, Carl J.
1995 Against the Current? Privatization, Markets, and the State in Water Rights: Chile, 1979–1993. Berkeley, CA: University of California.

1997 Bringing Water Markets Down to Earth: The Political Economy of Water Rights in Chile, 1976–1995. World Development 25,5:639–656.

1998 Against the Current: Privatization, Water Markets, and the State in Chile. Boston: Kluwer Academic Publishers.

Bauer, Dan
1987 The Dynamics of Communal and Hereditary Land Tenure among the Tigray of Ethiopia, In Bonnie McCay and James Acheson, eds. 1990 Question of the Commons: The Culture and Ecology of Communal Resources. Tucson: University of Arizona Press.

Bélidor, Bernard Forest de
1737–1753 Architecture hydraulique ou l'art de conduire, d'élever et de ménager les eaux pour les différents besoins de la vie. Paris.

Benderev, Aleksey, Deliana Gabeva, Petar Stefanov, and Borislav Velikov
1997 Characteristics of Large Springs in the Nastan-Trigrad Karst Basin by their Discharge Regimes. Review of the Bulgarian Geological Society (in Bulgarian) 58,2:115–121.

Benedict, Carol
1996 Bubonic Plague in Nineteenth Century China. Stanford: Stanford University Press.

Benoît, Philippe-Martin-Narcisse
1836 Manuel complet du boulanger, du négociant en grains, du meunier et du constructeur de moulins, 3rd ed. Paris: Encyclopédie Roret.

Berglund, Björn E.
1986 Handbook of Holocene Palaeocology and Palaeohydrology. Wiley, Chichester, UK.

Berrens, Robert P., David S. Brookshire, Michael McKee, and Christian Schmidt
1998 Implementing the Safe Minimum Standard Approach: Two Case Studies from the U.S. Endangered Species Act, Land Economics, 74,2:147–161.

Beschorner, Natasha
1992 Water and Instability in the Middle East. Adelphi Paper No. 273.

Bezzola, Gian-Reto
2000 Debris Flows. Lecture Notes of Postgraduate Studies in Hydraulic Structures. Lausanne: EPFL.

Binmore, Ken
1998 Game Theory and the Social Contract. Vol. 2. Just Playing. Cambridge, MA: MIT Press.

Binmore, Ken, Ariel Rubinstein, and Asher Wolinsky
1982 The Nash Bargaining Solution in Economic Modeling. Rand Journal of Economics 17:176–188.

Bishop, Richard
1978 Endangered Species and Uncertainty: The Economics of a Safe Minimum Standard, American Journal of Agricultural Economics 60,1:10–18.

Bjornlund, Henning and Jennifer McKay
2002 Aspects of Water Markets for Developing Countries: Experiences from Australia, Chile and the US. Environment and Development Economics 7:769–795.

Blake, David
2001 Proposed Mekong Dam Scheme in China Threatens Millions in Downstream Countries. World Rivers Review 16,3:4–5.

Bode, Barbara
1990 No Bells to Toll: Destruction and Creation in the Andes. New York: Paragon House.

Bojilova, Elena K.
1994 Karstic Springs in the Rhodopes Mountains. Proceeding of VITUKY, Budapest, Hungary.

2001 Stochastic Modelling of the Spring Discharges. 3rd International Conference Future Groundwater Resources at Risk, June 25–27; Lisbon, Portugal.

2004 Method of Composition Applied for Selected Karstic Springs in Bulgaria. BALWOIS Conference on Water Observation and Information System for Decision Support, Topic 5: Water Resources Protection and Ecohydrology, Ohrid, Republic of Macedonia (full text on CD-ROM).

Boulding, Kenneth E.
1991 What Is Evolutionary Economics? Journal of Evolutionary Economics 1:9–17.

Boyadjiev, Nikolaj
1964 The Karst Basins in Bulgaria and their Groundwaters. Bulletin of the Institute of Hydrology and Meteorology (in Bulgarian) 2:45–96.

Brentjes, Burchhard, Siegfried Richter, and Rolf Sonnemann
1978 Geschichte der Technik. Cologne: Aulis Deubner.

Brill, Eyal, Eithan Hochman, and David Zilberman
1997 Allocation and Pricing at the Water District Level. American Journal of Agricultural Economics 79,3:952–963.

Broggi, Jorge A.
1942 Letter to Sr. Director de Aguas e Irrigación. Lima: Biblioteca, Unidad de Glaciología y Recursos Hídricos.

Brokensha, David, D. Michael Warren, and Oswald Werner
1980 Indigenous Knowledge Systems and Development. Lanham, MD: University Press of America.

Bromley, David, ed.
1992 Making the Commons Work: Theory, Practice and Policy. San Francisco: Institute for Contemporary Studies.

Butcher, D. P. and Adrian Wood
1995 Sustainable Wetland Development in Highland Illubabor: A Preliminary Review of the Experience of Menschen fur Menschen. Unpublished report: The University of Huddersfield, Huddersfield.

Byers, Alton C.
2000 Contemporary Landscape Change in the Huascarán National Park and Buffer Zone, Cordillera Blanca, Peru. Mountain Research and Development 20,1:52–63.

Calder, Joshua and Jim Lee
1995 Aral Sea and Defense Issues. American University, Inventory of Conflict and Environment (ICE) Case Study No. 69.

Camerer, Colin F.
1989 An Experimental Test of Several Generalized Utility Theories. Journal of Risk and Uncertainty 2:61–104.

1995 Individual Decision Making. In The Handbook of Experimental Economics. Roth, J. H. Kagel and A., ed. Princeton: Princeton University Press. pp. 587–703.

Cannon, Terry
1994 Vulnerability Analysis and the Explanation of 'Natural' Disasters. In Disasters, Development and Environment. Varley, Ann, ed. New York: Wiley. pp. 13–30.

Capocaccia, A. Agostino, Alberto Mondini, and Umberto Forti
1973–1980 Storia della tecnica. Turin: Unione tipografico-editrice torinese.

Carey, Mark
2004 Glaciers. In Encyclopedia of World Environmental History. S. Krech III, J.R. McNeill and C. Merchant, ed. New York: Routledge. pp. 587–589.

2005 People and Glaciers in the Peruvian Andes: A History of Climate Change and Natural Disasters, 1941–1980. Ph.D. thesis. University of California.

Carlson, Lisa and Raymond Dacey
2006 Sequential Analysis of Deterrence Games with Declining Status Quo. Conflict Management and Peace Science 23,2:181–198.

Carrillo Ramírez, Alberto
1953 Ensayo monográfico de la Provincia de Bolognesi.

Carrión Vergara, Alberto
1959 Rasgos de pluma sobre Yungay. Forjando Ancash (Vocero del Club Ancash) Año 1, No. 2,30 Junio:19–21.

Castle, Emery N. and Robert P. Barrens
1993 Endangered Species, Economic Analysis, and the Safe Minimum Standard. Northwest Environmental Journal 9:108–130.

Castro-Lucic, Milka
2002 Local Norms and Competition for Water in Aymara and Atacama Communities (Northern Chile). In Water Rights and Empowerment. Boelens, Rutgerd and Paul Hoogendam, eds. The Netherlands. pp. 110–143.

Center for International Earth Science Information Network (CIESIN), Columbia University, International Food Policy Research Institute (IFPRI), and World Resources Institute (WRI)
2000 Gridded Population of the World. New York: CIESEN.

Central Statistics Authority (CSA), Ethiopia
1997 Ethiopia Statistical Abstract. Addis Ababa.

Chakravorty, Ujjayant, Eithan Hochman, and David Zilberman
1995 A Spatial Model of Optimal Water Conveyance. Journal of Environmental Economics and Management 28:25–41.

Chambers, Robert
1992 Rural Appraisal: Rapid, Relaxed and Participatory. In IDS Discussion Paper, No. 311. Brighton: Institute of Development Studies. Chambers, Robert.

1983 Rural Development: Putting the Last First. London: Longman.

Chichilnisky, Graciela
1994 North-South Trade and the Global Environment. The American Economic Review 84,4:851–874.

Chow, Ven Te, David R. Maidment, and Larry W. Mays
1988 Applied Hydrology. New York: McGraw-Hill.

Ciriacy-Wantrup, Siegfried von
1952 Resource Conservation: Economics and Policies. Berkeley: University of California Press.

Clague, John J. and Stephen G. Evans
2000 A Review of Catastrophic Drainage of Moraine-Dammed Lakes in British Columbia. Quaternary Science Reviews 19:1763–1783.

Clarke, Garry, David Leverington, James Teller and Arthur Dyke
2003 Superlakes, Megafloods, and Abrupt Climate Change. Science 301:922–923.

Clivaz, Christophe
1995 Tourisme et environnement dans l'espace alpin: L'exemple de Crans-Montana-Aminona. Collection Analyse des politiques publiques 6.

2001 Influence des réseaux d'action publique sur le changement politique. Le cas de l'écologisation du tourisme alpin en Suisse et dans le canton du Valais. Bâle/Genève/Munich: Helbing & Lichtenhahn.

Coase, Ronald
1960 The Problem of Social Cost. Journal of Law and Economics 3:1–44.

Cognat, Bruno
1973 La montagne colonisée. Paris: Ed. du Cerf.

Cohen, Jonathan E.
1992 International Law and the Water Politics of the Euphrates. New York University Journal of International Law and Politics 24:503–557.

Comisión de Control de Lagunas de la Cordillera Blanca (CCLCB)
1951–1954 Informes mensuales de los años 1951–1952-1953–1954. Huaraz: Biblioteca, Unidad de Glaciología y Recursos Hídricos.

Commission Internationale des Grands Barrages, (CIGB)
1974 Leçons tirées des accidents de barrages. Paris: CIGB.

Conklin, Harold C.
1980 Ethnographic Atlas of Ifugao: A Study of Environment, Culture, and Society in Northern Luzon. New Haven: Yale University Press.

Conseil fédéral (Switzerland)
1987 Message du Conseil fédéral concernant l'initiative populaire "pour la sauvegarde de nos eaux" et la révision de la loi fédérale sur la protection des eaux, 29.4.1987. Feuille Fédérale 139, II:1081–1231.

1995 Message relatif à la révision partielle de la loi fédérale sur l'utilisation des forces hydrauliques, 16.8.1995. Feuille Fédérale 147, IV:964–1001.

Coppey, Christian, Pierre-Antoine Masserey, and Gérard Rouvinez
1986 Crans-Montana. Un siècle de développement. Zürich: ETHZ, Travail de mémoire pour l'obtention du diplôme d'archictecture.

Coral Miranda, Reynaldo
1962 El aluvión de Huaraz: Relato de una tragedía. Lima: Litografia Universo, SA.

Corporación Peruana del Santa (CPS)
1944 Segunda Memoría. Lima: Biblioteca Nacional, Sala de Investigaciones.

Coussot, Philippe
1996 Les laves torrentielles, Connaissances à l'usage du praticien. Cemagref Editions.

Coward, Walter E.
1979 Principles of Social Organization in an Indigenous Irrigation System. Human Organization 38,1:28–36.

Crans-Montana Tourisme
2002 4e Rapport de gestion 2001. Crans Montana.

2006 6e Rapport de gestion 2003. Crans Montana.

Crook, Darren S. and Anne M. Jones
1999 Design Principles from Traditional Mountain Irrigation Systems (Bisses) in the Valais, Switzerland. Mountain Research and Development 19, 2:79–99.

Cropper, Maureen L. and Wallace E. Oates
1992 Environmental Economics: A Survey. Journal of Economic Literature 30:675–740.

Cubasch, Ulrich, Gerald Meehl, George Boer, Ronald Stouffer, Martin Dix, Akira Noda, Catherine Senior, Sarah Raper, and Kok-Seng Yap
2001 Projections of Future Climate Change. In Climate Change 2001: The Scientific Basis. Houghton, John, Y. Ding, D. Griggs, et al., eds. Cambridge: Cambridge University Press. pp. 526–582.

Cunnane, Conleth
1989 Statistical Distribution for Flood Frequency Analysis. Operational Hydrology Rep. 33 (WMO Publ. No. 718). Geneva: WMO.

Dacey, Raymond
1996a Dissatisfaction and the Illicit Response. Presented at the Fifth World Peace Science Society Congress, Amsterdam, June 3–6.

1996b Dissatisfaction and the Illicit Response 2. Presented at the Peace Science Society (International), Houston, October 25–27.

1996c The Maintenance of Authority: Risk Attitude and the Deterrence of Civil Disobedience. Presented at The American Economic Association, San Francisco, January 5–7, 1996.

1998 Risk Attitude, Punishment, and the Intifada. Conflict Management and Peace Science 16:77–78.

2002a Probability Weighting and the Persistence of Disagreements Among Constituents. Peace Economics, Peace Science and Public Policy 8,5–20.

2002b Policy Disputes over Water Development Projects. GAIA 11: 195–196.

2002c Decision Making under the ESA: A Decision–Theoretic Account of Disputes and their Resolution. Unpublished manuscript.

Dacey, Raymond and Kenneth Gallant
1997 Crime Control and Harassment of the Innocent. Journal of Criminal Justice 25:325–334.

Dali Environmental Protection Bureau
1997 Dali Bai Autonomous Prefecture People's Government's Decisions Regarding Protection and Control of Erhai's Ecology and Environment. In Dali Environmental Protection Magazine. Vol. 21.

Dali Gazetteer, (Jiajing Dai fuzhi (Jiajing reign-period Dali prefectural gazetteer)
1563 Microfilm 1055, from 'Rare Books, National Library, Peiping' in Menzies Library, Australian National University, roll 487, ed. Li Zhiyang, incomplete.

Dali County Gazetteer (Dali xianzhi)
1917 Rev. Zhang Peijue, ed. Zhou Zonglin, reprinted 1974, Chengwen, Tabei.

Dasgupta, Partha
1992 The Control of Resources. Cambridge, MA: Harvard University Press.

Dasgupta, Partha and Geoffrey Heal
1979 Economic Theory and Exhaustible Resources. Cambridge, UK: Cambridge University Press.

Daumas, Maurice
1962 Histoire générale des techniques. Paris: Presses Universitaires de France.

Davis, Mike
1998 Ecology of Fear: Los Angeles and the Imagination of Disaster. New York: Vintage Books.

de la Cadena, Marisol
2000 Indigenous Mestizos: The Politics of Race and Culture in Cuzco, Peru, 1919–1991. Durham: Duke University Press.

De Smedt, Florian H., Liu Yongbo, and Seifu Gebremeskel
2000 Hydrologic Modeling on a Catchment Scale using GIS and Remote Sensed Land Use Information. In Risk Analysis II. Brebbia, C.A., ed. Southampton, Boston: WIT Press. pp. 295–304.

Dearing, John Alfred
1999 Holocene Environmental Change from Magnetic Proxies in Lake Sediments. In Quaternary Climates, Environments and Magnetism. Maher, Barbara A. and Roy Thompson, eds. Cambridge: Cambridge University Press. pp. 231–278.

Demsetz, Harold
1967 Toward a Theory of Property Rights. American Economic Review 52,7:347–359.

Dengchuan, Zhouzhi
1854/5. In Gazetteer of Dengchuan Department. Vol. 13. pp. 166.

Denny, Patrick
1994 Biodiversity and Wetlands. Wetlands Ecology and Management 3,1:55–61.

Département Fédéral de l'Intérieur (DFI)
1991 Modifications du paysage en faveur de la pratique du ski. Directives pour la protection de la nature et du paysage. Berne.

Devarajan, Shantayanan and Anthony C. Fisher
1981 Hotelling's "Economics of Exhaustible Resources": Fifty Years Later. Journal of Economic Literature 19,1:65–73.

Diderot, Denis and Jean Le Rond d'Alembert
1762–1772 Recueil de planches sur les sciences, les arts libéraux, et les arts méchaniques, avec leur explication, appendix to the Encyclopédie.

Dinar, Ariel and Ashok Subramanian
1997 Water Pricing Experiences: An International Perspective. World Bank Technical Paper No. 386.

Dixit, Avinash K. and Robert S. Pindyck
1994 Investment under Uncertainty. Princeton: Princeton University Press.

Dixon, Alan B.
2002 The Hydrological Impacts and Sustainability of Wetland Drainage Cultivation in Illubabor, Ethiopia, Land Degradation and Development 13:17–31.

2003 Indigenous Management of Wetlands: Experiences in Ethiopia. Ashgate: Aldershot.

Döll Petra, Kaspar Frank and Bernhard Lehner
2003 A Global Hydrological Model for Deriving Water Availability Indicators: Model Tuning and Validation. Journal of Hydrology 270:105–134.

Dolsak, Nives and Elinor Ostrom
2003 The Commons in the New Millennium. Cambridge, MA: MIT Press.

Dorigo, Livio
1967 La Piena dell'Adige del novembre 1966 con brevi cenni alle precedenti piene, Ufficio Idrografico del Magistrato delle Acque. Ministero dei lavori pubblici-Servizio Idrografico.

Dourojeanni, Axel and Andrei Jouravlev
1999 El Codigo de Aguas de Chile: Entre la Ideologia y la Realidad. Debate Agrario 29–30:139–190.

Dregne, Harold E. and Nan-Ting Chou
1992 Global Desertification Dimensions and Costs. In The Degradation and Restoration of Arid Lands. Dregne, Harold E., ed. Lubbock: Texas Tech University.

Drochon, Rev. Père Jean-Emmanuel
1866 Un Chevalier apotre célestin. Augustienne.

Duan, Jinlu
2000 Dali lidai mingbei [Famous steles of successive dynasties in Dali]. Kunming: Yunnan minzu chubanshe.

Duan, Yanxue
1989 "Erhai-de bianyan" [The evolution of the Erhai]. In Yunnan Erhai kexue lunwenji [Collected Scientific Works on Erhai Lake in Yunnan]. Renxiang, Chen, ed. Kunming: Yunnan minzu chubanshe. pp. 212–222.

Dubach, Werner
1979 Die wohlerworbenen Rechte. Bundesamt für Wasserwirtschaft, Bern.

Dubas, Michel
2000 Roues hydrauliques, un regard d'ingénieur. Ingénieurs et Architectes Suisses 126,7:142–147.

2005 Design and efficiency of overshot water-wheels. International Journal on Hydropower & Dams 12,3:74–78.

Dubuis, Pierre
1990 Une économie alpine à la fin du Moyen Age. Orsières, l'Entremont et ses régions voisines, 1250–1500. 2 volumes. St. Maurice, Switzerland: Imprimerie St. Augustin.

1999 Le Bisse, témoin d'une civilisation alpine en mutation. In Le Rôle de l'eau dans le développement socio-économique des Alpes. Delaloye, Michel, ed. Sion: Institut universitaire Kurt Bösch. pp. 83–89.

Dugan, Patrick J.
1990 Wetland Conservation: A Review of Current Issues and Action. Gland: IUCN.

Duke Energy Perú
2002 Cañón del Pato: Hechos, datos y cifras. Lima: Duke Energy Perú.

Eagleson, Peter S.
1978 Climate, soil and vegetation. Water Resources. Research 14:705–776.

Easter, William K.
1986 Irrigation, Investment, Technology, and Management Strategies for Development. Boulder, CO: Westview Press.

Eckstein, Zvi, Dan Zackay, Yuval Nachtom, and Gideon Fishelson
1994 The Allocation of Water Resources between Israel, the West Bank and Gaza – An Economic Analysis (Hebrew). Economic Quarterly 41:331–369.

Electroperu
1975 Memoría bienal del Programa de Glaciología y Seguridad de Lagunas. Huaraz: Biblioteca, Unidad de Glaciología y Recursos Hídricos.

1989 Proyectos hidroeléctricos y de regulación de aguas en la cuenca del río Santa. Huaraz: Biblioteca, Unidad de Glaciología y Recursos Hídricos.

1995 Relación de Expedientes de Embalse 1995 solicitados por Electroperu S.A. Haraz. Huaraz: Biblioteca del Parque Nacional HuascaránArchive.

Electroperu, Glaciología y Recursos Hídricos
1997 Mapa indice de lagunas de la Cordillera Blanca. Huaraz: Electroperú (File Service), Lima.

Elvin, Mark and Darren Crook
2003 An Argument from Silence? The Implications of Xu Xiake's Description of the Miju River in 1639. In Collected Essays on Chinese History in Honour of Professor Li Yan on His Ninetieth Birthday and Sixtieth Year of Teaching. Xiaoliang, Wu, ed. Kunming: Yunnan University Press.

Elvin, Mark, Darren Crook, Shen Ji, Richard Jones, and John Dearing
2002 The Impact of Clearance and Irrigation on the Environment in the Lake Erhai Catchment from the Ninth to the Nineteenth Century. East Asian History 23:1–60.

Elvin, Mark, H. Nishioka, K. Tamura, and J. Kwek
1994 Japanese Studies on the History of Water Control in China. A Selected Bibliography. Institute of Advanced Studies, Australian National University, and the Centre for East Asian Cultural Studies for Unesco, Canberra and Tokyo.

Fagan, Brian
2000 The Little Ice Age: How Climate Made History, 1300–1850. New York: Basic Books.

Falkenmark, Malin and Cark Widstrand
1992 Population and Water Resources – A Delicate Balance. Population Bulletin 47,3:2–36.

Fan, Zhaoxin
late 1830sThe Three Rivers' Channel. In 1902, Langqiong xianzhi [Langqiong [Eryuan] County Gazetteer]. Vol. 11:60a–61b. Z. Kang, and Rev. L. Yingmei, ed. Chengwen: Taibei. pp. 481–484.

Fang, Guoyu
1998 Yunnan shiliao congkan.

Farmer, Darren L., Murugesu Sivapalan, and Chatchai Jothityangkoon
2003 Climate, Soil and Vegetation Controls upon the Variability of Water Balance in Temperate and Semi-Arid Landscapes: Downward Approach to Hydrological Prediction. Water Resources Research 39,2.

Fernández Concha, Jaime
1957 El problema de las lagunas de la Cordillera Blanca. *Boletín de la Sociedad Geológica del Perú* 32:87–95.

Fernández Concha, Jaime and Armin Hoempler
1953 Indice de lagunas y glaciares de la Cordillera Blanca. Huaraz: Biblioteca, Unidad de Glaciología y Recursos Hídricos.

Fernández, Justo
1942 13 de Diciembre de 1941: Crónicas completas de la tragedia. Editorial "Perú Libre".

Ferraz, Clarice and Franco Romerio
2004 Market Reorganisation, Supply Security and Hydro Resources. Oil, Gas and Energy Law (OGEL) 3.

Filippini, Massimo and Daniel Spreng
2001 Perspektiven der Wasserkraftwerke der Schweiz. Bundesamt für Energie, Bern.

Finon, Dominique
1996 Les coûts environnementaux de la production et de l'utilisation d'énergie: Méthodes d'évaluation et instruments d'internalisation. Revue de l'Energie 480:439–446.

Fiorentino, Mauro
1985 La valutazione dei volumi dei volumi di piena nelle reti di drenaggio urbano. Idrotecnica 3:141–152.

Fiorentino, Mauro and Maria Rosaria Margiotta
1998 La valutazione dei volumi di piena e il calcolo semplificato dell'effetto di laminazione dei grandi invasi. In Tecniche per la difesa dall'inquinamento. Frega, Giuseppe, ed: BIOS. pp. 203–222.

Fishburn, Peter C. and Gary Kochenberger
1979 Two-Piece von Neumann-Morgenstern Utility Functions. Decision Sciences 10:503–518.

Fisher, Franklin M.
2002 Water Value, Water Management, and Water Conflict: A Systematic Approach. GAIA 11:187–190.

Fisher, Franklin M., Shaul Arlosoroff, Zvi Eckstein, Salem Hamati Munther Haddadin, Annette Huber-Lee, Ammar Jarrar, Anan Jayyousi, Uri Shamir, and Hans Wesseling with a special contribution by Amer Salman and Emad Al-Karablieh
2005 Liquid Assets: An Economic Approach for Water Management and Conflict Resolution in the Middle East and Beyond. Washington: RFF Press.

Fisher, Franklin M., Shaul Arlosoroff, Zvi Eckstein, Munther Haddadin, Salem Hamati, and Ammar Jarrar Annette Huber-Lee, Anan Jayyousi, Uri Shamir, and Hans Wesseling
2002 Optimal Water Management and Conflict Resolution: The Middle East Water Project. Water Resources Research 10.1029/2001WR000943.

Fitzgerald, Charles Patrick
1941 The Tower of Five Glories, a Study of the Min-chia of Ta-li, Yunnan. London: Cresset Press.

Fliri, Franz
1998 Naturchronik von Tirol. Innsbruck: Universitätsverlag Wagner.

Frayer, Julia and Nazli Z. Uludere
2001 What is it Worth? Application of Real Options Theory to the Valuation of Generation Assets. The Electricity Journal October: 40–51.

Frey, Frederic
1993 The Political Context for Conflict and Cooperation over International River Basins. Water International 18,54–68.

Friedl, Gabriela and Alfred Wuest
2002 Disrupting Biogeochemical Cycles – Consequences of Damming. Aquatic Sciences – Research across Boundaries 64:55–65.

Friedman, Milton and Leonard J. Savage
1948 The Utility Analysis of Choices Involving Risk. Journal of Political Economy 56:279–304.

Fudenberg, Drew and David Levine
1998 The Theory of Learning in Games. Cambridge, MA: MIT Press.

Galli, Roberto
1996 Innalzamento diga Luzzone: Un parto difficile. Wasser, Energie, Luft 88:293–297.

Gardner, Roy
2003 Games for Business and Economics. New York: Wiley.

Garr, C.E. and Blair Fitzharris
1994 Sensitivity of Mountain Runoff and Hydro-electricity to Changing Climates. In Mountain Environments in Changing Climates. Beniston, Martin, ed. London and New York: Routledge.

Genev, Marin
2004 Tendencies of Annual River Discharge for Bulgaria during the XXth Century.

Genev, Marin, Strahil Gerassimov, Todorina Bojkova, et al.
1998 Tendencies of Multi-Annual Variations of Runoff of Bulgaria. Scientific Report, Library of the Ministry of Education and Science (in Bulgarian) 3,406.

Georges, Christian
2004 20th-Century Glacier Fluctuations in the Tropical Cordillera Blanca, Peru. Arctic, Antarctic, and Alpine Research 36,1:100–107.

Gerassimov, Strahil, Elena Bojilova, Tatiana Orehova, and Marin Genev
2001 Water Resources in Bulgaria During the Drought Period - Quantitative Investigations. In 29th IAHR Congress Proceedings. Development, Planning and Management of Surface and Ground Water Resources. Beijing, China: Tsinghua University Press.

Gerassimov, Strahil, Marin Genev, Elena Bojilova, and Tatiana Orehova
2004 Water Resources During the Drought, 85–100, Part IV: Drought Impact on Water Resources. In Drought in Bulgaria: a Contemporary Analogue for Climate Changes: Ashgate Studies in Environmental Policy and Practice.

Ghiglino, Luis A.
1950 De estudio a la Laguna de Jancarurush efectuada entre el 22 y 26 de Oct. de 1950. Lima: Biblioteca de Electroperú, Lima.

Gibbons, Robert
1992 Game Theory for Applied Economists. Princeton: Princeton University Press.

Giesecke, Alberto and Luke Lowther
1941 Informe sobre el aluvion de 13 de Diciembre de 1941. Lima: Biblioteca, Unidad de Glaciología y Recursos Hídricos, Huaraz.

Giovanola, Alain and Armin Karlen, eds.
1999 Les bisses du Valais. Sierre: Monographic.

Glachant, Jean-Michel and Dominique Finon
2003 Competition in European Electricity Markets: A Cross-Country Comparison. London: Edward Elgar.

Gleick, Peter H.
2000 The Changing Water Paradigm: A Look at Twenty-first Century Water Resources Development. Water International 25:127–138.

Glick, Thomas F.
1970 Irrigation and Society in Medieval Valencia. Cambridge, MA: Harvard University Press.

Global Runoff Data Centre (GRDC)
1999 Global Runoff Data: Global Runoff Data Centre, Koblenz, Germany.

Gostner, Walter
2002 Integrale Analyse eines murfähigen Wildbaches anhand einer Fallstudie: Master's Thesis for Postgraduate Studies in Hydraulic Structures. Zurich: Laboratory of Hydraulics, Hydrology and Glaciology, ETH-Zürich.

Gostner, Walter, Gian Reto Bezzola, Markus Schatzman, and Hans Erwin Minor
2003 Integral Analysis of Debris Flow in an Alpine Torrent – The Case Study of Tschengls. Rotterdam: Balkema.

Government of Ethiopia
1998 Annual Report of Illubabor Zone. Unpublished Report. Metu: Ministry of Agriculture, Government of Ethiopia.

Graves, Franz, Thomas Jenkin, and Dean Murphy
1999 Opportunities for Electricity Storage in Deregulating Markets. The Electricity Journal October:46–56.

Grenier, Louise
1998 Working with Indigenous Knowledge: A Guide for Researchers. Ottawa: IDRC.

Grove, Jean M.
1987 Glacier Fluctuations and Hazards. Geographical Journal 153,3:351–367.

1988 The Little Ice Age. London: Methuen.

Grupo Andinista Cordillera Blanca
1952 Estatutos y Reglamentos del Grupo Andinista "Cordillera Blanca" Aprobados en Asamblea de 9 Octubre [1952], Doc. # 14, Legajo 590. Lima: Ministerio del Interior, Prefectura, at AGN, Lima.

Guillet, David
1992 Covering Ground: Communal Water Management and the State in the Peruvian Highlands. Ann Arbor: University of Michigan Press.

1994 Canal Irrigation and the State: The 1969 Water Law and Irrigation Systems in the Colca Valley of Southern Peru. In Irrigation at High Altitudes: the Social Organization of Water Control Systems in the Andes. Mitchell, William P., David Guillet and Inge Bolin, eds. Washington: American Anthropological Association. pp. 167–188.

Güner, Serdar
1997 The Syrian-Turkish War of Attrition: the Water Dispute. Studies in Conflict and Terrorism 20,105–116.

1998 Signalling in the Syrian-Turkish Water Conflict. Conflict Management and Peace Science 16:185–206.

1999 Water Alliances in the Euphrates-Tigris Basin. In Environmental Change Adaptation, and Security. Lonergan, Steve C., ed. Dordrecht: Kluwer. pp. 301–316.

Handmer, John, Edmund Penning-Rowsell, and SueTapsell
1999 Flooding in a Warmer World: The View from Europe. In Climate Change and Risk. Downing, Thomas E., Alexander A. Olsthoorn and Richard S.J. Tol, eds. London: Routledge.

Hardin, Garret
1968 The Tragedy of the Commons. Science, 162:1243–1248.

Hargreaves Heap, Shaun P. and Yanis Varoufakis
1995 Game Theory: A Critical Introduction. London: Routledge.

Harrison, Gareth P. and Whittington H. (Bert)
2001 Impact of Climatic Change on Hydropower Investment. Hydropower in the New Millennium, Proceedings of the 4th International Conference on Hydropower Development, Bergen, 20–22 June 2001:257–261.

Harsanyi, John C.
1977 Rational Behavior and Bargaining Equilibrium in Games and Social Situations. Cambridge, UK: Cambridge University Press.

Harsanyi, John C. and Reinhart Selten
1988 A General Theory of Equilibrium Selection in Games. Cambridge, MA: MIT Press.

Hastenrath, Stefan and Alcides Ames
1995 Recession of Yanamarey Glacier in Cordillera Blanca, Peru, during the Twentieth Century. Journal of Glaciology 41,127:191–196.

Haverkort, Bertus and Wim Hiemstra (eds.)
1999 Food for Thought: Ancient Visions and New Experiments of Rural People. London: Zed Books.

Heller, Michael
2000 The Three Faces of Private Property. 79 Oregon Law Review 417.

Hendriks, Jan
1998 Water as Private Property: Notes on the Case of Chile. In Searching for Equity: Conceptions of Justice and Equity in Peasant Irrigation. Boelens, Rutgerd, Gloria Davila and Rigoberta Menchu, eds: Van Gorcum Press. pp. 297–309.

Hewitt, Kenneth
1997 Regions of Risk: A Geographical Introduction to Disaster. Essex, UK: Longman.

Hirschmann, Alfred O.
1970 Exit, Voice, and Loyalty: Responses to Decline in Firms, Organizations and States. Cambridge, MA: Harvard University Press.

Hirshleifer, Jack
1983 From Weakest Link to Best Shot: The Voluntary Provision of Public Goods. Public Choice 41,371–386.

Hofer, Thomas
1998 Floods in Bangladesh: A Highland–Lowland Interaction? University of Berne.

Hollis, George E.
1990 Environmental Impacts of Development on Wetlands in Arid and Semi-Arid Lands. Hydrological Sciences Journal 35,4:411–428.

Horsman, Stuart
2001 Water in Central Asia: Regional Cooperation or Conflict? In Central Asian Security. Allison, Roy and Leana Jonson, eds. London: Royal Institute of International Affairs. pp. 70–94.

Hotelling, Harold
1931 The Economics of Exhaustible Resources. Journal of Political Economy 39:137–175.

Hou, Yunqin
No date. In Dengchuan zhouzhi. pp. 46–48.

Howe, Charles
2002 Policy Issues and Institutional Impediments in the Management of Groundwater: Lessons from Case Studies. Environment and Development Economics 7,4:625–641.

Hristo, Antonov and Dancho Danchev
1980 Groundwater in the Republic Bulgaria (in Bulgarian). Sofia: Technika.

Hunt, Robert C.
1988 Size and Structure of Authority in Canal Irrigation Systems. Journal of Anthropological Research 44,4:335–355.

1992 Inequality and Equity in Irrigation Communities. Third Common Property Conference of the International Association for the Study of Common Property, September 17–20, 1992.

Hunt, Robert C. and Eva Hunt
1976 Canal Irrigation and Local Social Organization. Current Anthropology 17:129–157.

Huth, Paul
2001 Standing Your Ground: Territorial Disputes and International Conflict. Ann Arbor: University of Michigan Press.

Ibérico, Mariano
1954 Los Nevados: Notas sobre el paisaje de la Sierra. La Hora Año 1, 89, 27 Enero, 3.

Instituto Geológico Minero y Metalúrgico (INGEMMET)
1979 Boletín Informativo, Oficina Regional Huaraz, Programa: Glaciología y Seguridad de Lagunas. Harat: Litho Offset Impresores.

International Institute of Rural Reconstruction
1996 Recording and Using Indigenous Knowledge: A Manual. Cavite, Philippines: International Institute of Rural Reconstruction.

Intergovernmental Panel on Climate Change (IPCC)
2001 Climate Change 2001 – The Scientific basis. Contribution of Working Group I to the Third Assessment Report of the Intergovernmental Panel on Climate Change (IPCC). Houghton, John T., Y. Ding, D. J. Griggs, et al., eds. Cambridge, UK and New York: Cambridge University Press.

Irving
1952 Lo que el pueblo de Huaraz debe conocer en torno al plano regulador. El Departamento 10 Noviembre, 3.

Iverson, Richard M.
1997 The Physics of Debris Flows. Reviews of Geophysics 35,3:245–296.

Iverson, Richard M. and James W. Vallance
2001 New Views of Granular Mass Flows. Geology 29,2:115–118.

Ives, Jack, Bruno Messerli, and Ernst Spiess
1997 Introduction. In Mountains of the World: A Global Priority. Bruno Messerli, Jack Ives, ed. New York and London: Parthenon. pp. 1–15.

Izaguirre, Isaías L.
1954 Llanganuco. La Hora Año1, No. 6, 15 Julio, 2.

Jan, Ch.D. and H.W. Shen
1993 Review Dynamic Modeling of Debris Flows. International Workshop on Debris Flows, Kagoshima, Japan, September 6–8:33–42.

Jaranoff, Dimitar
1959 Karst Hydrogeology in the Upper Krichim River Basin. In Karst Groundwater in Bulgaria (in Bulgarian). Sofia: Tehnika. pp. 52–65.

Kahneman, Daniel and Amos Tversky
1979 Prospect Theory: An Analysis of Decisions Under Risk. Econometrica 47:263–291.

2000 Choices, Values, and Frames. Cambridge, UK: Cambridge University Press.

Kandori, Michihiro, George Mailath, and Rafael Rob
1993 Learning, Mutation, and Long Run. Econometrica 61:29–56.

Kaser, Georg, Alcides Ames, and Marino Zamora
1990 Glacier Fluctuations and Climate in the Cordillera Blanca, Peru. Annals of Glaciology 14:136–140.

Kaser, Georg and Christian Georges
1997 Changes of the Equilibrium-Line Altitude in the Tropical Cordillera Blanca, Peru, and Their Spatial Variations. Annals of Glaciology 24:344–349.

Kaser, Georg and Henry Osmaston
2002 Tropical Glaciers. New York: Cambridge University Press.

Kattelmann, Richard
2003 Glacial Lake Outburst Floods in the Nepal Himalaya: A Manageable Hazard? Natural Hazards 28:145–154.

Kebede Tato
1993 Evaluation of the Environmental Components of the Menschen für Menschen Projects in Ethiopia, Unpublished report. Addis Ababa: Menschen für Menschen.

Kerr, John
2002 Sharing the Benefits of Watershed Management in Sukhomajri, India. In Selling Forest Environmental Services. Pagiola, Stefano, Josh Bishop and Natasha Landell-Mills, eds. Sterling, VA: Earthscan.

Khanna, Madhu, Katrin Millock, and David Zilberman
1999 Sustainability, Technology, and Incentives. In Flexible Incentives for the Adoption of Environmental Technologies in Agriculture. Casey, Frank, Andrew Schmitz, Scott Swinton, et al., eds. Norwell, MA: Kluwer Academic.

Kienholz, H.
1995 Gefahrenbeurteilung und –bewertung – Auf dem Weg zu einem Gesamtkonzept. Schweiz. Z. Forstw. Jg 146,9:701–725.

Kolars, John F. and William A. Mitchell
1991 The Euphrates River and the Southeastern Anatolia Development Project. Carbondale: Southern Illinois University Press.

Koleva, Ekaterina and Raina Peneva
1990 Climatic Reference Book. Precipitation in Bulgaria (in Bulgarian). Sofia: Bulgarian Academy of Sciences Press.

Kreps, David M.
1993 Game Theory and Economic Modelling. Oxford: Oxford University Press (The Clarendon Press).

Krewitt, Wolfram, Thomas Heck, Alfred Trukenmüller, and Friedrich Rainer
1999 Environmental Damage Costs from Fossil Electricity Generation in Germany and Europe. Energy Policy 27:173–183.

Larson, Brooke
2004 Trials of Nation Making: Liberalism, Race, and Ethnicity in the Andes, 1810–1910. New York: Cambridge University Press.

Last, William M. and John P. Smol, eds.
2002a Tracking Environmental Change Using Lake Sediments, Volume 1: Basin Analysis, Coring, and Chronological Techniques. Dordrecht: Kluwer Academic.

2002b Tracking Environmental Change Using Lake Sediments, Volume 2: Physical and Geochemical Methods. Dordrecht: Kluwer Academic.

Lave, Lester B. and Tunde Bayvanyos
1998 Risk Analysis and Management of Dam Safety. Risk Analysis 18,4:455–462.

Le Roy Ladurie, Emmanuel
1971 Times of Feast, Times of Famine: A History of Climate since the Year 1000. Garden City, NY: Doubleday.

Leach, Edmund
1961 Pul Eliya. Oxford: Oxford University Press.

LeMarquand, David G.
1977 International Rivers: The Politics of Co-operation. University of British Columbia and Waterloo Research Center.

Lenhard, Vera Christine and Raimund Rodewald
2000 Die Allmende als Chance: Nachhaltige Landschaftsentwicklung mit Hilfe von institutionellen Ressourcenregimen. GAIA 9,1: 50–57.

Leung, L. Ruby, Yun Qian, Xindi Bian, Warren M. Washington, Jongil Han, and John O. Roads
2004 Mid-Century Ensemble Regional Climate Change Scenarios for the Western United States. Climate Change 62:75–113.

Li, Hao
2002 Ming. Sanyi suibi [Literary jottings by Suibi or 'From Western Yunnan'].

Li Yiheng, ['Yudi shanren']
Middle MingHuaicheng yeyu [Record of night-time conversations from Huai City].

Li, Zhengqing
1998 Dali Xizhou wenhua shi [History of the Culture of Xizhou in Dali]. Kunming: Yunnan minzu chubanshe.

Li, Zhiyang
No date Dali fushi.

Lihui, Chen
2004 Dam-Building Decision Institution in China and Integrated Management of Watershed. Paper presented at the International Conference on Advances in Integrated Mekong River Management, Australian National University, Canberra. http://hdl.handle.net/1885/43119.

Liniger, Hanspeter, Rolf Weingartner, and Martin Grosjean, eds.
1998 Mountains of the World: Water Towers for the 21st Century – A Contribution to Global Freshwater Management. Berne.

Litton, Consul George
1903 Commercial Report on the Province of Yunnan, with Special Reference to the Burma-Têngyüeh Trade: Rangoon Office of the Superintendant, Government Printing Burma.

Liu, Wenzheng
1991 (reprint of original 1621–1627) Dianzhi. Kunming: Yunnan jiaoyu chubanshe.

Lliboutry, Louis, Benjamín Morales A., André Pautre, and Bernard Schneider
1977 Glaciological Problems Set by the Control of Dangerous Lakes in Cordillera Blanca, Peru. I. Historical Failures of Morainic Dams, Their Causes and Prevention. Journal of Glaciology 18,79:239–254.

Lowi, Miriam R.
1993 Water and Power: The Politics of a Scarce Resource in the Jordan Basin. Cambridge: Cambridge University Press.

Luterbacher, Urs and Ellen Wiegandt
1995 Social Dimensions of Resource Use: The Freshwater Case. Swiss Political Science Review 1,4:131–142.

2002 Water Control and Property Rights: An Analysis of the Middle East Situation. In Climatic Change: Implications for the Hydrological Cycle. Beniston, Martin, ed. Dordrecht/Boston/London: Kluwer Academic. pp. 379–410.

2005 Cooperation or Confrontation: Sustainable Water Use in an International Context. In Freshwater and International Economic Law. Weiss, Edith Brown, Laurence Boisson de Chazournes and Nathalie Bernasconi-Osterwalder, eds. Cambridge, UK: Cambridge University Press. pp. 11–34.

Luterbacher, Urs, Valery Kuzmichenok, Gulnara Shalpykova, and Ellen Wiegandt
2008 Glaciers and Efficient Water Use in Central Asia. In Darkening Peaks. Orlove, Benjamin, Ellen Wiegandt and Brian Luckman, eds. Berkeley: University of California Press. (forthcoming)

Maass, Arthur and Raymond L. Anderson
1978 And the Desert Shall Rejoice: Conflict, Growth and Justice in Arid Environments. Cambridge Raymond: The MIT Press.

1986 And The Desert Shall Rejoice: Conflict, Growth and Justice in Arid Environments. Malabar, FL: R.E. Krieger.

Mabry, Jonathan and David Cleveland
1996 The Relevance of Indigenous Irrigation: A Comparative Analysis of Sustainability. In Canals and Communities: Small-Scale Irrigation Systems. Mabry, Jonathon, ed. Tuscon: University of Arizona Press. pp. 227–260.

Mainali, Arbind P. and Nallamuthu Rajaratnam
1991 Hydraulics of debris flows. In Water Resources Engineering Report (WRE 91–2): Department of Civil Engineering, University of Alberta, Edmonton, Canada, T6G 2G7.

Manabe, Syukuro
1969 Climate and the Ocean Circulation 1. Atmospheric Circulation and the Hydrology of the Earths Surface. Monthy Weather Review 97,11: 739–774.

Manfreda, Salvatore
2002 An analysis of risk vulnerability in the Province of Trento. Conference "Mountains: Sources of Water, Sources of Knowledge". IUKB, Sion.

Manfreda, Salvatore, Mauro Fiorentino, and Vito Iacobellis
2005 DREAM: A Distributed Model for Runoff, Evapotranspiration, and Antecedent Soil Moisture Simulation. Advances in Geosciences 2:31–39.

Mangin, William P.
1955 Estratificación social en el Callejón de Huaylas. Revista del Museo Nacional (Lima-Perú) XXIV, pp. 174–189.

Margolis, Howard
1996 Dealing with Risk. Chicago: University of Chicago Press.

Maskrey, Andrew
1994 Disaster Mitigation as a Crisis of Paradigms: Reconstructing after the Alto Mayo Earthquake, Peru. In Disasters, Development and Environment. Varley, Ann, ed. New York: Wiley. pp. 109–123.

Maynard Smith, John
1982 Evolution and the Theory of Games. Cambridge: Cambridge University Press.

Maynard Smith, John and George Price
1973 The Logic of Animal Conflicts. Nature 246:15–18.

McCann, James C.
1995 People of the Plow: An Agricultural History of Ethiopia 1800–1990. Madison, WI: University of Wisconsin Press.

McCay, Bonnie and James Acheson, eds.
1987 Question of the Commons: The Culture and Ecology of Communal Resources. Tucson: University of Arizona Press.

McCorkle, Constance M. and G. D. McClure
1995 Farmer Know-How and Communication for Technology Transfer: CTTA in Niger. In The Cultural Dimension of Development: Indigenous Knowledge Systems. Warren, D. Michael, L. Jan Slikkerveer and David Brokensha, eds. Bourton-on-Dunsmore, UK: Practical Action Publishing. pp. 323–332.

Mekong News
2004 Chinese Data Brings Early Flood Warning. In Mekong News. Vol. Aug–Oct 2004. p. 3.

Menna, Federico
1998 Della possibilità di utilizzare in via parziale e temporanea la capacità d'invaso dei serbatoi idroelettrici del bacino idrografico del'Adige per la laminazione delle piene. Trento.

Messerli, Bruno, Daniel Viviroli, and Rolf Weingartner
2004 Mountains of the World – Vulnerable Water Towers for the 21st Century. Contribution to the Royal Colloquium on Mountain Areas: A Global Resource, Abisko 2003. Ambio, Special Report 13:32–40.

Meybeck, Michel, Pamela Green, and Charles J. Vörösmarty
2001 A New Typology for Mountains and other Relief Classes: An Application to Global Continental Water Resources and Population Distribution. Mountain Research and Development 21,1:34–45.

Millar, David
1993 Farmer Experimentation and the Cosmovision Paradigm. In Cultivating Knowledge: Genetic Diversity, Farmer Experimentation and Crop Research. Boef, Walter de, Kojo Amanor, Kate Wellard, et al., eds. London: Intermediate Technology Publications. pp. 44–50.

Mitchel, John G.
2002 Down the Drain? In National Geographic. Vol. 202. pp. 34–51.

Montaigne, Fen
2002 Water Pressure. In National Geographic. Vol. 202. pp. 2–33.

Montes, Emilio
1945 Catásrofe de Chavín. Uncited newspaper clipping in Doc. # 182, Legajo 465, Ministerio del Interior, Prefectura, at AGN, Lima.

Montginoul Marielle, Benoît Rossignol, and Patrice Garin
2000 Une résolution des conflits d'usages par des actions sur la coordination et la demande en eau: le cas français. Territoires en mutation 7:77–89.

Mool, Pradeep
2001 Glacial Lakes and Glacial Lake Outburst Floods, Mountain Development Profile 2.

Morales Arnao, César
2001 Las Cordilleras del Perú. Lima: Banco Central de Reserva del Perú.

Morales, Benjamín
1969 Las lagunas y glaciares de la Cordillera Blanca y su control. Boletín del Instituto Nacional de Glaciología (Peru) 1:14–17.

Müller, Gerald and Klemens Kaupert
2004 Performance Characteristics of Water Wheels. Journal of Hydraulic Research 42,5:1–10.

Müller, Hansruedi, Bernhard Kramer, and Jost Krippendorf
1993 Freizeit und Tourismus. Eine Einführung in Theorie und Politik. Berner Studien zu Freizeit und Tourismus 28,5.

Müller, Wilhelm
1939 Die Wasserräder, 2nd ed. Leipzig: Moritz Schäfer.

Mundy, Paul A. and J. Lin Compton
1995 Indigenous Communication and Indigenous Knowledge. In The Cultural Dimension of Development: Indigenous Knowledge Systems. Warren, D. Michael, L. Jan Slikkerveer and David Brokensha, eds. Bourton-on-Dunsmore, UK: Practical Action Publishing. pp. 112–123.

Murra, John
1972 El "Control Vertical" de un Maximo de Pisos Ecologicos en la Economia de las Societades Andinas. In Visita de la Provincia de Léon de Huanaco (1562). Vol. 2. Huanuco, Peru: Universidad Hermilio Valdizán. pp. 429–476.

Naff, Thomas and Ruth C. Matson
1984 Water in the Middle East: Conflict or Cooperation. Boulder, CO: Westview Press.

Nakarado, Gary L.
1996 A Marketing Orientation is the Key to a Sustainable Energy Future. Energy Policy 24:187–193.

Nash, John
1950 The Bargaining Problem. Econometrica 18:155–162.

1953 Two Person Cooperative Games. Econometrica 21:128–140.

National Environmental Research Council (NERC)
1975 Flood Studies Report, Vol. I. London.

National Research Council (NRC)
1986 Proceedings of the Conference on Common Property Management.

Neilson, William S.
1993 An Expected Utility-User's Guide to Nonexpected Utility Experiments. Eastern Economics Journal 19:257–274.

Neogodo
1954 A trece años del aluvion de Huaraz. La Hora Año 1, 54, 14 Diciembre, 3.

Netting, Robert McC.
1981 Balancing on an Alp. Cambridge, UK: Cambridge University Press.

Nijssen, Bart, Greg M. O'Donnell, Dennis P. Lettenmaier, Dag Lohmann, and Eric F. Wood
2001 Predicting the Discharge of Global Rivers, Journal of Climate 14,15:307–323.

O'Brien, James S., Pierre Y. Julien, and W.T. Fullerton
1993 Two-Dimensional Water Flood and Mudflow Simulation. Journal of Hydraulic Engineering, ASCE 119,2:244–261.

Office fédéral de l'économie des eaux (OFEE)
2001 Rapport concernant une loi fédérale sur les ouvrages d'accumulation. Berne, 29.8.2001.

Oliver-Smith, Anthony

1977 Traditional Agriculture, Central Places, and Postdisaster Urban Relocation in Peru. American Ethnologist 4,1:102–116.

1986 The Martyred City: Death and Rebirth in the Andes. Albuquerque: University of New Mexico Press.

1999 Peru's Five-Hundred-Year Earthquake: Vulnerability in Historical Context. In The Angry Earth: Disaster in Anthropological Perspective. Oliver-Smith, Anthony and Susannah M. Hoffman, eds. New York: Routledge. pp. 74–88.

Oliver-Smith, Anthony and Susanna M. Hoffman

1999 The Angry Earth: Disaster in Anthropological Perspective. New York: Routledge.

Olson, Robert

1992 The Kurdish Question in the Aftermath of the Gulf War: Geopolitical and Geostrategic Changes in the Middle East. Third World Quarterly 13:475–499.

Orehova, Tatiana

2001 Analysis of Water Temperature Variability for some Bulgarian Springs. 3rd International Conference Future Groundwater Resources at Risk, Lisbon, Portugal:681–688.

2002a A Comparison between Water Years 2000 and 2001 for Groundwater in Bulgaria. Proceedings of the 22nd Annual AGU Hydrology Days. Colorado State University. Fort Collins. CO, April 1–4:216–223.

2002b Seasonal Variability of the Groundwater Regime for Several Aquifers in Bulgaria. ICHE-2002 Conference, Warsaw, Poland:15.

Orehova, Tatiana and Elena Bojilova

2001a Impact of the Recent Drought Period on Groundwater in Bulgaria. In 29th IAHR Congress Proceedings. Theme A Development, Planning and Management of Surface and Ground Water Resources. Beijing, China: Tsinghua University Press.

2001b Some Investigations Concerning Groundwater Regime in the Mediterranean and Black Sea zones in Bulgaria. 3rd International Conference Future Groundwater Resources at Risk, Lisbon, Portugal:689–696.

Organisation for Economic Co-operation and Development (OECD)
2001 Freshwater. OECD Environmental Outlook:97–107.

Orlove, Benjamin
1993 Putting Race in Its Place: Order in Colonial and Postcolonial Peruvian Geography. Social Research 60:301–336.

Osborne, Martin and Ariel Rubinstein
1990 Bargaining and Markets. San Diego, CA: Academic Press.

Ostrom, Elinor
1987 Institutional Arrangements for Resolving the Commons Dilemma. In The Culture and Ecology of Communal Resources. Bonnie McCay and James Acheson, eds. Tucson, AZ: University of Arizona Press. pp. 250–265.

1990 Governing the Commons: The Evolution of Institutions for Collective Action. New York: Cambridge University Press.

1992 Crafting Institutions for Self-Governing Irrigation Systems. San Francisco: Institute for Contemporary Studies Press.

1998 Reformulating the Commons. In The Commons Revisited: An Americas Perspective. Burger, Joanna, Richard Norgaard, Elinor Ostrom, et al., eds. Washington: Island Press. pp. 1–26.

Ostrom, Elinor, ed.
1986 Issues of Definition and Theory: Some Conclusions and Hypotheses, Proceedings of the Conference on Common Property Management. Washington: National Academy Press.

Ostrom, Elinor, Joanna Burger, Christopher Field, Richard Norgaard, and David Policansky
1999 Revisiting the Commons: Local Lessons, Global Challenges. Science 284,5412:1–10.

Ostrom, Elinor, Thomas Dietz, Nives Dolsak, Paul C. Stern, Susan Stonich, and Elke U. Weber, eds.
2002 The Drama of the Commons. Washington: National Academy Press.

Ostrom, Elinor and Roy Gardner
1993 Coping with Asymmetries in the Commons: Self-Governing Irrigation Systems Can Work. Journal of Economic Perspectives 7,4:93–112

Ouarda, Taha B.M.J., Y. Hamdi, and Bernard Bobée
2002 A General System for Frequency Estimation in Hydrology (FRESH) with Historical Data. Proceedings of the PHEFRA Workshop, Barcelona, 16–19 October.

Oud, Engelbertus
2002 The Evolving Context of Hydropower Development. Energy Policy 30:1215–1223.

Pagiola, Stefano
2002 Paying for Water Services in Central America: Learning from Costa Rica. In Selling Forest Environmental Services. Pagiola, Stefano, Joshua Bishop and Natasha Landell-Mills, eds. Sterling, VA: Earthscan.

Pajuelo Prieto, Rómulo
2002 Vida, muerte y resurrección: Testimonios sobre el Sismo-Alud 1970. Yungay: Ediciones Elinca.

Paul, Frank, Andreas Kääb, Max Maisch, Tobias Kellenberger, and Wilfried Haeberli
2004 Rapid Disintegration of Alpine Glaciers Observed with Satellite Data. Geophysical Research Letters 31, L21402, doi:10.1029/2004GL020816.

Pelda, Kurt
2004 Streit ums Wasser des Nils Ägypten pocht auf seine "historischen Rechte". Neue Zürcher Zeitung.

Pelet, Paul-Louis
1988 Turbit et turbine, les roues hydrauliques horizontales du Valais. Vallesia 43:125–164.

1991 Des Rois Mages à la dynamo, les roues hydrauliques verticales en Valais. Vallesia 46:245–276.

1998 A la force de l'eau, les turbines de bois du Valais. Sierre: Monographic.

Perú, Asociación Automotriz del
1963 El Perú y sus rutas. Lima: SESATOR.

Plate, Erich J.
2002 Flood Risk and Flood Management. 267:2–11.

Postel, Sandra
1992 Last Oasis: Facing Water Scarcity. New York: W.W. Norton.

1999 Pillar of Sand: Can the Irrigation Miracle Last? New York: W.W. Norton.

Putz, Charlotte
2003 La gestion de l'eau potable sur le Haut Plateau. Conception, élaboration et mise en oeuvre d'un SIG prototype pour atteindre une gestion durable de l'eau. Mémoire de licence (non publié). Université de Lausanne.

Rabin, Matthew
1998 Psychology and Economics. Journal of Economic Literature 36:11–46.

Raffeiner, H.
1990 Mit Gefahren leben. Die freiwillige Feuerwehr in Tschengls 1890–1990, Meran.

Reichenbach, Paola, Fausto Guzzetti, and Mauro Cardinali
1998 Carta delle Aree Colpite da Movimenti Franosi e da Inondazioni – 2a edizione. CNR – IRPI Perugia, U.O. 3.1. GNDCI.

Ren, Yin
1902 In Langqiong xian zhi (Langqiong [Eryuan] County Gazetteer). Vol. 11:60a–61b. Kang Zhou, revised by Yingmei Luo, ed: reprinted Tapei: Chengwen Chubanshe, 1975. pp. 481–484.

Repetto, Robert
1986 Skimming the Water: Rent-seeking and the Performance of Public Irrigation Systems. World Resources Institute Research Report No. 4.

Retlinger, Gerald
1939 South of the Clouds: A Winter Ride Through Yün-nan. London: Faber & Faber.

Reynard, Denis
2002 Histoires d'eau. Bisses et irrigation en Valais au XVe siècle. Lausanne: Cahiers lausannois d'histoire médiévale, no. 30.

Reynard, Emmanuel
2000a Gestion patrimoniale et intégrée des ressources en eau dans les stations touristiques de montagne. Les cas de Crans-Montana-Aminona et Nendaz (Valais). Travaux et Recherches, n. 17, Institut de Géographie, Université de Lausanne 17.

2000b Cadre institutionnel et gestion des ressources en eau dans les Alpes: deux études de cas dans des stations touristiques valaisannes. Revue Suisse de Science Politique 6,1:53–85.

2001a Aménagement du territoire et gestion de l'eau dans les stations touristiques alpines. Le cas de Crans-Montana-Aminona (Valais, Suisse). Rev. Géogr. Alpine 89,3:7–19.

2001b GIS and Water Resource Management in Crans-Montana-Aminona (Switzerland). Unpublished research report. University of Lausanne.

Reynard Emmanuel, Adèle Thorens, and Corine Mauch
2001 Développement historique des régimes institutionnels de l'eau en Suisse entre 1870 et 2000. In Institutionelle Regime für natürliche Ressourcen: Boden, Wasser und Wald im Vergleich. Knoepfel Peter, Kissling-Näf Ingrid und Frédéric Varone (Hrsg.), ed. Basel/Genf/München: Helbing & Lichtenhahn. pp. 101–139.

Reynolds, John M.
1993 The Development of a Combined Regional Strategy for Power Generation and Natural Hazard Risk Assessment in a High-Altitude Glacial Environment: An Example from the Cordillera Blanca, Peru. In Natural Disasters: Protecting Vulnerable Communities. Merriman, P.A. and C.W.A. Browitt, eds. London: Thomas Telford. pp. 38–50.

Rhoades, Robert E. and Anthony Bebbington
1995 Farmers Who Experiment: An Untapped Resource for Agricultural Research and Development. In The Cultural Dimension of Development: Indigenous Knowledge Systems. Warren, D. Michael, L. Jan Slikkerveer and David Brokensha, eds. Bourton-on-Dunsmore, UK: Practical Action Publishers. pp. 296–307.

Richards, Paul

1985 Indigenous Agricultural Revolution: Ecology and Food Production in West Africa. London: Hutchinson.

Richardson, Shaun D. and John M. Reynolds

2000a Degradation of Ice-Cored Moraine Dams: Implications for Hazard Development. In Debris-Covered Glaciers. Nakawo, C.F.R.M. and A. Fountain, eds. Oxfordshire, UK: International Association of Hydrological Sciences. pp. 187–198.

2000b An Overview of Glacial Hazards in the Himalayas. Quaternary International 65–66:31–47.

Robins, Philip

1991 Turkey and the Middle East. London: Pinter.

Rocher, Emile

1904 Histoire des princes de Yun-nan et leurs relations avec la China d'après des documents historiques chinois. Bulletin de l'École française d'Extrême Orient 4.

Rodda, John C.

1994 Mountains–A Hydrological Paradox or Paradise? Beiträge zur Hydrologie der Schweiz 35:41–51.

Roggeri, Henri

1998 Tropical Freshwater Wetlands: A Guide to Current Knowledge and Sustainable Management. Dordrecht: Kluwer Academic.

Romerio, Franco

1994 Il settore elettrico in Ticino, i suoi legami con il mercato svizzero e la politica cantonale in materia di energia elettrica, 1894-1994-2044. CUEPE, Université de Genève (with a French summary).

1999 L'hydroélectrique entre Etat et marché. In Quels Systèmes énergétiques pour le XXIe siècle? Hollmuller, Pierre, Bernard Lachal, Franco Romerio and Willi Weber, eds. Geneva: CUEPE, University of Geneva, pp. 107–122.

2002a Which Paradigm for Managing the Risk of Ionising Radiation? Risk Analysis 22,1:59–66.

2002b European Electrical Systems and Alpine Hydro Resources. GAIA 3:200–202.

2003a Ouverture des marchés de l'électricité à la concurrence: problèmes, perspectives et études de cas. In L'énergie, controverses et perspectives. Lachal, Bernard and Franco Romerio, eds: CUEPE, Université de Genève. pp. 231–248.

2003b Opening the Swiss Electricity Market to Competition. In Competition in European Electricity Markets: A Cross-Country Comparison. Glachant, Jean-M. and Dominique Finon, eds. pp. 241–254.

Rossi, Fabio, Mauro Fiorentino, and Pasquale Versace
1984 Two Component Extreme Value Distribution for Flood Frequency Analysis. Water Resources. Research. 20,7:847–856.

Rothman, Mitchell
2000 Measuring and Apportioning Rents from Hydroelectric Power Developments. World Bank Discussion Paper No. 419.

Rubinstein, Ariel
1982 Perfect Equilibrium in a Bargaining Model. Econometrica 50,1:97–109.

Rudoy, Alexei N.
2002 Glacier-Dammed Lakes and Geological Work of Glacial Superfloods in the Late Pleistocene, Southern Siberia, Altai Mountains. Quaternary International 87,1:119–140.

Saitzew, Manuel
1950 Die Partnerwerke in der schweizerischen Elektrizitätswirtschaft, Ihr Wesen, Ihre Verbreitung und die Motive Ihrer Gründung. Gutachten erstattet der Etzelwerk AG.

Salazar Bondy, Augusto
1970 La técnica y la reubicación de Huaraz. El Diario de Huaraz 29 Agosto, 2.

Saleh, Maria and Ariel Dinar
2004 The Institutional Economics of Water: A Cross-country Analysis of Institutions and Performance. Cheltenham, UK; Northampton, MA: Edward Elgar.

Sarıibrahimoğlu, Lale
1995 PKK-Su Çemberi Kiriliyor. Cumhuriyet, 19 February.

Schlager, Edella and Elinor Ostrom
1992 Property-Rights Regimes and Natural Resources: A Conceptual Analysis. Land Economics 68,3:249–262.

Scoones, Ian
1990 Wetlands in Drylands: Key Resources for Agricultural and Pastoral Development in Africa. Ambio 20,8:366–371.

Selten, Reinhart
1975 Reexamination of the Perfectness Concept for Equilibrium Points in Extensive Games. International Journal of Game Theory 4: 25–55.

1978 The Chain Store Paradox. Theory and Decision 9:127–159.

1991 Evolution, Learning, and Economic Behavior. Games and Economic Behavior 3:3–24.

Service de l'Aménagement du Territoire (SAT)
1998 Gestion de l'eau. Département de l'environnement et de l'aménagement du territoire, Etude de base pour le Plan Directeur Cantonal. Sion.

Shah, Farhed A., David Zilberman, and Ujjayant Chakravorty
1993 Water Rights Doctrines and Technology Adoption. In The Economics of Rural Organization: Theory, Practice, and Policy. Hoff, Karla, Avishay Braverman and Joseph Stiglitz, eds. New York: Oxford University Press.

Shalpykova, Gulnara
2002 Water Disputes in Central Asia: The Syr Darya River Basin. Unpublished Master's Thesis in International Relations, The International University of Japan.

Shapiro, Judith
2001 Mao's War against Nature: Politics and the Environment in Revolutionary China. Cambridge, UK: Cambridge University Press.

Shimshack, Jay P. and Michael B. Ward
2005 Regulator Reputation, Enforcement, and Environmental Compliance, Journal of Environmental Economics and Management 50:519–540.

Shiva, Vandana
2002 Water Wars. Cambridge, MA: South End Press.

Shortle, James S. and Richard D. Horan
2001 The Economics of Nonpoint Pollution Control. Journal of Economic Surveys 15,3:255–289.

Silvius, Marcel, Michael Oneka, and Jan Verhagen
2000 Wetlands: Lifelines for People at the Edge. Phys. Chem. Earth (B) 15:645–652.

Singer, Charles Joseph, Eric John Holmyard, and Alfred Rupert Hall, eds.
1954–1984 A History of Technology. Oxford: Clarendon Press.

Siy, Robert Y. Jr.
1982 Community Resource Management: Lessons from the Zanjera. Quezon City: University of the Philippines Press.

Slaymaker, Curry, Joel Albrecht
1967 Proyecto: 'Parque Nacional Huascarán': Informe elevado al Servicio Forestal y de Caza del Ministerio de Agricultura por voluntarios del Cuerpo de Paz de Los Estados Unidos de Norte América. Lima, Noviembre. Doc #. I-VARIOS-008 at Biblioteca, Unidad de Glaciología y Recursos Hídricos, Huaraz.

Smith, Tony
2003 Farm Exports Boom in Argentina. New York Times, March 26, 2003.

Smol, John P., H. John, B. Birks and William M. Last, eds.
2002 Tracking Environmental Change Using Lake Sediments Volume 3: Terrestrial, Algal, and Siliceous Indicators. Dordrecht: Kluwer Academic.

Snorrason, Árni, Helga P. Finnsdóttir, and Marshall Moss, eds.
2002 The Extremes of the Extremes: Extraordinary Floods, 271. Oxfordshire, UK: International Association of Hydrological Sciences.

Société d'Histoire du Valais Romand (SHVR)
1995 Les bisses: Actes du colloque international, Sion, 15–18 septembre 1994. Sion: Annales valaisannes.

Solanes, Miguel
1996 Water Rights: Institutional Elements. CEPAL Review 59:83–96.

Solomon Abate
1994 Land Use Dynamics, Soil Degradation and Potential for Sustainable Use in Metu Area, Illubabor Region, Ethiopia. African Studies Series A13.

Solomon Mulugeta
2004 Socio-economic Determinants of Wetland Cultivation in Kemise, Illubabor Zone, Southwestern Ethiopia. Eastern Africa Social Science Research Review, 20,1:93–114.

Song Yingxing, [Wung Ying-hsing]
1637 Tiangong kaiwu [T'ien-kung k'ai-wu: Heaven-Nature and the Artificer Develop Commodities], reprinted 1966 In Chinese Technology in the Seventeenth Century, E-T. Z. Sun and S-C. Sun, eds.: Pennsylvania State University Press.

Spann, Hans J. and Jaime Fernández Concha
1950 Informe sobre el origen del aluvion de la quebrada de Los Cedros. Lima, 3 Noviembre. Doc #. I-GEOL-003 at Biblioteca, Unidad de Glaciología y Recursos Hídricos, Huaraz.

Stacul, P.
1979 Wildbachverbauung in Südtirol gestern und heute. Sonderbetrieb für Bodenschutz. Wildbach- und Lawinenverbauung, Autonome Provinz Bozen – Südtirol.

Starr, Joyce
1991 Water Wars. Foreign Policy 82:17–36.

Stauffer, Julie
1999 The Water Crisis. Montreal: Black Rose Books.

Stein, William W.
1974 Countrymen and Townsmen in the Callejón de Huaylas, Peru: Two Views of Andean Social Structure. Buffalo: Council on International Studies, State University of New York at Buffalo.

Steinberg, Ted
2000 Acts of God: The Unnatural History of Natural Disaster in America. New York: Oxford University Press.

Stephens, David W.
1990 Risk and Incomplete Information in Behavioral Ecology. In Risk and Uncertainty in Tribal and Peasant Economies. Cashdan, Elizabeth, ed. Boulder, CO: Westview Press.

Sturm, Leonhard-Christoph
1815 Vollständige Mühlen-Baukunst, 5th ed. Nuremberg: Augsburg.

Subprefecto Lucar, Caraz
1942 Telegram to Mendez Muñoz, Director Gobierno duputado Capitán. Caraz, 16 de enero de 1942. Doc. # 11, Legajo 425, Ministerio del Interior, Prefectura, at AGN, Lima.

Sullivan, Caroline
2002 Calculating a Water Poverty Index. World Development 30,7:1195–1210.

Swift, Jeremy J.
1979 Notes on Traditional Knowledge, Modern Knowledge and Rural Development. In Institute for Development Studies Bulletin. Vol. 10. pp. 41–43. University of Sussex, UK.

Tafesse Asres
1996 Agro-Ecological Zones of South-West Ethiopia. Unpublished MSc. Thesis. University of Trier.

Takahashi, T.
1991 Debris Flow. Rotterdam: Balkema.

Tang, Shui Yan
1992 Institutions and Collective Action: Self-Governance in Irrigation. San Francisco: ICS Press.

Tello, Julio C.
1960 Chavín. Lima: Universidad Nacional Mayor de San Marcos.

Tercier, Pierre and Christian Roten
2000 La responsabilité civile imposée aux exploitants d'ouvrages d'accumulation selon la Loi sur les ouvrages à accumulation. In Risques majeurs: Perception, globalisation et management. Wagner, Jean-Jacques and Jérôme Faessler, eds: Université de Genève.

Thobani, Mateen, ed.
1995 Peru: A User-Based Approach to Water Management and Irrigation Development. Washington DC: World Bank

Thompson, Roy and Frank Oldfield
1986 Environmental Magnetism. London: Allen & Unwin.

Tognacca, C.
1999 Beitrag zur Untersuchung der Entstehungsmechanismen von Murgängen. Laboratory of Hydraulics, Hydrology and Glaciology, ETH-Zürich Report 164.

Trask, Parker D.
1953 El problema de los aluviones de la Cordillera Blanca. Boletín de la Sociedad Geográfica de Lima LXX:5–75.

Trawick, Paul B.
1995 Water Reform and Poverty Alleviation in the Highlands. In Peru: A User-Based Approach to Water Management and Irrigation Development. World Bank Report # 13642-PE. Thobani, Mateen, ed. Washington D.C.: World Bank.

2001a Successfully Governing the Commons: Principles of Social Organization in an Andean Irrigation System. Human Ecology 29,1:1–25.

2001b The Moral Economy of Water: Equity and Antiquity in the Andean Commons. American Anthropologist 103,2:361–379.

2002a The Moral Economy of Water: General Principles for Successfully Governing the Commons. GAIA: Ecological Perspectives in Science, the Humanities and Economics 11:191–194.

2002b Trickle-down Theory, Andean Style: Indigenous Practices Provide a Lesson in Sharing. Natural History 111,8:60–65.

2003a Against the Privatization of Water: An Indigenous Model for Improving Existing Laws and Successfully Governing the Commons. World Development 31,6:977–996.

2003b The Struggle for Water in Peru: Comedy and Tragedy in the Andean Commons. Palo Alto: California: Stanford University Press.

2005 The Moral Economy of Water: A Cross-Cultural Study of Principles for Successfully Governing the Commons: Final Report to the John D. and Catherine T. MacArthur Foundation on a Research and Writing Grant in their *Program on Global Security and Sustainability*. [archived as a working paper on the Digital Library of the Commons of the International Asssociation for the Study of Common Property (IASCP)].

Treacy, John
1994a Teaching Water: Hydraulic Management and Terracing in Corporaque, the Colca Valley, Peru. In Irrigation at High Altitudes: the Social Organization of Water Control Systems in the Andes. Vol. 12. Mitchell, William P. and David Guillet, eds. Society for Latin American Anthropology Publication Series: American Anthropological Association. pp. 99–114.

1994b Las Chacras de Corporaque: Andeneria y Riego en el Valle del Colca. Lima: Instituto de Estudios Peruanos.

Tsur, Yacov
1997 The Economics of Conjunctive Ground and Surface Water Irrigation Systems: Basic Principles and Empirical Evidence from Southern California. In Decentralization and Coordination of Water Resource Management. Parker, Douglas D. and Yacov Tsur, eds. Norwell, MA: Kluwer Academic.

Tversky, Amos and Daniel Kahneman
1992 Advances in Prospect Theory: Cumulative Representation of Uncertainty. Journal of Risk and Uncertainty 5:297–323.

Tversky, Amos and Craig R. Fox
1995 Weighing Risk and Uncertainty. Psychology Review 102:269–283.

UNCRD
1994 Regional Development and Environmental Management for the Dali-Lake Erhai Area, Yunnan Provincial People’s Government, China. Nagoya, Japan: UNCRD.

Unidad de Glaciología e Hidrología (UGH)
1990 Afianzamiento hídrico del Río Santa, I Etapa. Informe, Huaraz, Enero. Doc #. I-HIDRO-006 at Biblioteca, Unidad de Glaciología y Recursos Hídricos, Huaraz.

United Nations Development Program (UNDP)
1999 Prosperity Fades: Jimma and Illubabor Zones of Oromia Region. Addis Ababa: Emergency Unit for Ethiopia Assessment Mission, EUE, UNDP.

Varone, Frédéric, Emmanuel Reynard, Ingrid Kissling-Näf, and Corine Mauch
2002 Institutional Resource Regimes. The Case of Water in Switzerland. Integrated Assessment 3:78–94.

Vasquez, John
1996 Distinguishing Rivals that Go to War from Those that Do Not: A Quantitative Comparative Case Study of the Two Paths to War. International Studies Quarterly 40:531–558.

Veiga da Cunha, Luis.
1994 The Aral Sea Crisis: A Great Challenge in Transboundary Water Resources Management. Presented at NATO Advanced Research Workshop on Transboundary Water Resources Management: Technical and Institutional Issues, Skopelos, Greece, May 1994.

Vickers, Amy
2001 Water Use and Conservation. Amherst, MA: Water Plow Press.

Villani, Paolo
1993 Extreme Flood Estimation Using Power Extreme Value (PEV) Distribution. In Proceedings of the IASTED International Conference, Modeling and Simulation. Hamza, M. H., ed. Pittsburgh: University. of Pittsburgh. pp. 470–476.

Villón, Pedro Cristóbal
1959 El Huascarán. Forjando Ancash Año 1, No. 2, 30 de Junio, 11.

Vischer, Daniel and Gian Reto Bezzola
2000 Bewegliche Wildbachbrücken und alternative Lösungen. Wasserwirtschaft, Zeitschrift für Wasser und Umwelt 90,7/8:394–398.

Viviroli, Daniel
2001 Zur hydrologischen Bedeutung der Gebirge, Publikationen Gewässerkunde, No. 265. Department of Geography, University of Berne.

Viviroli, Daniel and Rolf Weingartner
2002 The Significance of Mountains as Sources of the World's Fresh Water. GAIA 11,3:182–186.

Viviroli, Daniel, Rolf Weingartner, and Bruno Messerli
2003 Assessing the Hydrological Significance of the World's Mountains. Mountain Research and Development 23,1:32–40.

Von Braun, Joachim
1997 The Links between Agricultural Growth, Environmental Degradation, and Nutrition and Health: Implications for Policy and Research. In Sustainability, Growth, and Poverty Alleviation. Vosti, Stephan A. and Thomas Reardon, eds. Baltimore: Johns Hopkins University.

Von Neumann, John and Oscar Morgenstern
1944 Game Theory and Economic Behavior. Princeton, NJ: Princeton University Press.

Vörösmarty, Charles J., Pamela Green, Joseph Salisbury, and Richard B. Lammers
2000 Global Water Resources: Vulnerability from Climate Change and Population Growth. Science 289:284–288.

Wade, Robert
1986 Common Property Resource Management in South Indian Villages. In Proceedings of the Conference on Common Property Resource Management: National Academy Press, Washington DC. pp. 231–258.

1988 Village Republics. Economic Conditions for Collective Action in South India. Cambridge, UK: Cambridge University Press.

1992 Common Property Resource Management in South Indian Villages. In Making the Commons Work: Theory, Policy and Practice. Bromley, Daniel, ed. San Francisco: Institute for Contemporary Studies Press. pp. 207–228.

Walt, Stephen
1987 The Origin of Alliances. Ithaca, NY: Cornell University Press.

Warren, D. Michael
1991 Using Indigenous Knowledge in Agricultural Development. World Bank Discussion Paper 127.

Weitzman, Martin L.
1974 Prices vs. Quantities. Review of Economic Studies 41,4:477–491.

Wiegandt, Ellen
1977 Inheritance and Demography in the Swiss Alps. Ethnohistory 24,2:133–148.

1980 Un Village en Transition. Ethnologica helvetica 4:63–93.

2004 Losing Ground: Challenges to Mountain Agriculture in Switzerland. Culture and Agriculture 26,1,2:110–123.

Wiens, Harold. J
1967 Han Chinese Expansion in South China, A revised edition of his earlier China's March toward the Tropics. Shoe String Press.

Wigley, Thomas M. L. and Sarah C. B. Raper
2005 Extended Scenarios for Glacier Melt due to Anthropogenic Forcing. Geophysical Research Letters 32, L05704, doi:10.1029/2004GL021238.

Wisner, Ben, Piers Blaikie, Terry Cannon, and Ian Davis
2004 At Risk: Natural Hazards, People's Vulnerability and Disasters. New York: Routledge.

Wolf, Aaron
1994 A Hydropolitical History of the Nile, Jordan, and Euphrates River Basins. In International Waters of the Middle East from Euphrates-Tigris to Nile. Biswas, Asit, ed. Oxford: Oxford University Press.

Wood, Adrian P.
1996 Wetland Drainage and Management in South-West Ethiopia: Some Environmental Experiences of an NGO. In The Sahel Workshop 1996. Reenburg, Annette, Henrik Secher Marcussen and Ivan Nielsen, eds. Copenhagen: SEREIN, Institute of Geography, University of Copenhagen.

Wood, Adrian P., Afework Hailu, Patrick G. Abbot, and Alan B. Dixon
2002 Sustainable Management of Wetlands in Ethiopia: Local Knowledge versus Government Policy. In Strategies for Wise Use of Wetlands: Best Practices in Participatory Management, Proceedings of a Workshop held at the 2nd International Conference on Wetlands and Development, November 1998, Dakar, Senegal. Gawler, Meg, ed. Gland: IUCN. Pp. 81–88.

World Bank
1995 Peru: A User-based Approach to Water Management and Irrigation Development. Report No. 13642-PE. Washington, D.C.: World Bank.

World Commission on Dams (WCD)
2000 Dams and Development: A New Framework for Decision Making. The Report of the World Commission on Dams.

World Data Centre
1994 Trends '93, A Compendium of Data on Global Change. World Data Centre.

World Meterological Organisation (WMO)
2004 Unusual Weather Causes Widespread Flood, Heavy Casualties and Damages in Asia. WMO Press Release No. 710.

Wüstenhagen, Rolf, Jochen Markard, and Bernhard Truffer
2003 Diffusion of Green Power Products in Switzerland. Energy Policy 31:621–632.

Xiangcan, Jin
2003 Analysis of Eutrophication State and Trend for Lakes in China, Papers from Bolsena Conference (2002), Residence Time in Lakes: Science, Management, Education. Journal of Limnology 62, Supplement 1:60–66.

Xue, Lin, ed.
1999 Dali fengwu zhi [The scenery of Dali]. Kunming: Yunnan Renmin Chubanshe.

Yang, Dewen
1988 Yunnan Dali Shi Dazhantun erhao Hanmu. 5:449–456.

Yang, Hong and Alexander J.B. Zehnder
2002 Water Scarcity and Food Import: A Case Study for Southern Mediterranean Countries. World Development 30,8:1413–1430.

Yauri Montero, Marcos
1961 Ganchiscocha: Leyendas, cuentos y mitos de Ancash. Lima: Ediciones Piedra y Nieve.

2000 Leyendas ancashinas. Lima: Lerma Gómez.

Young, Peyton
1993 The Evolution of Conventions. Econometrica 61:57–84.

Zagorchev, Ivan
1995 Pirin. Geological Guidebook. Acad. Pub. House "Prof. Marin Drinov".

2001 Southwest Bulgaria. Geological Guidebook. International Conference Geodynamic Hazards, Late Alpine Tectonics and Neotectonics in the Rhodope Regions, 5FP, Sofia, Bulgaria. Geological Institute.

Zamora Cobos, Marino
1983 Inventario y seguridad de lagunas en la Cordillera Blanca.Trabajo presentado al Forum Recursos Hídricos del Rio Santa, Chimbote, 18–20 Noviembre 1983, Huaraz, Noviembre. Doc #. I-INVEN-010 at Biblioteca, Unidad de Glaciología y Recursos Hídricos, Huaraz.

Zapata Luyo, Marco
2002 La dinámica glaciar en lagunas de la Cordillera Blanc. Acta Montana (Czech Republic) 19,123:37–60.

Zappa, Massimiliano
2002 Multiple-Response Verification of a Distributed Hydrological Model at Different Spatial Scales.

Zegarra, M.
1941 Sobre la catástrofe de Huaraz: Informe técnico del origen de ésta. El Comercio 21 Diciembre, 3.

Zerihun Woldu
2000 Plant Biodiversity in the Wetlands of Illubabor Zone, Unpublished report, Report 3 for Objective 2. Huddersfield, UK: University of Huddersfield.

Zeuthen, Frederik
1930 Problems of Monopoly and Economic Warfare. London: Routledge and Kegan Paul.

Zhao, Jinhua and David Zilberman
1999 Irreversibility and Restoration in Natural Resource Development. Oxford Economic Papers 51:559–573.

1999 Irreversibility and Restoration in Natural Resource Development. Oxford Economic Papers 51:559–573.

Zhou Kang, comp. and rev. Luo Yingmei
1902/ 1975 Langqiong Xianzhi [Langqiong [Eryuan] County Gazetteer], repr. Chengwen chubanshe: Taipei, 1975 11,60a–61b:481–484.

Zilberman, David
1997 Water Marketing in California and the West. Gianni Foundation Paper. Department of Agricultural and Resource Economics, University of California, Berkeley.

INDEX

Advances in Global Change Research

1. P. Martens and J. Rotmans (eds.): *Climate Change: An Integrated Perspective.* 1999 ISBN 0-7923-5996-8
2. A. Gillespie and W.C.G. Burns (eds.): *Climate Change in the South Pacific: Impacts and Responses in Australia, New Zealand, and Small Island States.* 2000 ISBN 0-7923-6077-X
3. J.L. Innes, M. Beniston and M.M. Verstraete (eds.): *Biomass Burning and Its Inter-Relationships with the Climate Systems.* 2000 ISBN 0-7923-6107-5
4. M.M. Verstraete, M. Menenti and J. Peltoniemi (eds.): *Observing Land from Space: Science, Customers and Technology.* 2000 ISBN 0-7923-6503-8
5. T. Skodvin: *Structure and Agent in the Scientific Diplomacy of Climate Change.* An Empirical Case Study of Science-Policy Interaction in the Intergovernmental Panel on Climate Change. 2000 ISBN 0-7923-6637-9
6. S. McLaren and D. Kniveton: *Linking Climate Change to Land Surface Change.* 2000 ISBN 0-7923-6638-7
7. M. Beniston and M.M. Verstraete (eds.): *Remote Sensing and Climate Modeling: Synergies and Limitations.* 2001 ISBN 0-7923-6801-0
8. E. Jochem, J. Sathaye and D. Bouille (eds.): *Society, Behaviour, and Climate Change Mitigation.* 2000 ISBN 0-7923-6802-9
9. G. Visconti, M. Beniston, E.D. Iannorelli and D. Barba (eds.): *Global Change and Protected Areas.* 2001 ISBN 0-7923-6818-1
10. M. Beniston (ed.): *Climatic Change: Implications for the Hydrological Cycle and for Water Management.* 2002 ISBN 1-4020-0444-3
11. N.H. Ravindranath and J.A. Sathaye: *Climatic Change and Developing Countries.* 2002 ISBN 1-4020-0104-5; Pb 1-4020-0771-X
12. E.O. Odada and D.O. Olaga: *The East African Great Lakes: Limnology, Palaeolimnology and Biodiversity.* 2002 ISBN 1-4020-0772-8
13. F.S. Marzano and G. Visconti: *Remote Sensing of Atmosphere and Ocean from Space: Models, Instruments and Techniques.* 2002 ISBN 1-4020-0943-7
14. F.-K. Holtmeier: *Mountain Timberlines.* Ecology, Patchiness, and Dynamics. 2003 ISBN 1-4020-1356-6
15. H.F. Diaz (ed.): *Climate Variability and Change in High Elevation Regions: Past, Present & Future.* 2003 ISBN 1-4020-1386-8
16. H.F. Diaz and B.J. Morehouse (eds.): *Climate and Water: Transboundary Challenges in the Americas.* 2003 ISBN 1-4020-1529-1
17. A.V. Parisi, J. Sabburg and M.G. Kimlin: *Scattered and Filtered Solar UV Measurements.* 2004 ISBN 1-4020-1819-3
18. C. Granier, P. Artaxo and C.E. Reeves (eds.): *Emissions of Atmospheric Trace Compounds.* 2004 ISBN 1-4020-2166-6
19. M. Beniston: *Climatic Change and its Impacts.* An Overview Focusing on Switzerland. 2004 ISBN 1-4020-2345-6
20. J.D. Unruh, M.S. Krol and N. Kliot (eds.): *Environmental Change and its Implications for Population Migration.* 2004 ISBN 1-4020-2868-7
21. H.F. Diaz and R.S. Bradley (eds.): *The Hadley Circulation: Present, Past and Future.* 2004 ISBN 1-4020-2943-8
22. A. Haurie and L. Viguier (eds.): *The Coupling of Climate and Economic Dynamics.* Essays on Integrated Assessment. 2005 ISBN 1-4020-3424-5

Advances in Global Change Research

23. U.M. Huber, H.K.M. Bugmann and M.A. Reasoner (eds.): *Global Change and Mountain Regions.* An Overview of Current Knowledge. 2005
ISBN 1-4020-3506-3; Pb 1-4020-3507-1
24. A.A. Chukhlantsev: *Microwave Radiometry of Vegetation Canopies.* 2006
ISBN 1-4020-4681-2
25. J. McBeath, J. Rosenbery: *Comparative Environmental Politics.* 2006
ISBN 1-4020-4762-2
26. M.E. Ibarrarán and R. Boyd: *Hacia el Futuro.* Energy, Economics, and the Environment in 21st Century Mexico. 2006 ISBN 1-4020-4770-3
27. N.J. Roserberg: *A Biomass Future for the North American Great Plains:* Toward Sustainable Land Use and Mitigation of Greenhouse Warming. 2006 ISBN 1-4020-5600-1
28. V. Levizzani, P. Bauer and F.J. Turk (eds.): *Measuring Precipitation from Space.* EURAINSAT and the Future. 2007 ISBN 978-1-4020-5834-9
29. N.H. Ravindranath and M. Ostwald: *Carbon Inventory Methods.* Handbook for Greenhouse Gas Inventory, Carbon Mitigation and Roundwood Production Projects. 2007
ISBN 978-1-4020-6546-0
30. To be Published.
31. E. Wiegandt (ed.): *Mountains: Sources of Water, Sources of Knowledge.* 2008
ISBN 978-1-4020-6747-1